Nanobioelectronics - for Electronics, Biology, and Medicine

Nanostructure Science and Technology

Series Editor: David J. Lockwood, FRSC
National Research Council of Canada
Ottawa, Ontario, Canada

Current volumes in this series:

Alternative Lithography: Unleashing the Potentials of Nanotechnology
Edited by Clivia M. Sotomayor Torres

Nanoparticles: Building Blocks for Nanotechnology
Edited by Vincent Rotello

Nanostructured Catalysts
Edited by Susannah L. Scott, Cathleen M. Crudden, and Christopher W. Jones

Nanotechnology in Catalysis, Volumes 1 and 2
Edited by Bing Zhou, Sophie Hermans, and Gabor A. Somarjai

Polyoxometalate Chemistry for Nano-Composite Design
Edited by Toshihiro Yamase and Michael T. Pope

Self-Assembled Nanostructures
Edited by Jin Z. Zhang, Zhong-lin Wang, Jun Liu, Shaowei Chen, and Gang-yu Liu

Semiconductor Nanocrystals: From Basic Principles to Applications
Edited by Alexamder L. Efros, David J. Lockwood, and Lenoid Tsybekov

---

A Continuation Order Plan is available for this series. A continuation order will bring delivery of each new volume immediately upon publication. Volumes are billed only upon actual shipment. For further imformation please contact the publisher.

Nanobioelectronics - for Electronics, Biology, and Medicine

Edited by

Andreas Offenhäusser
Forschungszentrum Jülich
Germany

Ross Rinaldi
University of Lecce
Italy

Editors
Andreas Offenhäusser
Institute of Bio- and Nanosystems
Forschungszentrum Jülich
D-52425 Jülich
Germany
a.offenhaeusser@fz-jeulich.de

Ross Rinaldi
CNR Lecce
Ist. Nazionale di Fisica della Materia
National Nanotechnology Lab. (NNL)
Via Arnesano, 5
73100 Lecce
Italy
ross.rinaldi@unile.it

ISBN: 978-1-4419-1857-4 e-ISBN: 978-0-387-09459-5
DOI: 10.1007/978-0-387-09459-5

Printed on acid-free paper

springer.com

Contents

Contributors

Allison J. Beattie
Centre for Cell Engineering
University of Glasgow
Glasgow G12 8QQ
United Kingdom

Alessandro Bramanti
STMicroelectronics, Research Unit of Lecce,
c/o Distretto Tecnologico ISUFI,
Via per Arnesano, km.5, I-73100 Lecce, Italy

Franco Calabi
National Nanotechnology Laboratory of CNR-INFM,
Unità di Ricerca IIT, Distretto Tecnologico ISUFI,
Via per Arnesano, km.5, I-73100 Lecce, Italy

Qijin Chi
Technical University of Denmark,
Department of Chemistry and NanoDTU
2800 Kgs. Lyngby, Denmark

Adam S.G. Curtis
Centre for Cell Engineering
University of Glasgow
Glasgow G12 8QQ
United Kingdom

Rosa Di Felice
National Center on nanoStructures and bioSystems at Surfaces
of INFM-CNR, Center for NanoBiotechnology,
Modena, Italy

Grit Festtag
Institut of Physical High Technology,
P.O.B.100 239; D-07702 Jena, Germany

Monika Fischler
Institute of Inorganic Chemistry,
Rheinisch-Westfälisch Technische Hochschule Aachen
Landoltweg 1, Aachen, Germany

Wolfgang Fritzsche
Institut of Physical High Technology,
P.O.B.100 239; D-07702 Jena, Germany

Melanie Homberger
Institute of Inorganic Chemistry,
Rheinisch-Westfälisch Technische Hochschule Aachen
Landoltweg 1, Aachen, Germany

Sven Ingebrandt
Institute of Bio- and Nanosystems,
Forschungszentrum Jülich, D-52425 Jülich, Germany

Palle S. Jensen
Technical University of Denmark,
Department of Chemistry and NanoDTU
2800 Kgs. Lyngby, Denmark

Ingo Köper
Max Planck Institute for Polymer Research,
Ackermannweg 10, 55128 Mainz, Germany

Wolfgang Knoll
Max Planck Institute for Polymer Research,
Ackermannweg 10, 55128 Mainz, Germany

Giuseppe Maluccio
National Nanotechnology Laboratory of CNR-INFM,
Unità di Ricerca IIT, Distretto Tecnologico ISUFI,
Via per Arnesano, km.5, I-73100 Lecce, Italy

Liberato Manna
National Nanotechnology Laboratory of CNR-INFM,
Unità di Ricerca IIT, Distretto Tecnologico ISUFI,
Via per Arnesano, km.5, I-73100 Lecce, Italy

Robert Möller
Institut of Physical High Technology,
P.O.B.100 239; D-07702 Jena, Germany

Andreas Offenhäusser
Institute of Bio- and Nanosystems,
Forschungszentrum Jülich, D-52425 Jülich, Germany

Michael Pabst
Institute of Bio- and Nanosystems,
Forschungszentrum Jülich, D-52425 Jülich, Germany

Teresa Pellegrino
National Nanotechnology Laboratory of CNR-INFM,
Unità di Ricerca IIT, Distretto Tecnologico ISUFI,
Via per Arnesano, km.5, I-73100 Lecce, Italy

Pier Paolo Pompa
National Nanotechnology Laboratory of CNR-INFM,
Unità di Ricerca IIT, Distretto Tecnologico ISUFI,
Via per Arnesano, km.5, I-73100 Lecce, Italy

Dietmar Pum
Center for NanoBiotechnology,
University of Natural Resources and Applied Life Sciences Vienna,
Gregor Mendelstr. 33, A-1180 Vienna, Austria

Mathis Riehle
Centre for Cell Engineering
University of Glasgow
Glasgow G12 8QQ
United Kingdom

B.H. Roelofsen
Biomedical Signals and Systems Department,
Faculty of Electrical Engineering,
Mathematics and Computer Science/Institute for Biomedical Technology,
University of Twente, The Netherlands

T.G. Ruardij
Biomedical Signals and Systems Department,
Faculty of Electrical Engineering,
Mathematics and Computer Science/Institute for Biomedical Technology,
University of Twente, The Netherlands

Wim Rutten
Biomedical Signals and Systems Department,
Faculty of Electrical Engineering,
Mathematics and Computer Science/Institute for Biomedical Technology,
University of Twente, The Netherlands

Ulrich Simon
Institute of Inorganic Chemistry,
Rheinisch-Westfälisch Technische Hochschule Aachen
Landoltweg 1, Aachen, Germany

Uwe B. Sleytr
Center for NanoBiotechnology,
University of Natural Resources and Applied Life Sciences Vienna,
Gregor Mendelstr. 33, A-1180 Vienna, Austria

G. van Staveren
Biomedical Signals and Systems Department,
Faculty of Electrical Engineering,
Mathematics and Computer Science/Institute for Biomedical Technology,
University of Twente, The Netherlands

Jens Ulstrup
Technical University of Denmark,
Department of Chemistry and NanoDTU
2800 Kgs. Lyngby, Denmark

Inga K. Vockenroth
Max Planck Institute for Polymer Research,
Ackermannweg 10, 55128 Mainz, Germany

Chris D.W. Wilkinson
Department of Electronics and Electrical Engineering
University of Glasgow
Glasgow G12 8LT
United Kingdom

Introduction

The combination of biological elements with electronics is of great interest for many research areas. Inspired by biological signal processes, scientists and engineers are exploring ways of manipulating, assembling, and applying biomolecules and cells on integrated circuits, joining biology with electronic devices. The overall goal is to create bioelectronic devices for biosensing, drug discovery, and curing diseases, but also to build new electronic systems based on biologically inspired concepts. This research area called bioelectronics requires a broad interdisciplinary and transdisciplinary approach to biology and material science. Even though at the frontier of life science and material science, bioelectronics has achieved in the last years many objectives of scientific and industrial relevance, including aspects of electronics and biotechnology. Although the first steps in this field combined biological and electronic units for sensor applications (e.g., glucose oxidase on an oxygen electrode), we see now many applications in the fields of genomics, proteomics, and celomics as well as electronics. This approach challenges both the researcher and the student to learn and think outside of their zones of comfort and training.

Today, one can fabricate electrically active structures that are commensurate in size with biomolecules. The advancement of nanotechnology has influenced bioelectronics to a large extent. New inspection tools, such as scanning probe microscopy, developed in the last two decades have become ubiquitous systems to image nanoscale structure and estimate certain structural, mechanical, and functional characteristics of biological entities, ranging from proteins and DNA to cells and tissues. Various modes of imaging and SPM-based spectroscopy have been developed to correlate structure, properties, and chemomechanical interactions between biological units in different environments. This has induced rapid improvement in control, localization, handling, assembling, and subsequent modification of these biological entities. New understanding of properties of interfaces and binding mechanisms has been achieved. In particular, the detailed investigation of self-assembling processes at the base of protein and DNA formation and ligand–receptor interactions has opened new routes to the design and engineering of hybrid systems, comprising inorganic nanostructures and biological "smart" matter. In parallel different technologies have been developed to produce structures below 100 nm with nanometer control: first, electron beam lithography, which is most often employed, but also ion beam lithography, X-ray lithography, scanning probe

A. Offenhäusser and R. Rinaldi (eds.), *Nanobioelectronics - for Electronics, Biology, and Medicine*,
DOI: 10.1007/978-0-387-09459-5_1, © Springer Science+Business Media, LLC 2009

lithography, and alternative techniques such as soft lithography. The latter in particular has been demonstrated to be compatible with the handling and modification of organic and biological materials, and there exist already in literature various examples of protein patterning realized by means of this technique. Finally, it is worth mentioning the progress made by chemistry in the production of colloidal nano-objects such as spherical particles, rods, tetrapods and combinations, characterized by wide tunability in sizes and emission wavelengths, along with the development of the biochemical ability to join them to biological entities.

Having tools similar in size to biomolecules enables us to manipulate, measure, and (in the future) control them with electronics, ultimately connecting their unique functions. The combination of inorganic nano-objects with biological molecules leads to hybrid systems with special properties that provide fascinating scientific and technological opportunities. A bioelectronic interface joins structured, functional surfaces, and circuits to nucleic acids (e.g., DNA), proteins at the single molecule level. The need of development of new strategies for the functional integration of biological units and electronic systems or nanostructured materials were also facilitated by the parallel progress in biochemistry and molecular biology, namely, advances in protein engineering, with the ability to make "designer" proteins and peptides with specific functions or combinations of functions; and the establishment of surface display technologies, with the ability to generate and screen large repertoires of peptides and nucleic acids for high-affinity binding to potentially any structure (organic or inorganic). New nanostructured sensors, electronic nanocircuitries based on biomolecules, and biomolecular templates are a few examples in which biology meets nanoelectronics.

Moreover, the similar dimensions of biomolecules and electronic nanostructures have opened the way for fabrication of bioelectronic hybrid systems of novel functions. In the last years, considerable research was focused on understanding transport phenomena between biological materials and electronic systems. Recent advances in the field have demonstrated electrical contacting of redox proteins with electrodes—the use of DNA or proteins as templates to assemble nanoparticles and nanowires. This combination of biomolecules with nano-objects will find applications in various disciplines. In turn, recent studies have opened the way to the use of nanoelectrodes, nano-objects, and nanotools in living cells and tissue, for both fundamental biophysical studies and cellular signaling detection. Another research direction is based on the functional connection of neuronal signal processing elements and electronics in order to build brain–machine interfaces and future information systems.

The different aspects of bioelectronics reviewed in this book emphasize the immense developments in the field of bioelectronics and nanobioelectronics. These technological and scientific advancements show that bioelectronics is a ripe discipline based on solid ground. The range of themes addressed emphasizes key aspects and future perspectives of nanobioelectronics. The book discusses the electronic coupling of DNA and proteins with electronic devices to build new information systems and apply the systems as biosensors. The exploitation of networks of neurons connected with electronic devices in future information processing systems and the use of nano-objectes to assess cellular function is also discussed in detail.

The topics of these hybrid nanobioelectronic systems are both interesting for fundamental research and to enhance industrial competitiveness through research, education, and transfer of technology. Applications of these technologies include:

- *Nanoelectronics for the future*. The fascinating world of the bio–self-assembly provides new opportunities and directions for future electronics, opening the way to a new generation of computational systems based on biomolecules and biostructures at the nanoscale.
- *Life sciences*. Rapid pharmaceutical discovery and toxicity screening using arrays of receptors on an integrated circuit, with the potential to develop targeted "smart drugs."
- *Medical diagnostics*. Rapid, inexpensive, and broad-spectrum point-of-use human and animal screening for antibodies specific to infections
- *Environmental quality*. Distinguishing dioxin isomers for cleaning up polluted sites, improving production efficiency of naturally derived polysaccharides such as pectin and cellulose, and measuring indoor air quality for "sick" buildings.
- *Food safety*. Array sensors for quality control and for sensing bacterial toxins.
- *Crop protection*. High-throughput screening of pesticide and herbicide candidates.
- *Military and civilian defense*. Ultrasensitive, broad-spectrum detection of biological warfare agents and chemical detection of antipersonnel land mines, screening passengers and baggage at airports, and providing early warning for toxins from virulent bacterial strains.

Therefore, the different topics addressed in this book will be of interest to the interdisciplinary research community. We hope that this collection of chapters will provide physics, chemists, biologists, material scientists, and engineers with a comprehensive perspective of the field. Furthermore, the book is aimed to attract young researchers and introduce them to the field, while providing newcomers with an enormous collection of literature references.

The book is organized into three sections: The first is on nanobioelectronics and DNA, the second is on nanobioelectronics and proteins, and the third is on nanobioelectronics and cells. In each section there is a preface describing the key properties of the basic bio-units on which the sections have been focused. The sections are in turn divided in two parts: The first presents the biological element as a part of a (possible) nanoelectronic device, and the second highlights how the recent and fast progress (development) of nanothechnologies can meet the life science world to explore, understand, and possibly control mechanisms that have not been explored up to now. We hope that from the conjunction of the two ways, bio-to-nano and nano-to-bio, a new broad discipline could come up, aimed to increase the scientific progress of the whole scientific community and everyone's wellness in the near future.

Part A
DNA-Based Nanobioelectronics

Deoxyribonucleic acid (DNA) is a nucleic acid that contains the genetic instructions for the development and function of living organisms. The main role of DNA in the cell is the long-term storage of information. It is often compared to a blueprint, since it contains the instructions to construct other components of the cell, such as proteins and RNA molecules. The DNA segments that carry genetic information are called genes, but other DNA sequences have structural purposes or are involved in regulating the expression of genetic information.

DNA is a long polymer made from repeating units called nucleotides. The DNA chain is 22 to 24 Å wide and one nucleotide unit is 3.3 Å long. Although these repeating units are very small, DNA polymers can be enormous molecules containing millions of nucleotides. For instance, the largest human chromosome is 220 million base pairs long.

In living organisms, DNA does not usually exist as a single molecule, but instead as a tightly associated pair of molecules. These two long strands entwine like vines in the shape of a double helix. The nucleotide repeats contain both the backbone of the molecule, which holds the chain together, and a base, which interacts with the other DNA strand in the helix. In general, a base linked to a sugar is called a nucleoside and a base linked to a sugar and one or more phosphate groups is called a nucleotide. If multiple nucleotides are linked together, as in DNA, this polymer is referred to as a polynucleotide.

The backbone of the DNA strand is made from alternating phosphate and sugar residues. The sugar in DNA is the pentose (five-carbon) sugar 2-deoxyribose. The sugars are joined together by phosphate groups that form phosphodiester bonds between the third and fifth carbon atoms in the sugar rings. These asymmetric bonds mean a strand of DNA has a direction. In a double helix the direction of the nucleotides in one strand is opposite to their direction in the other strand. This arrangement of DNA strands is called antiparallel. The asymmetric ends of a strand of DNA bases are referred to as the 5′ (*five prime*) and 3′ (*three prime*) ends. One of the major differences between DNA and RNA is the sugar, with 2-deoxyribose being replaced by the alternative pentose sugar ribose in RNA.

The DNA double helix is held together by hydrogen bonds between the bases attached to the two strands. The four bases found in DNA are adenine (abbreviated A), cytosine (C), guanine (G), and thymine (T). These four bases are attached to the sugar/phosphate to form the complete nucleotide.

These bases are classified into two types, adenine and guanine, which are fused five- and six-membered heterocyclic compounds called purines, whereas cytosine and thymine are six-membered rings called pyrimidines. A fifth pyrimidine base, called uracil (U), replaces thymine in RNA and differs from thymine by lacking a methyl group on its ring. Uracil is normally only found in DNA as a breakdown product of cytosine, but a very rare exception to this rule is a bacterial virus called PBS1 that contains uracil in its DNA.

The double helix is a right-handed spiral. As the DNA strands wind around each other, they leave gaps between each set of phosphate backbones, revealing the sides of the bases inside. There are two of these grooves twisting around the surface of the double helix: one groove is 22 Å wide and the other 12 Å wide. The larger groove is called the major groove, while the smaller, narrower groove is called the minor groove. The narrowness of the minor groove means that the edges of the bases are more accessible in the major groove. As a result, proteins like that can bind to specific sequences in double-stranded DNA usually read the sequence by making contacts to the sides of the bases exposed in the major groove.

Each type of base on one strand forms a bond with just one type of base on the other strand. This is called complementary base pairing. Here, purines form hydrogen bonds to pyrimidines, with A bonding only to T, and C bonding only to G. This arrangement of two nucleotides joined together across the double helix is called a base pair. In a double helix, the two strands are also held together by forces generated by the hydrophobic effect and pi stacking, but these forces are not affected by the sequence of the DNA. As hydrogen bonds are not covalent, they can be broken and rejoined relatively easily. The two strands of DNA in a double helix can therefore be pulled apart like a zipper, either by a mechanical force or high temperature. As a result of this complementarity, all the information in the double-stranded sequence of a DNA helix is duplicated on each strand, which is vital in DNA replication. Indeed, this reversible and specific interaction between complementary base pairs is critical for all the functions of DNA in living organisms.

The two types of base pairs form different numbers of hydrogen bonds, AT forming two hydrogen bonds, and GC forming three hydrogen bonds. The GC base pair is therefore stronger than the AT base pair. As a result, it is both the percentage of GC base pairs and the overall length of a DNA double helix that determine the strength of the association between the two strands of DNA. Long DNA helices with a high GC content have strongly interacting strands, whereas short helices with high AT content have weakly interacting strands. The strength of this interaction can be measured by finding the temperature required to break the hydrogen bonds, their melting temperature (also called T_m value). When all the base pairs in a DNA double helix melt, the strands separate and exist in solution as two entirely independent molecules.

Based on these properties DNA is of great interest for applications in bioelectronics. This is in the focus of the first part which is divided into two sections: The first focuses on the use of DNA for future nanoelectronic devices, whereas the second relates to recent developments in the fields of biodiagnostics and genomincs.

"DNA-Mediated Assembly of Metal Nanoparticles: Fabrication, Structural Features, and Electrical Properties" is the title of the first chapter of the first section. It is a great challenge to organize nanoparticles in one to three dimensions in order to study the electronic and optical coupling between the particles, and to even use these coupling effects for the set-up of novel nanoelectronic, diagnostic or nanomechanical devices. Here the authors describe the principles of DNA-based assembly of metal nanoparticles in one, two, and three dimensions together with structural features, and summarize different methods of liquid-phase synthesis of metal nanoparticles as well as their functionalization with DNA. Concepts, which have been developed up to now for the assembly are explained, whereas selected examples illustrate the electrical properties of these assemblies as well as potential applications.

The second chapter, "DNA-Based Nanoelectronics" reports about the exploration of DNA to implement nanoelectronics based on molecules. The unique properties in self-assembling and recognition in combination with well established biotechnological methods makes DNA very attractive for concepts of auto-organizing nanocircuits. Nevertheless, the conductivity of DNA is still under debate. Here the author briefly reviews the state-of-the-art knowledge on this topic.

The first chapter of the second section, entitled "DNA Detection with Metallic Nanoparticles" draws attention to the development of detection schemes with high specificity and selectivity needed for the detection of biomolecules. Here, the authors describe the use of metal nanoparticles as markers to overcome some of the obstacles of the classical DNA labeling techniques. The unique properties of nanoparticles can be used for a variety of detection methods such as optical, electrochemical, electromechanical, or electrical detection methods. In this chapter the authors give an overview of the use of metal nanoparticles as labels for DNA detection in solution and in surface-bound assays.

Finally, the last chapter of this part of the book, "Label-Free, Fully Electronic Detection of DNA with a Field-Effect Transistor Array," gives an introduction into label-free detection of DNA with an electronic device. Electronic biosensors based on field-effect transistors (FET), offer an alternative approach for the direct and time-resolved detection of biomolecular binding events, without the need to label the target molecules. These semiconductor devices are sensitive to electrical charge variations that occur at the surface/electrolyte interface and on changes of the interface impedance. Using the highly specific hybridization reaction of DNA molecules, which carry an intrinsic charge in liquid environments, unknown—so-called target—DNA sequences can be identified.

DNA for Electronics

1

DNA-Mediated Assembly of Metal Nanoparticles: Fabrication, Structural Features, and Electrical Properties

Monika Fischler, Melanie Homberger, and Ulrich Simon

1 INTRODUCTION

Many different synthetic routes have been developed in order to obtain metal nanoparticles of different sizes and shapes. The evolution of high-resolution physical measurements together with the elaboration of theoretical methods applicable to mesoscopic systems inspired many scientists to create fascinating ideas about how these nanoparticles can provide new technological breakthroughs; for example, in nanoelectronic, diagnostic, or sensing devices (de Jongh 1994; Schön and Simon 1995; Simon 1998; Feldheim and Foss 2002; Schmid 2004; Willner and Katz 2004; Rosi and Mirkin 2005). Nanoparticles with a diameter between one and several tens of nanometres possess an electronic structure that is an intermediate of the discrete electronic levels of an atom or molecule and the band structure of a bulk material. The resulting size-dependent change of physical properties is called the *quantum size effect* (QSE) or *size quantization effect* (Halperin 1986).

A. Offenhäusser and R. Rinaldi (eds.), *Nanobioelectronics - for Electronics, Biology, and Medicine*,
DOI: 10.1007/978-0-387-09459-5_2,

This behavior raises fundamental questions about the design of "artificial molecules" or "artificial solids" built up from nanoscale subunits which finally lead to a new state of matter. Therefore, ordered assemblies of uniform nanoparticles in one, two, or three dimensions are required. Such arrays of nanoparticles exhibit delocalized electron states that depend on the strength of the electronic coupling between the neighboring nanoparticles, whereas the electronic coupling depends mainly on the particle size, the particle spacing, the packing symmetry, and the nature and covering density of the stabilizing organic ligands (Remacle and Levine 2001).

Thus, it is a great challenge to organize nanoparticles in one to three dimensions in order to study the electronic and optical coupling between the particles, and to even utilize these coupling effects for the set-up of novel nanoelectronic, diagnostic, or nanomechanical devices (Willner and Katz 2004).

This chapter focuses on how DNA can be used as a construction material for the controlled assembly of metal nanoparticles. The enormous specificity of Watson-Crick base-pairing together with the chemists ability to synthesize virtually any DNA sequence by automated methods allow the convenient programming of artificial DNA architectures. Furthermore, short DNA fragments (up to approximately 100 nm) possess great mechanical rigidity. Thus, upon using short DNA fragments the DNA effectively behaves like rigid rod spacers between two tethered functional molecular components (e.g., nanoparticles). Moreover, DNA displays a relatively high physicochemical stability. Hence, DNA holds the promise of allowing the bottom-up self-assembly of complex nanodevices, where, for example, in the course of further miniaturization, conductive DNA-based structures could reduce time and costs in future nanofabrication (Stoltenberg and Woolley 2004).

We aim to acquaint the reader with the principles of DNA-based assembly of metal nanoparticles. Starting with a brief introduction into the different methods of liquid-phase synthesis of metal nanoparticles and their functionalization with DNA, we give an overview on the assembly of nanoparticles in one, two, and three dimensions. The structural features and electrical properties will be exemplarily described together with emerging applications.

2 MATERIALS SYNTHESIS

2.1 LIQUID PHASE SYNTHESIS OF METAL NANOPARTICLES

The common way for the synthesis of metal nanoparticles is the reduction of soluble metal salts in the presence of stabilizing ligand molecules (typically in excess) in solution (Fig. 1.1).

The reduction is achieved either by suitable reducing agents (e.g., hydrogen, boron hydride, methanol, citric acid, and others) or electrochemically. In order to stabilize the formed nanoparticles it is necessary to perform the reduction in the presence of molecules that are able to bind to the nanoparticles surface. These are all molecules with electron donor functionalities (e.g., carboxylates, amines, phosphines, thiols). The stabilization effect refers to sterical and electrostatic effects. Sterical stabilization means that the protecting molecules surround the nanoparticles comparable to a protective shield due to the required space of the molecules. Electrostatic stabilization refers to coulombic repulsion between the particles caused by the charge introduced by the ligand.

The protected metal nanoparticles synthesized this way can be further modified by ligand exchange reactions. This allows varying the nanoparticle properties (e.g., solubility) or the chemical functionality of the nanoparticle system.

The great variety of different reducing agents together with the great variety of different types of stabilizing molecules has led to a huge diversity of metal nanoparticles with different sizes, shapes, and ligand molecules. In the following, the preparation of selected metal nanoparticles is exemplarily described. For detailed overviews on the synthetic routes and surface modification methods one could refer to Schmid, Daniel and Astruc, and Richards and Boennemann (Daniel and Astruc 2004; Schmid 2004; Richards and Boennemann 2005).

2.1.1 REDUCTION OF SOLUBLE METAL SALTS WITH REDUCING AGENTS

For a long time the most popular route for synthesizing metal nanoparticles in the liquid phase was the reduction of $HAuCl_4$ with sodium citrate in aqueous solution, a route that was first reported in 1951 (Turkevitch et al. 1951). This route allowed the preparation of gold nanoparticles with sizes ranging from 14.5 ± 1.4 to 24 ± 2.9 nm. Thereby, the sizes of the formed nanoparticles could be controlled by the ratio of the gold precursor and the citrate. This method is still often used due to the fact that the citrate ligand can easily be exchanged and, thus, further modifications of the nanoparticle surface are enabled (see Chapter 2.2).

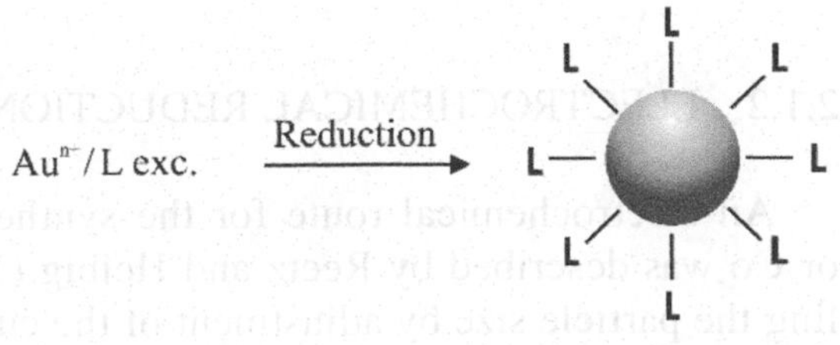

Au^{n+}/L exc. —Reduction→

FIG. 1.1. General reaction scheme for the preparation of metal nanoparticles via reduction of a metal salt in the presence of stabilizing ligand molecules (L).

In 1995 Möller and co-workers introduced an approach utilizing amphiphilic block copolymers as templates for the preparation of small gold nanoparticles of a diameter of 2.5, 4, and 6 nm (Spatz et al. 1995; Spatz, et al. 1996). Amphiphilic block copolymers tend to form micelles in solvents that dissolve only one block of the co-polymer well. The shape and stability of the micelles depend on the solvent (polar or non-polar), the relative composition of the block co-polymer, and the concentration. In their approach Möller and co-workers used symmetrical polystyrene-b-polyethylene oxide (PS-b-PSO). Under the conditions employed, this block co-polymer assembled to spherical micelles in toluene. Upon addition of the metal salt precursor $LiAuCl_4$, the Li^+ ions formed a complex with the polyethylene oxide block, whereas the tetrachloroaurate ions were bound as counter-ions within the core of the micelles. After reduction of the metal ions either by adding hydrazine or initiating the electron beam of the TEM, nanoparticles were formed inside the micellar core. The size of the formed nanoparticles depended on the size of the micelle and the loading ratio $LiAuCl_4$/PS-b-PSO. This approach provides a good tool for the formation of polymer films containing gold nanoparticles of defined size.

The most prominent example for the synthesis of gold nanoparticles with a narrow size-distribution or even uniformity is the preparation of the so-called Schmid-cluster: $Au_{55}(PPh_3)_{12}Cl_6$ (Schmid et al. 1981). The prominence rises from the quantum size behavior of the cluster, a fact that makes these clusters promising particles for future nanoelectronic applications (Schmid 2004). The cluster was prepared by the reduction of the metal salt $Au(PPh_3)Cl$ with in situ formed B_2H_6 and could be isolated as black microcrystalline solid, and characterized by TEM and small-angle X-ray diffraction (Schmid et al. 1999). This cluster is an example of a so-called full-shell cluster. Full-shell clusters are considered to be constructed by shells, each having $10\,n^2 + 2$ atoms (n = number of shells) (Schmid et al. 1990; Schmid 2004). Further examples for full-shell clusters are $[Pt_{309}phen^*_{36}O_{30}]$ and $[Pd_{561}phen_{36}O_{200}]$ (phen* = bathophenantroline and phen = 1,10-phenantroline) (Vargaftik et al. 1985; Schmid et al. 1989; Moiseev et al. 1996). The Pt_{309} cluster is synthesized by the reduction of Pt(II)acetate with hydrogen in the presence of phenantroline and following oxidation with O_2. The Pd_{561} cluster is one product of the analogous reduction of Pd(II) acetate with hydrogen in the presence of phenantroline or bathophenantroline, respectively.

2.1.2 ELECTROCHEMICAL REDUCTION OF METAL SALTS

An electrochemical route for the synthesis of nanoparticles from Pd, Ni, or Co was described by Reetz and Helbig (1994). This route allowed controlling the particle size by adjustment of the current density. The electrochemical setup was a two-electrode one, in which the anode consisted of the bulk metal and the supporting electrolyte contained tetraalkylammonium salts, which served as ligand molecules. The process itself can be described as follows: The bulk metal is oxidized at the anode, the metal cations migrate to the cathode, and a consecutive reduction takes place, resulting in the formation of the tetra-alkylammonium-stabilized nanoparticles. Using this technique, particle sizes in the range of 1.4 to 4.8 nm with a narrow size distribution could be obtained. One advantage seems to be the broad variation range of the corresponding ligand shell. Since the tetraalkylamonium ions are added to the reaction mixture, the thickness of the ligand shell can be varied by changing the length of the alkyl chain.

2.2 PREPARATION OF DNA-FUNCTIONALIZED METAL NANOPARTICLES

For the construction of nanoparticle assemblies in one, two, and three dimensions DNA oligomers as ligand molecules have become an important tool. The design and synthesis of the DNA oligomers as well as of other native and non-natural nucleic acid derivatives is a routine technology today. DNA sequences up to 120 nucleotides in length, modified with a large variety of chemical substituents, such as amino- and thiol groups attached to the 3′ or 5′-terminus, are readily available by a multitude of commercial suppliers. This chapter outlines a selection of methods to functionalize nanoparticles with DNA oligomers. An overview on general procedures is given in Fig. 1.2.

Gearheart and co-workers described the functionalization of citrate stabilized gold nanoparticles with unmodified DNA oligomers by ligand exchange (Fig. 1.2A). Here, the oligomer with the negatively charged phosphate backbone binds electrostatically to the gold nanoparticles and thereby replaces citrate ligands on the nanoparticles' surface (Gearheart et al. 2001). Thus, the binding ability of the DNA depends on the nanoparticle size and curvature as well as on the kinked, bent, or straight DNA morphology.

Numerous other protocols for the synthesis of DNA-modified gold nanoparticles are related to the initial description of these materials by Mirkin and co-workers who used the chemisorption of thiol- or amino-functionalized oligomers on the gold nanoparticle (Mirkin et al. 1996). This method is displayed in Fig. 1.2B. Briefly, citrate-stabilized gold nanoparticles are mixed with DNA oligomers derivatized with alkylthiolgroups at the 3′- or 5′-terminus, incubated for prolonged times up to several days, and purified by repeated centrifugation to remove unbound oligomers in the supernatant. The resulting DNA-functionalized gold nanoparticles are water soluble and stable for months.

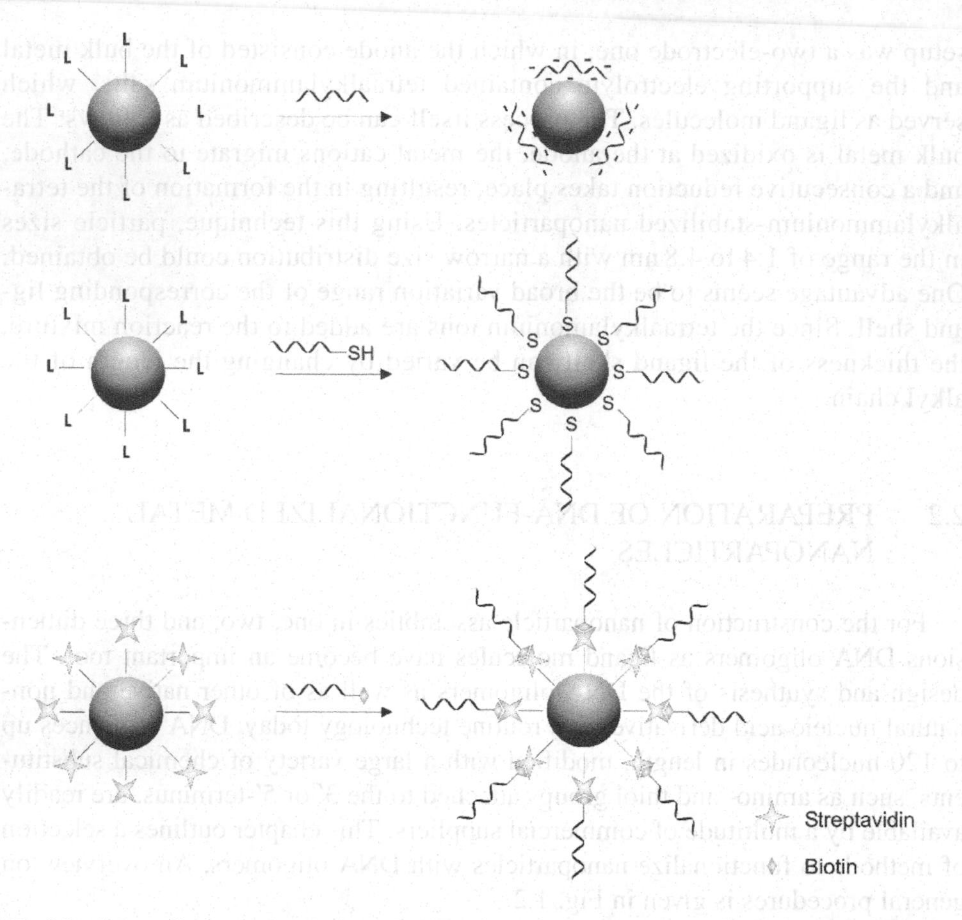

FIG. 1.2. Schematic presentation of three different principles for the preparation of DNA-oligomer functionalized nanoparticles. **A.** Via ligand exchange: Oligomers are attached electrostatically to the nanoparticle surface. **B.** Via ligand exchange: Thiol-terminated oligomers are attached covalently to the nanoparticle surface. **C.** Via streptavidin/biotin system: Biotinylated oligomers bind specifically to streptavidin-modified particles.

Slight variations of this protocol have been reported again by Mirkin and co-workers and Niemeyer and co-workers (Storhoff et al. 1998; Niemeyer et al. 2003; Hazarika et al. 2004). The latter reports the synthesis of DNA-modified gold nanoparticles that contain more than one single-stranded oligo sequence (Fig. 1.3). The number of different sequences attached to the nano-particle ranges from two (difunctional) up to seven (heptafunctional). These oligofunctional gold nanoparticles reveal almost unaltered hybridization capabilities compared with conventional monofunctional conjugates. Because of the extraordinary specificity of Watson-Crick base pairing, the various

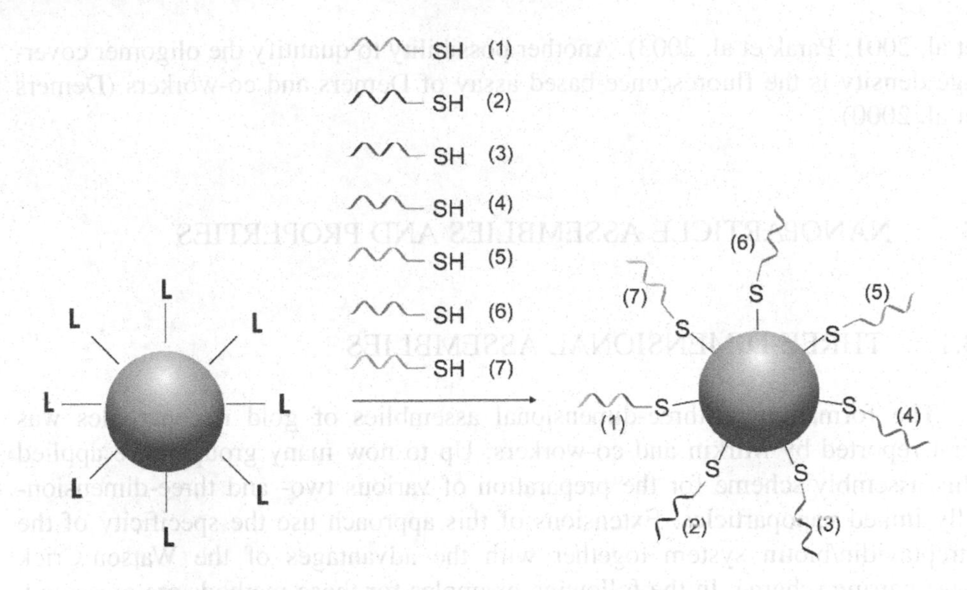

FIG. 1.3. Synthesis of oligofunctional DNA-gold nanoparticles by ligand exchange with different thiolated oligos (1–7) yielding up to heptafunctional particles.

oligonucleotide sequences can therefore be individually and selectively addressed as members of an orthogonal coupling system present at the particle's surface. Applications of such oligofunctional DNA gold nanoparticles are reported later in this chapter.

Further approaches for the preparation of DNA-functionalized gold nanoparticles including the use of polymers (Chen et al. 2004), dithiol- (Letsinger et al. 2000), and trithiol-linkers (Li et al. 2002) in between the DNA moiety and the gold particle have been reported.

Other methods for the coupling of DNA oligomers and a large variety of other biomolecules to the nanoparticle surfaces take advantage of the highly specific binding affinity of the streptavidin (STV)/biotin system (Niemeyer 2001a; Cobbe et al. 2003; Willner and Katz 2004). Streptavidin offers four native binding sites for biotin and therefore serves as an ideal linker between nanoparticles and biotinylated DNA oligomers or other biomolecules that are modified with biotin moieties (Fig. 1.2C).

For many applications it is essential to purify the DNA-gold nanoparticles and quantify the density of oligomer coverage on their surface. This can be achieved by gel electrophoresis for example, as described by Alivisatos and co-workers (Zanchet et al. 2001; Parak et al. 2003). Another possibility to quantify the oligomer coverage density is the fluorescence-based assay of Demers and co-workers (Demers et al. 2000).

3 NANOPARTICLE ASSEMBLIES AND PROPERTIES

3.1 THREE-DIMENSIONAL ASSEMBLIES

The formation of three-dimensional assemblies of gold nanoparticles was first reported by Mirkin and co-workers. Up to now many groups have applied this assembly scheme for the preparation of various two- and three-dimensionally linked nanoparticles. Extensions of this approach use the specificity of the streptavidin/biotin system together with the advantages of the Watson-Crick base pairing scheme. In the following examples for these methods are presented. Furthermore, some properties of the resulting DNA-based networks are summarized exemplarily.

In the original work Mirkin and co-workers used 13 nm gold nanoparticles that were modified according to the method described in Chapter 2.2 with non-complementary thiol-terminated oligonucleotides. Upon the addition of a double-stranded DNA containing two single-stranded ends ("sticky ends"), complementary to the particle-bound DNA, aggregation due to DNA hybridization occurred (Fig. 1.4A) (Mirkin et al. 1996). This aggregation process became visible in the slow precipitation of the macroscopic DNA-nanoparticle network and was shown to be reversible. Because in this route the nanoparticle surface is covered with multiple DNA molecules, the aggregates are two- or three-dimensionally linked. Evidence of the assembly process can be given by TEM images (Fig. 1.4B and C). Typical images reveal close-packed assemblies of the colloids with uniform particle separations of about 6 nm, corresponding to the length of the double-stranded DNA linker.

In a published extension of their work Mirkin and Li showed that care has to be taken if nanoparticles modified with deoxyguanosin-rich DNA strands are used for the assembly process. Within this study they showed that in the case of deoxyguanosin-rich DNA-modified nanoparticles self-assembly already occurs upon increasing the buffer salt concentration, and stable networks are formed in the presence of potassium (Li and Mirkin 2005).

The DNA hybridization scheme developed by Mirkin and co-workers was also applied for the preparation of nanoparticle networks comprised of different types of nanoparticles. For example gold nanoparticles of either 31 or 8 nm diameter

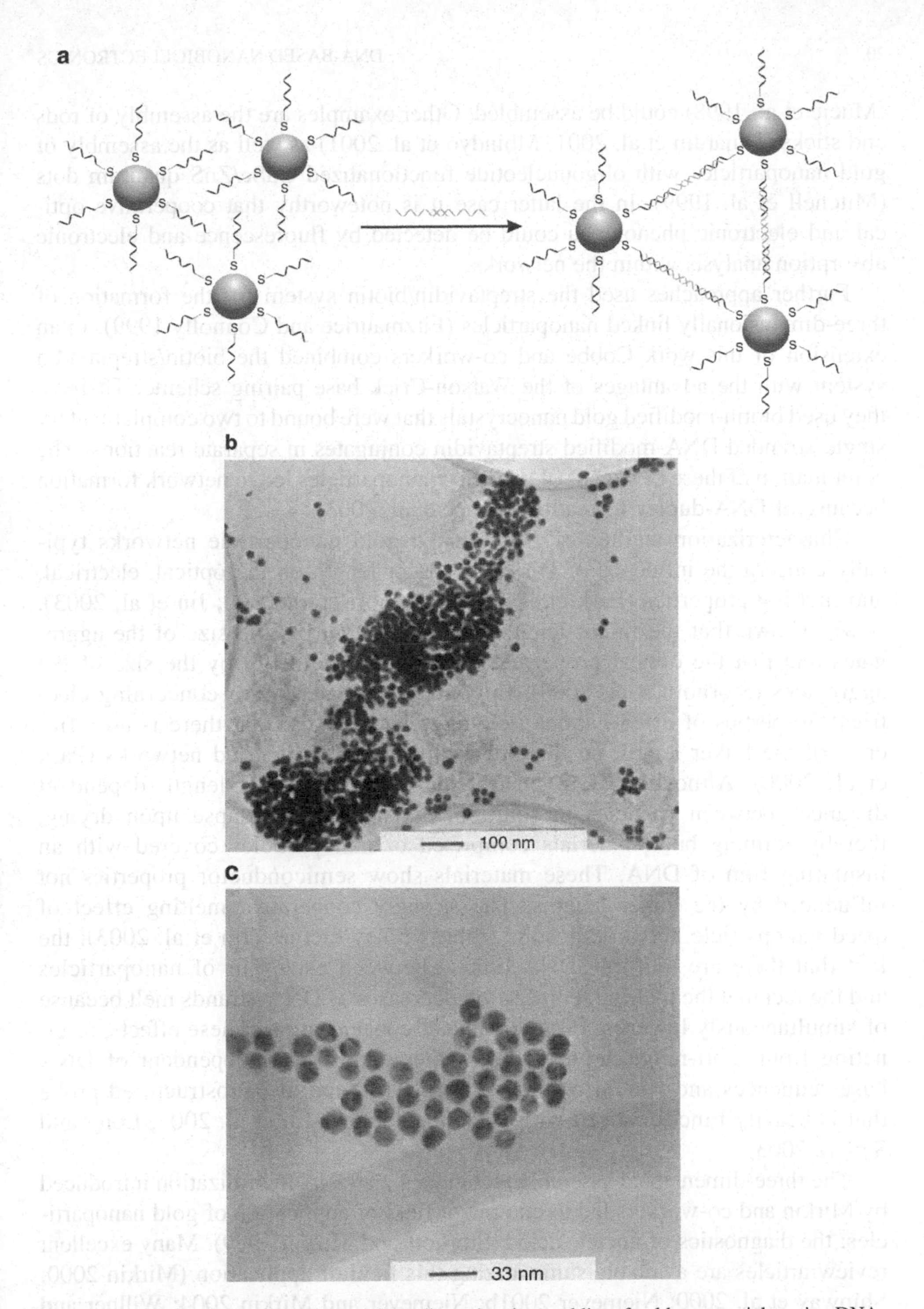

FIG. 1.4. **A.** Scheme for the formation of three-dimensional assemblies of gold nanoparticles using DNA hybridization. **B.** TEM image of an aggregated DNA/colloid hybrid material. **C.** TEM image of a two-dimensional colloid aggregate showing the ordering of the DNA linked gold nanoparticles (**B** and **C** reprinted from Mirkin et al. 1996 with permission of the Nature Publishing Group.).

(Mucic et al. 1998) could be assembled. Other examples are the assembly of rods and sticks (Dujardin et al. 2001; Mbindyo et al. 2001) as well as the assembly of gold nanoparticles with oligonucleotide functionalized CdSe/ZnS quantum dots (Mitchell et al. 1999). In the latter case it is noteworthy that cooperative optical and electronic phenomena could be detected by fluorescence and electronic absorption analysis within the networks.

Further approaches used the streptavidin/biotin system for the formation of three-dimensionally linked nanoparticles (Fitzmaurice and Connolly 1999). In an extension of this work Cobbe and co-workers combined the biotin/streptavidin system with the advantages of the Watson-Crick base pairing scheme. Thereby, they used biotin-modified gold nanocrystals that were bound to two complementary single-stranded DNA-modified streptavidin conjugates in separate reactions. The combination of these two types of modified nanoparticles led to network formation because of DNA-duplex formation (Cobbe et al. 2003).

Characterization studies of DNA-linked gold nanoparticle networks typically concern the influence of the DNA spacer length on the optical, electrical, and melting properties (Park et al. 2000; Storhoff et al. 2000; Jin et al. 2003). It was shown that the linker length kinetically controls the size of the aggregates and that the optical properties are influenced directly by the size of the aggregates (Storhoff et al. 2000). In contrast to that, studies concerning electrical properties of dried nanoparticle aggregates reveal that there is no influence of the linker length on the electrical properties of dried networks (Park et al. 2000). Although SAXS clearly indicated the linker length–dependent distances between particles in solution, the networks collapse upon drying, thereby forming bulk materials comprised of nanoparticles covered with an insulating film of DNA. These materials show semiconductor properties not influenced by the linker lengths. The strongly cooperative melting effect of dried nanoparticle networks results from two key factors (Jin et al. 2003): the fact that there are multiple DNA linkers between each pair of nanoparticles and the fact that the melting temperature decreases as DNA-strands melt because of simultaneously lowering of the local salt concentration. These effects, originating from short-range duplex–duplex interactions, are independent of DNA base sequences and should be universal for any type of nanostructured probe that is heavily functionalized with oligonucleotides (Jin et al. 2003; Long and Schatz 2003).

The three-dimensional assembly techniques via DNA hybridization introduced by Mirkin and co-workers led to one major field of application of gold nanoparticles: the diagnostics of nucleic acids (Storhoff and Mirkin 1999). Many excellent review articles are available summarizing this field of application (Mirkin 2000; Shipway et al. 2000; Niemeyer 2001b; Niemeyer and Mirkin 2004; Willner and Katz 2004; Rosi and Mirkin 2005).

3.2 TWO-DIMENSIONAL ASSEMBLIES

The DNA-based assembly scheme for the formation of three-dimensionally linked gold nanoparticles was also used for the formation of ordered two-dimensional nanoparticle networks. In general, the two-dimensional structures are achieved by binding the metal nanoparticles to a substrate surface via DNA hybridization. Thereby, one challenging goal is the formation of conducting two-dimensional metal nanoparticle arrays. Various approaches have been developed to achieve this. Some examples are presented in the following.

In 2002 Mirkin and co-workers presented a setup for the electrical detection of specific DNA-strands (Fig. 1.5) (Park et al. 2002). The principle of this method is analogous to the previously described formation of three-dimensional assemblies: A silicon surface between gold microelectrodes is modified with oligonucleotides. DNA-functionalized gold nanoparticles are added and in the presence of a complementary target DNA-strand the particles are immobilized on the surface due to hybridization. Thereby, particle densities of ≥420 particles per μm^2 were achieved. It turned out that these particle densities were too small to directly obtain conducting structures. So to close the gap the nanoparticles were plated with silver. Using this method Mirkin and co-workers were able to detect specific DNA-strands down to concentrations of $0.5 \cdot 10^{-12}$ mol·L^{-1} (see Fig. 1.5).

Koplin et al. introduced a method for the formation of electrically conducting monolayers of DNA-modified 15-nm gold nanoparticles (Fig. 1.6) (Koplin et al. 2006). The method is described as follows: First, an aminosilylated surface is formed via condensation of the hydroxy groups at the silicon substrate surface with 3-aminopropyltrimethoxysilane. Following this the surfaces are modified with the homobifunctional linker reagent disuccinimidylglutarate in order to bound dendritic hyper-branched poly(amidoamine) starburst monomers in the following step. The so formed polymeric dendrimer thin layer is further activated with the linker reagent disuccinimidylglutarate for the covalent attachment of 5′-aminofunctionalized DNA oligomers. Because of specific Watson-Crick base pairing, gold nanoparticles, functionalized with complementary oligonucleotides, are immobilized on substrate surfaces. By using this system an increase of the particle density on

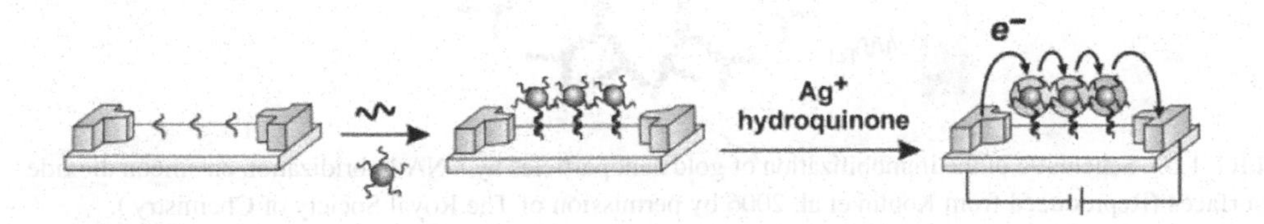

FIG. 1.5. Scheme of the array-based electrical detection of DNA (Reprinted with permission from Park et al. 2002, copyright 2002 AAAS.).

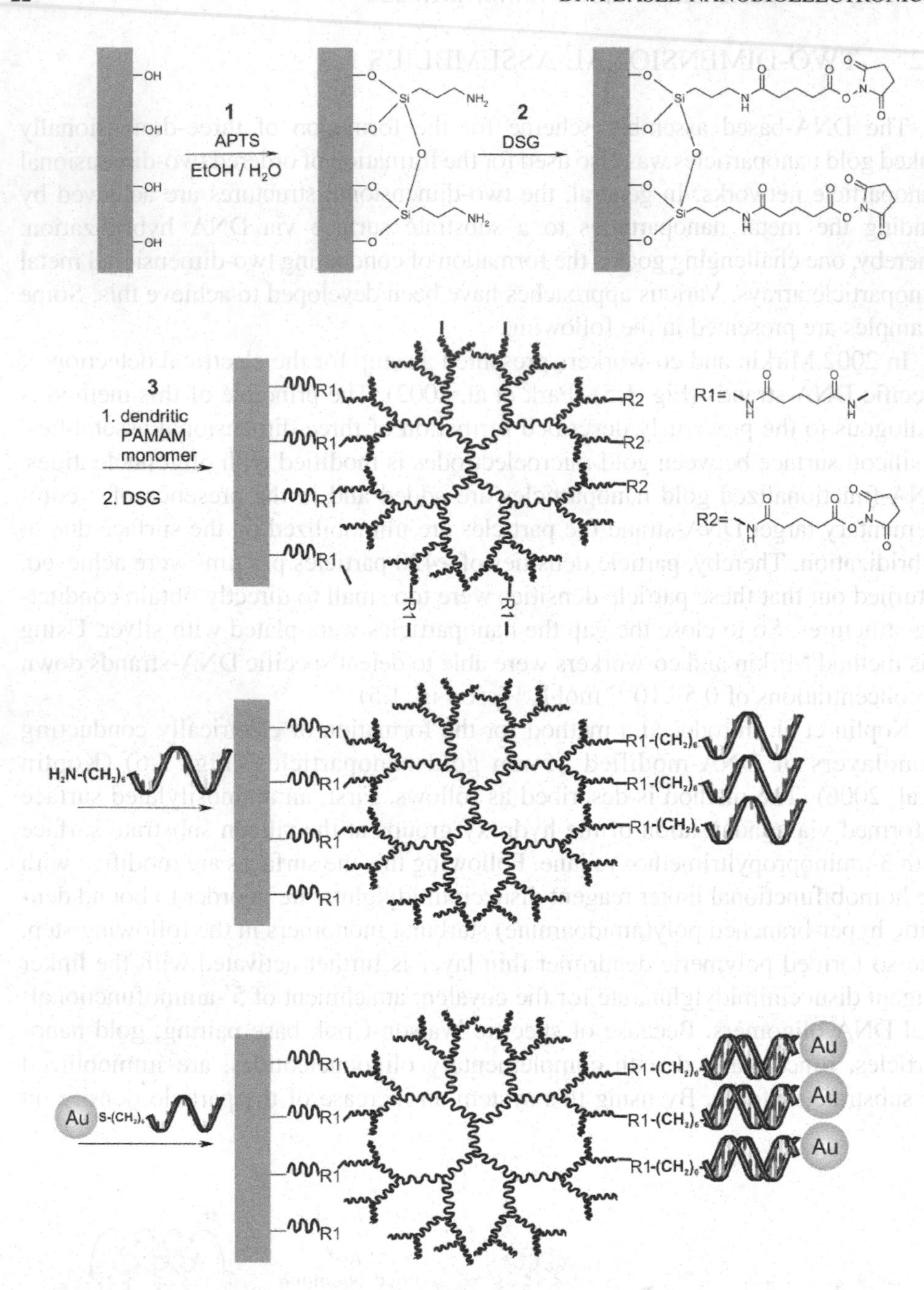

FIG. 1.6. Schematic of the immobilization of gold nanoparticles by DNA-hybridization on silicon dioxide surfaces (Reproduced from Koplin et al. 2006 by permission of The Royal Society of Chemistry.).

the substrate surfaces up to ≥850 particles per μm^2 was achieved (Fig. 1.7). The arrays were characterized by I–V measurements and temperature dependent impedance spectroscopy (IS) (Fig. 1.8). The electrical features of these layers showed pronounced field dependence as well as a thermal activation of the conductivity, reflecting classical hopping transport.

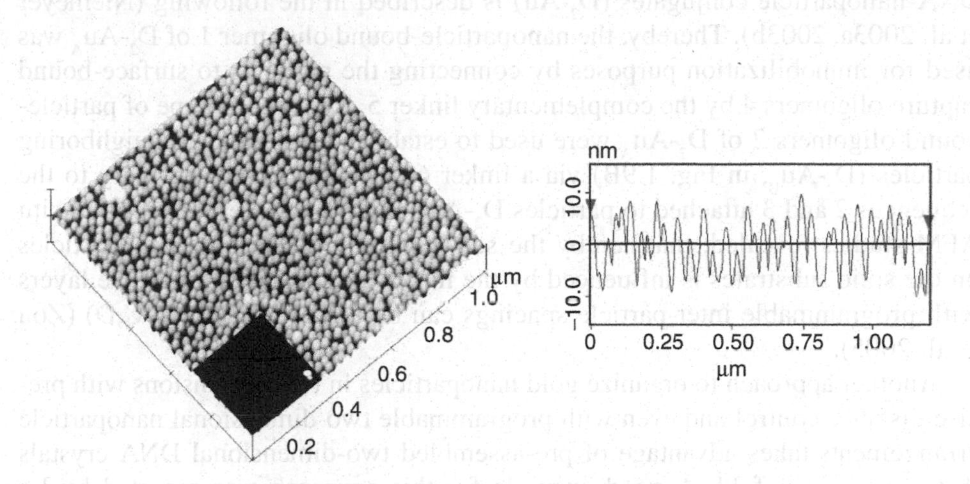

FIG. 1.7. **Left:** AFM-image of a gold nanoparticle monolayer, immobilized by DNA base pairing (≥850 particles/nm^2). The small inset shows a height image of a substrate that was modified with a non-complementary oligonucleotide (negative control). **Right:** Height profile of the surface (Reproduced from Koplin et al. 2006 by permission of The Royal Society of Chemistry.).

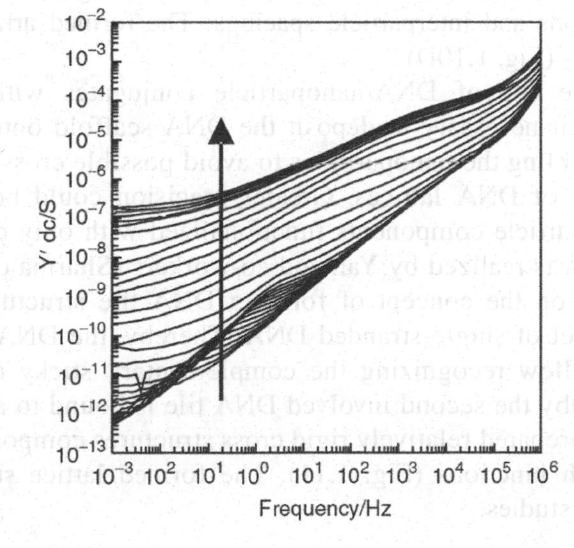

FIG. 1.8. Admittance spectra (plot of Y' vs. υ) of gold nanoparticle monolayers, immobilized on silicon substrates via specific DNA-hybridization for temperatures in between 75 and 300 K. The arrow indicates increasing temperature) (Reproduced from Koplin et al. 2006 by permission of The Royal Society of Chemistry.).

Another approach to obtain ordered two-dimensional networks with defined particle spacings utilizes oligofunctional gold nanoparticles containing different DNA sequences (Fig. 1.9A), which were introduced in Chapter 2.2. Exemplary for this approach the formation of surface-bound layers comprised of closely packed, cross-linked nanoparticles via self-assembly of difunctional DNA-nanoparticle conjugates (D_2-Au) is described in the following (Niemeyer et al. 2003a, 2003b). Thereby, the nanoparticle-bound oligomer 1 of D_2-Au_A was used for immobilization purposes by connecting the particles to surface-bound capture oligomers 4 by the complementary linker 5. The second type of particle-bound oligomers 2 of D_2-Au_A were used to establish cross-links to neighboring particles (D_2-Au_B, in Fig. 1.9B) via a linker 6, which is complementary to the sequences 2 and 3 attached to particles D_2-Au_A and D_2-Au_B, respectively. In-situ AFM studies revealed, that hereby the self-assembly of the gold nanoparticles on the solid substrates is influenced by the linker length and that particle layers with programmable inter-particle spacings can be achieved (Fig. 1.9C,D) (Zou et al. 2005).

Another approach to organize gold nanoparticles in two-dimensions with precise distance control and even with programmable two-dimensional nanoparticle arrangements takes advantage of pre-assembled two-dimensional DNA crystals that serve as scaffold. A good example for this approach was reported by Le and co-workers (Le et al. 2004). Thereby, 6 nm gold nanoparticles, functionalized with multiple single-stranded DNA oligomers, were self-assembled into high-density two-dimensional arrays by hybridization to a pre-assembled DNA scaffolding (Fig. 1.10). In this way, many thousands of DNA sequence encoded gold nanoparticles could be organized into regular arrays with defined particle locations and interparticle spacings. The formed arrays were proven by AFM studies (Fig. 1.10D).

The use of DNA/nanoparticle conjugates with multiple DNA strands made it necessary to deposit the DNA scaffold onto a solid substrate before assembling the nanoparticles to avoid possible cross-linking between multiple layers of DNA lattices. Greater precision could be achieved by the use of nanoparticle components functionalized with only one single-stranded DNA. This was realized by Yan and co-workers (Sharma et al. 2006). Their method bases on the concept of forming DNA tile structures via the self-assembly of a set of single-stranded DNA. Thereby, the DNA tiles carry "sticky ends" that allow recognizing the complementary sticky ends of other DNA tiles, whereby the second involved DNA tile is bound to a gold nanoparticle. Thus, they prepared relatively rigid cross structures composed of four four-arm DNA branch junctions (Fig. 1.11). The formed lattice structures were proven by AFM studies.

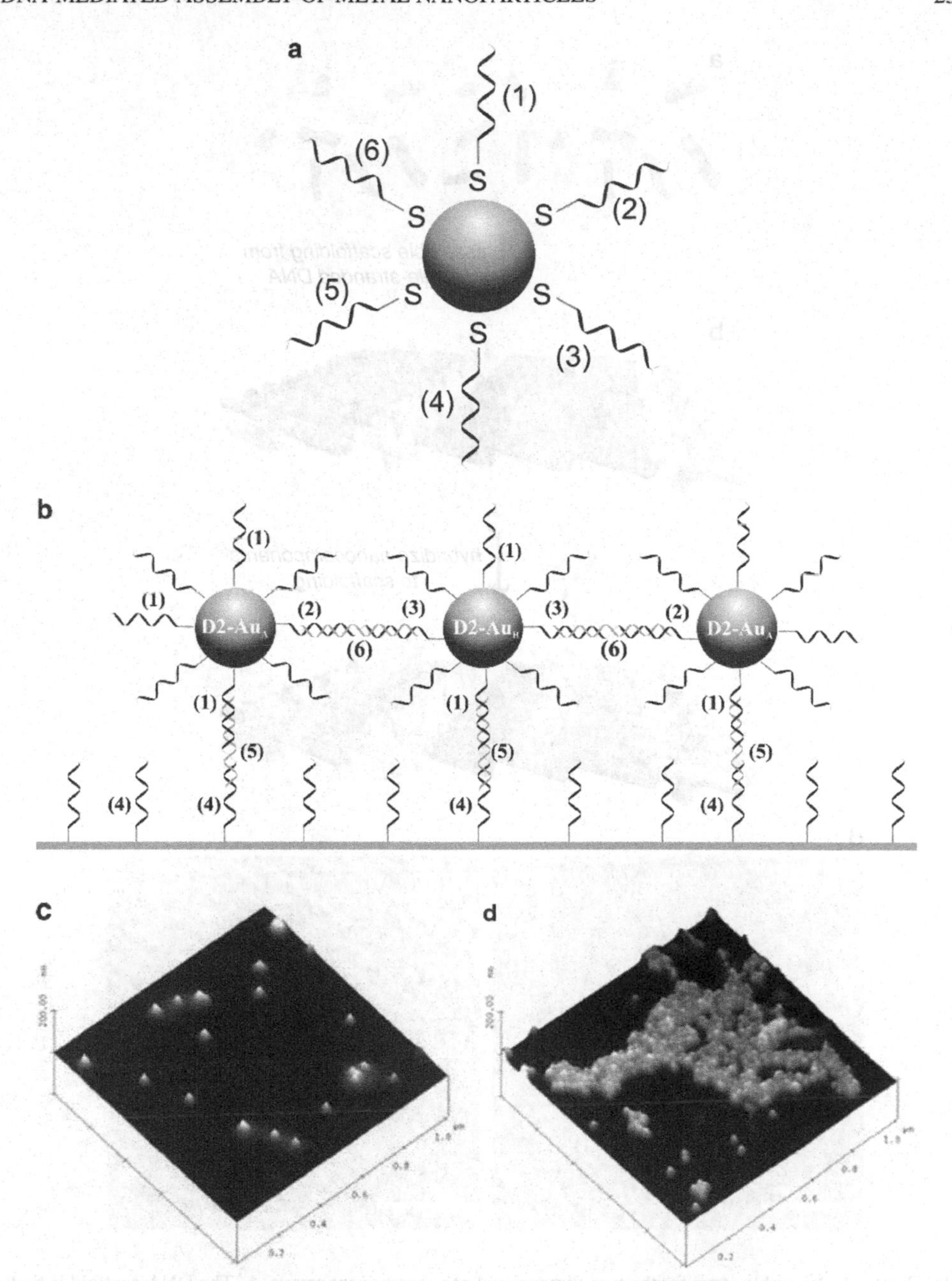

FIG. 1.9. **A.** Schematic drawing of hexafunctional DNA-gold nanoparticle conjugate D_6-Au, containing six coding oligonucleotides. **B.** Schematic drawing of DNA-directed immobilization of cross-linked difunctional DNA-gold nanoparticles D_2-Au. **C,D.** AFM images of layer assemblies prepared in the absence *(C)* or presence *(D)* of cross-linking oligonucleotide (**C** and **D** reprinted from Niemeyer et al. 2003 with permission from Elsevier.).

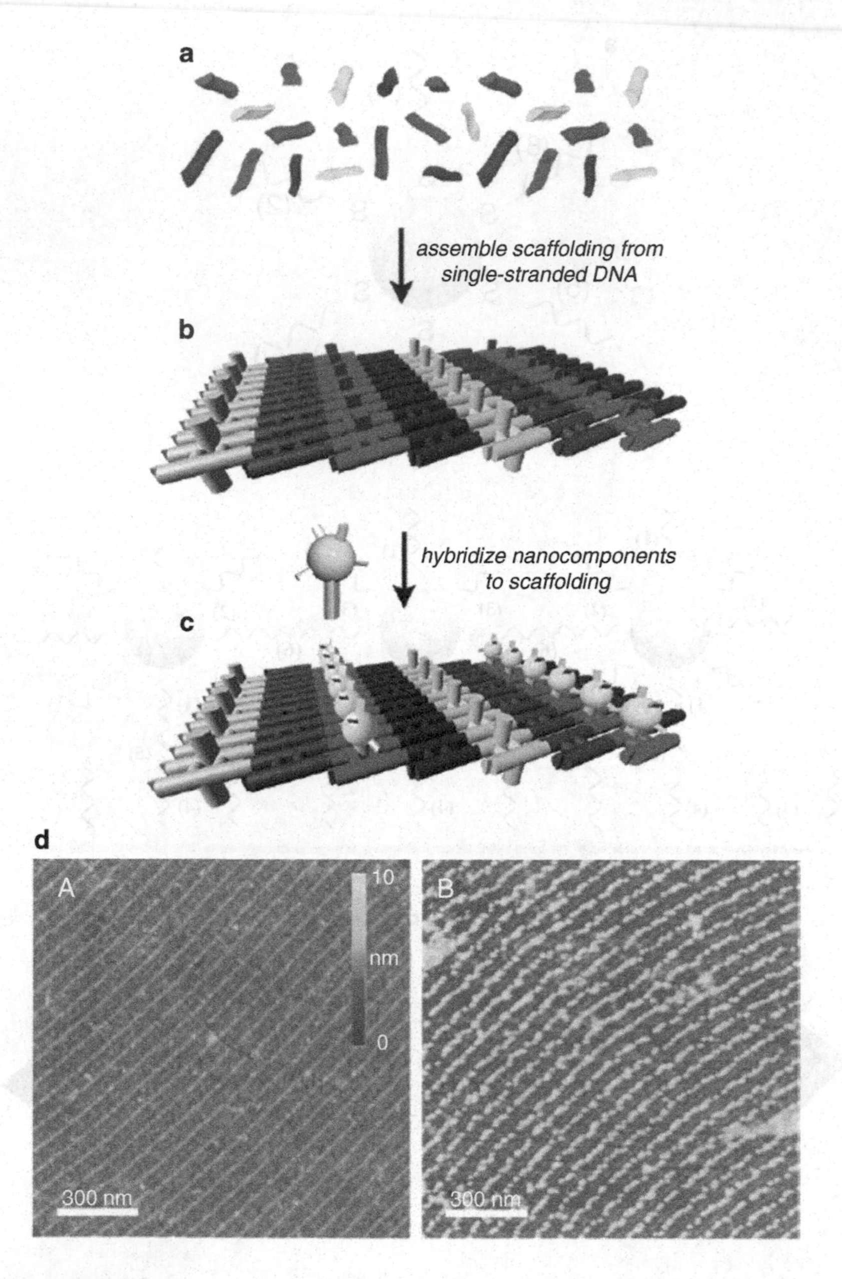

FIG. 1.10. **A.** Assembly steps for the two-dimensional nanocomponent arrays. **A.** The DNA-scaffold is first assembled in solution from a set of 21 strands. **B.** A suspension of the DNA scaffolding is deposited on mica. **C.** The scaffold is combined with DNA-encoded nanocomponents, which attach to the open hybridization sites. (Note: Although this diagram shows one nanocomponent occupying each site, single nanoparticles can also attach to multiple sites via hybridization of multiple, nanoparticle-bound strands.) **A.** Topographical AFM image of the DNA scaffolding. **D.** AFM image of the DNA scaffolding after nanocomponent assembly (Reprinted with permission from Le et al. 2004, Copyright 2004 American Chemical Society.).

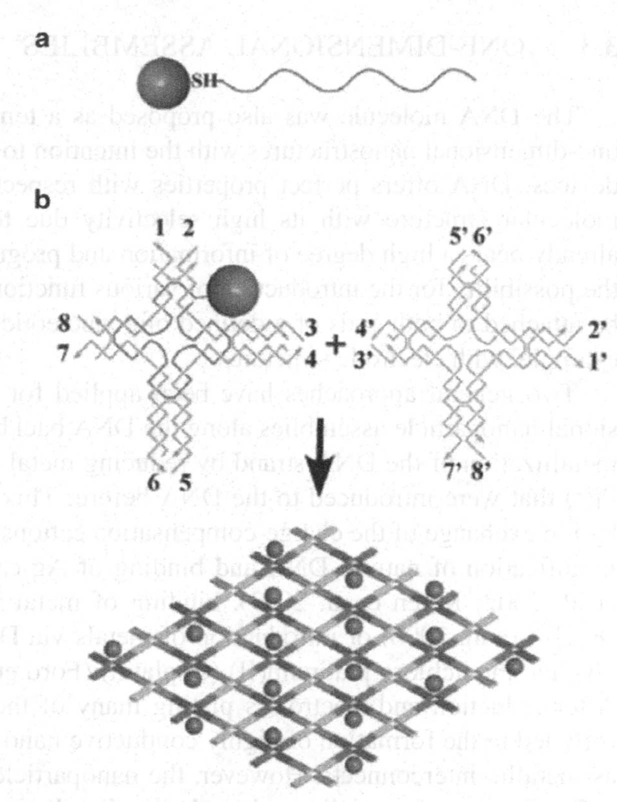

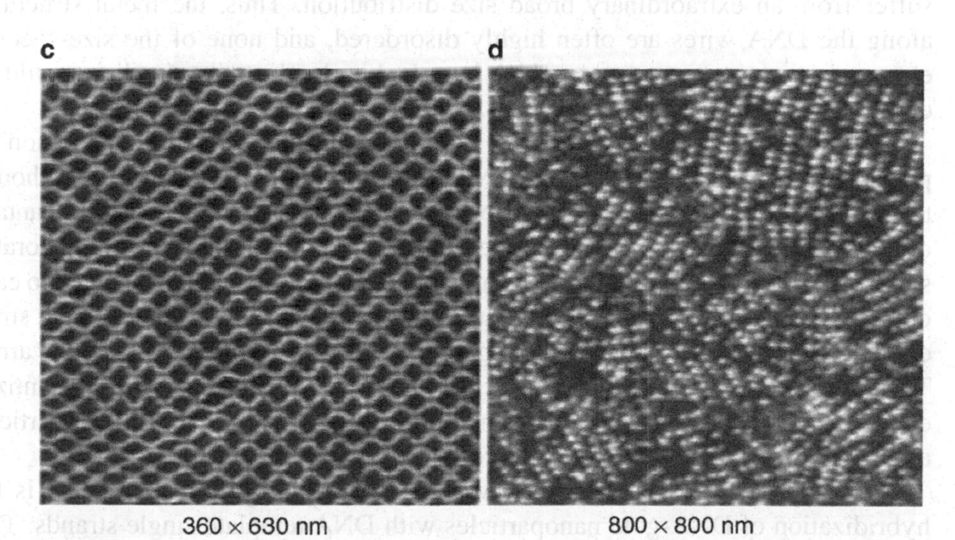

FIG. 1.11. Reaction scheme for the DNA templated assembly of periodical gold nanoparticle arrays. **A.** A 1:1 conjugate of a gold nanoparticle with a thiolated DNA-strand (isolated by electrophoresis). **B.** Hybridization of two DNA tiles via "sticky end." **C.** AFM image of the DNA scaffold without gold nanoparticles. **D.** AFM image showing the patterning of gold nanoparticles on the self-assembled DNA tile structure (A–D reprinted from Sharma et al. 2006 with permission of Wiley-VCH Verlag GmbH&Co. KGA, Weinheim, copyright 2006.).

3.3 ONE-DIMENSIONAL ASSEMBLIES

The DNA molecule was also proposed as a template for the construction of one-dimensional nanostructures with the intention to generate nanoscale electronic devices. DNA offers perfect properties with respect to this goal. The chain-like molecular structure with its high selectivity due to Watson-Crick base pairing already bears a high degree of information and programmability and further offers the possibility for the introduction of various functional groups. Again, groups can be attached to both ends of a defined oligonucleotide to contact the molecule, for example, with electrode structures.

Two general approaches have been applied for the formation of one-dimensional nanoparticle assemblies along the DNA backbone. One describes the direct metallization of the DNA-strand by reducing metal cations (e.g., Ag^{+}, Au^{3+}, Pd^{2+}, Pt^{2+}) that were introduced to the DNA before. This can be achieved for example by ion exchange of the charge-compensation cations (Braun et al. 1998), aldehyde modification of natural DNA and binding of Ag-cations to these groups (Keren et al. 2002; Keren et al. 2003), binding of metal ions to modified DNA bases (Burley et al. 2006) or introduction of metals via DNA binding Pt-comlexes like *cis*-diaminodichloro platinum(II) (cisplatin) (Ford et al. 2001; Seidel et al. 2004). After reduction and electroless plating many of these approaches have successfully led to the formation of highly conductive nanowires, which could be applied as metallic interconnects. However, the nanoparticles formed during this process suffer from an extraordinary broad size distribution. Thus, the metal structures along the DNA wires are often highly disordered, and none of the size-specific electronic transport properties, which base on single-electron tunneling, could be observed or even used in nanodevices.

For that reason only the second approach, the selective immobilization of preformed metal nanoparticles to the DNA-strand is described in detail here. Although the site selective immobilization of preformed nanoparticles can take advantage of the precise size control and defined surface chemistry of the well-elaborated synthetic routes, it is a difficult task to assemble the particles in direct contact to each other over extended domains, where the inter-particle spacing is identical and small enough to allow direct dipolar coupling or even electronic transport along the array. The approach to fulfill the requirements of one-dimensional assembly for quantized electronically transport follows different binding mechanisms between nanoparticles and DNA, which are described in the following.

One approach for the site-selective binding of nanoparticles to DNA is the hybridization of DNA-gold nanoparticles with DNA template single strands. This method allows a spatially defined immobilization of single even different nano-objects along DNA template strands benefiting from the specificity of Watson-Crick base pairing (Fig. 1.12A).

Alivisatos and co-workers showed for the first time, that a discrete number of water-soluble Au_{55}-clusters with one N-propylmaleimide ligand per cluster can couple selectively to a sulforyl-group incorporated into single-stranded DNA oligomers (Alivisatos et al. 1996). Single-stranded oligonucleotides, modified at either the 3′ or 5′ termini with a free sulforyl group, were coupled with an excess of the nanoparticles. By combination of these oligomer-functionalized nanoparticles with suitable oligonucleotide single strand templates, parallel (head-to-tail) and antiparallel dimers (head-to-head) were obtained (Fig. 1.12B). The linear alignment of the clusters and the center-to-center distance, which ranged from 2 to 6 nm, respectively, were shown by means of TEM. UV/Vis absorbance measurements indicated changes in the spectral properties of the nanoparticles as a consequence of the supramolecular organization (Loweth et al. 1999). Deng and co-workers demonstrated the formation of cluster chains with several hundreds of nanometer in length by this method (Deng et al. 2005). A template DNA single strand containing one repeated sequence was prepared by rolling circle polymerization. Gold nanoparticles

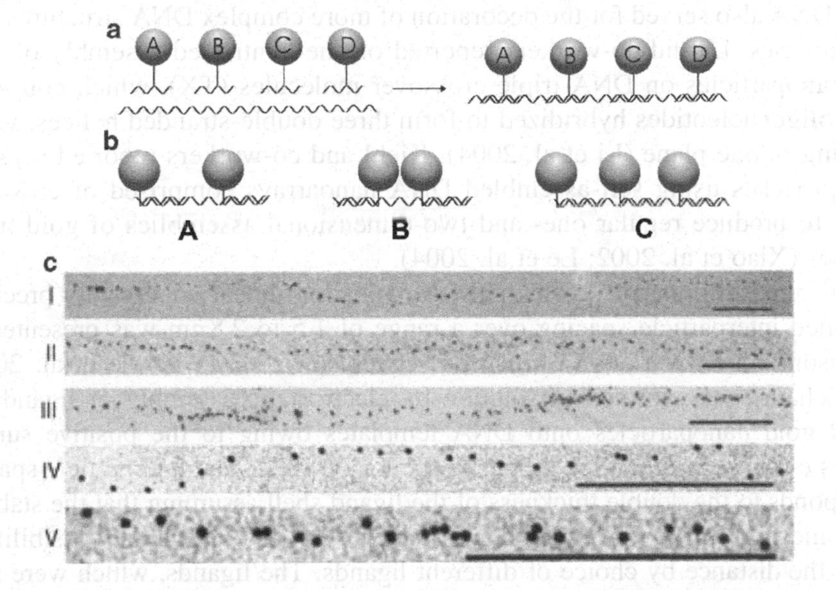

Fig. 1.12. **A.** Schematic representation of DNA-directed one-dimensional assembly of four different nanoscale building blocks to form a stoichiometrically and spatially defined supramolecular aggregate. **B.** Conjugates from gold particles (represented as shaded spheres) and 3′- or 5′-thiolated oligonucleotides allow the fabrication of head-to-head *(A)* or head-to-tail *(B)* homodimers. A template containing the complementary sequence in triplicate effects the formation of the trimer *(C)*. **C.** Long cluster chains prepared by the hybridization method. Scale bar in images I–V: 200 nm (Reprinted from Deng et al. 2005 with permission of Wiley-VCH Verlag GmbH&Co. KGA, Weinheim, copyright 2005.).

connected to single sequence containing oligomers were attached to this template strand by hybridization. The resulting chain-like nanoparticle formations are shown in Fig. 1.12C.

A variety of essential basic studies on the DNA-directed assembly of macromolecules were carried out with the biotin/streptavidin system (Niemeyer 2001). The covalently attached oligonucleotide moiety supplements the four native biotin-binding sites of streptavidin with a specific recognition domain for a complementary nucleic acid sequence. This bispecificity allows using the DNA-streptavidin conjugates as adapters to assemble basically any biotinylated compound along a nucleic acid template. It also opens the possibility to organize gold nanoparticles along DNA stands as shown by Niemeyer and co-workers (Niemeyer et al. 1998). Gold particles (1.4 nm) containing a single amino-substituent were derivatized with a biotin group, and subsequently the biotin moiety was used to organize the nanoparticles into the tetrahedral superstructure defined by the biotin-binding sites of the streptavidin. In the following, the nanoparticle-loaded proteins self-assemble in the presence of a complementary single-stranded nucleic acid carrier molecule. The strong biotin-streptavidin affinities in combination with the specific nucleic acid hybridization capabilities of the DNA also served for the decoration of more complex DNA structures with nanoparticles. Li and co-workers reported on the controlled assembly of 5 nm gold nanoparticles on DNA-triple crossover molecules (TX), which consist of seven oligonucleotides hybridized to form three double-stranded helices, which are lying in one plane (Li et al. 2004). Kiehl and co-workers reported on similar approaches using self-assembled DNA nanoarrays comprised of crossover motifs to produce regular one- and two-dimensional assemblies of gold nanoparticles (Xiao et al. 2002; Le et al. 2004).

A method to organize 1.5-nm gold particles into linear chains with precisely controlled interparticle spacing over a range of 1.5 to 2.8 nm was presented by Hutchison and co-workers (Warner and Hutchison 2003; Woehrle et al. 2004). These chains were formed in solution by electrostatic assembly of ligand stabilized gold nanoparticles onto DNA templates owing to the positive surface charges of quarternary amine groups in the ligand shell. The interparticle spacing corresponds to the double thickness of the ligand shell assuming that the stabilizing ligands are in fully extended conformation. This fact opens the possibility of tuning the distance by choice of different ligands. The ligands, which were used for this immobilization process are (2-mercaptoethyl)trimethylammonium iodide (TMAT, 1), [2-(2-mercaptoethoxy)ethyl]trimethyl-ammonium toluene-4-sulfonate (MEMA, 6), and {2-[2-(2-mercaptoethoxy)ethoxy]ethyl}-trimethylammonium toluene-4-sulfonate (PEGNME,11), resulting in interparticle distances of 1.5, 2.1, and 2.8 nm, respectively. Because the interparticle spacing is enforced by the ligand shell rather than by the scaffold, the spacing is uniform even in nonlinear sections of the chain. This method allows assembling the particles over extended areas up to 1 μm in length and a total coverage of >90%. The immobilization scheme and representative TEM images of linear nanoparticle chains of Au_n-TMAT, Au_n-MEMA, and Au_n-PEGNME, respectively, assembled on λ-DNA are displayed in Fig. 1.13.

Another example for the electrostatic-binding of gold nanoparticle to the DNA backbone has been reported by Braun and co-workers (Braun et al. 2005). Briefly,

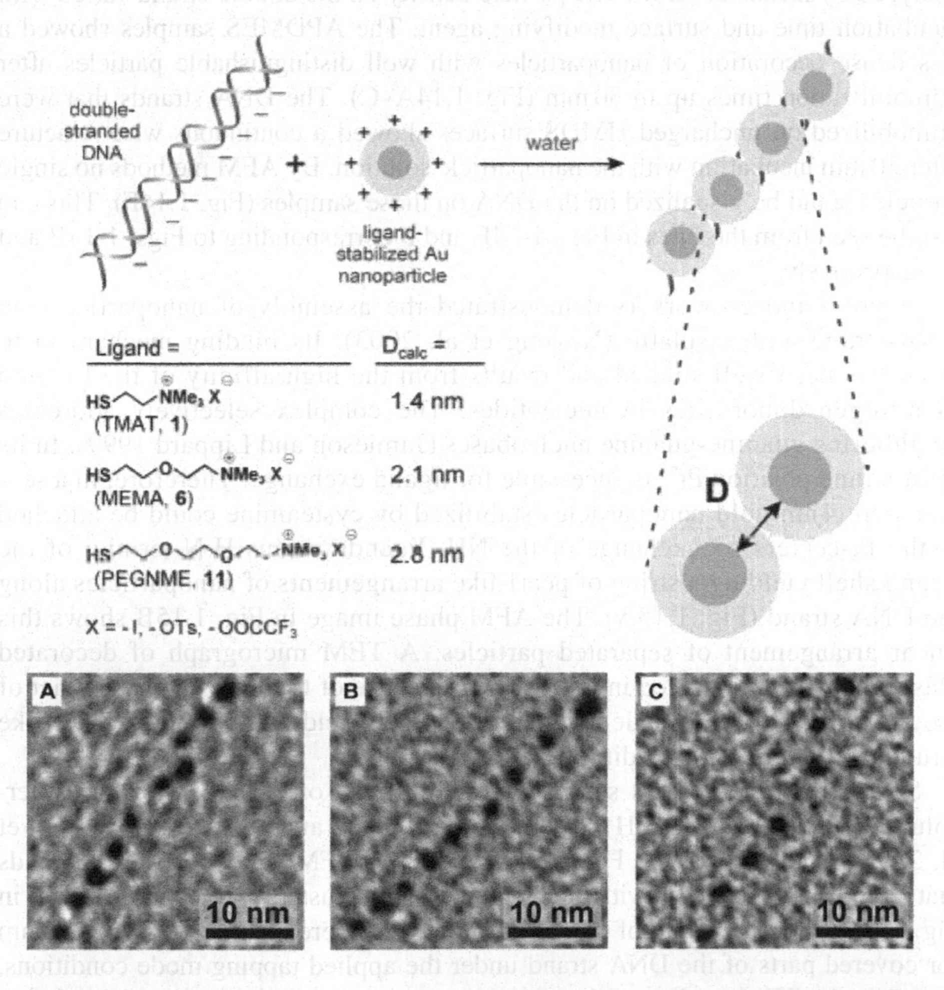

FIG. 1.13. Schematic representation of solution phase formation of linear close-packed DNA-nanoparticle assemblies. The three ligands used are shown in the figure. TEM images of linear, close-packed nanoparticle-DNA assemblies are depicted in the figure. The TEM micrographs demonstrate the correlation between the size of the three different ligands and the resulting interparticle distance (Reprinted with permission from Wöhrle et al. 2004, copyright 2004 American Chemical Society.).

3.5-nm gold nanoparticles were decorated with thiocholine by ligand exchange and furthermore bound to plasmid DNA, which was combed on aminopropyldimethylethoxysilane (APDMES) or hexamethyldisilazane (HMDS)-modified Si/SiO_2 surfaces before. Immobilization of the nanoparticles to the surface bound DNA was accomplished by placing a drop of aqueous nanoparticle solution on the substrate. The nanoparticles were attached to the DNA through the high number of positive quaternary amine surface charges in the thiocholine ligand shell, which are independent of solution pH. After washing and drying the substrates were analyzed by means of AFM. The particle density on the double-strand varied with incubation time and surface modifying agent. The APDMES samples showed a less dense decoration of nanoparticles with well distinguishable particles after immobilization times up to 30 min (Fig. 1.14A–C). The DNA strands that were immobilized on uncharged HMDS surfaces showed a continuous wire structure after 30 min incubation with the nanoparticle solution. By AFM methods no single particles could be visualized on the DNA on those samples (Fig. 1.14F). This can also be seen from the lines in Figs. 1-14E and F corresponding to Figs. 1-14B and D, respectively.

Noyong and co-workers demonstrated the assembly of nanoparticles on DNA-strand with cisplatin (Noyong et al. 2003). Its binding mechanism to DNA has been well studied and results from the high affinity of the Pt^{2+}-ion to nitrogen donor sites in nucleotides. The complex selectively addresses neighboring guanine-guanine nucleobases (Jamieson and Lippard 1999). In its intra-strand position Pt^{2+} is accessible for ligand exchange. Therefore, in a second step, 4 nm gold nanoparticles stabilized by cysteamine could be attached to the Pt-centers by exchange of the NH_3 ligands versus H_2N termini of the ligand shell yielding a string of pearl-like arrangements of nanoparticles along the DNA strand (Fig. 1.15A). The AFM phase image in Fig. 1.15B shows this linear arrangement of separated particles. A TEM micrograph of decorated plasmid DNA is displayed in Fig. 1.15C. Because of the high energy impact of the electron beam the particles melted together yielding continuous wire-like structures under TEM conditions.

Schmid and co-workers studied the interactions of λ-DNA with the water-soluble $[Au_{55}(Ph_2C_6H_4SO_3H)_{12}Cl_6]$ cluster by TEM and AFM methods (Liu et al. 2003; Tsoli et al. 2005). Fig. 1.16A depicts an AFM study of λ-DNA strands that were partly decorated with the mentioned gold cluster. The section analysis in Fig. 1.16A shows a height of 0.5 to 0.6 nm for uncovered parts and 2.05 ± 0.2 nm for covered parts of the DNA strand under the applied tapping mode conditions. Surprisingly, TEM studies of these structures revealed dramatic size degradation in the linear arrangement of the clusters. Modeling experiments showed that the degradation of cluster size is due to a conformational change of the DNA structure from the B-DNA to the A-DNA formed under ultra–high vacuum conditions.

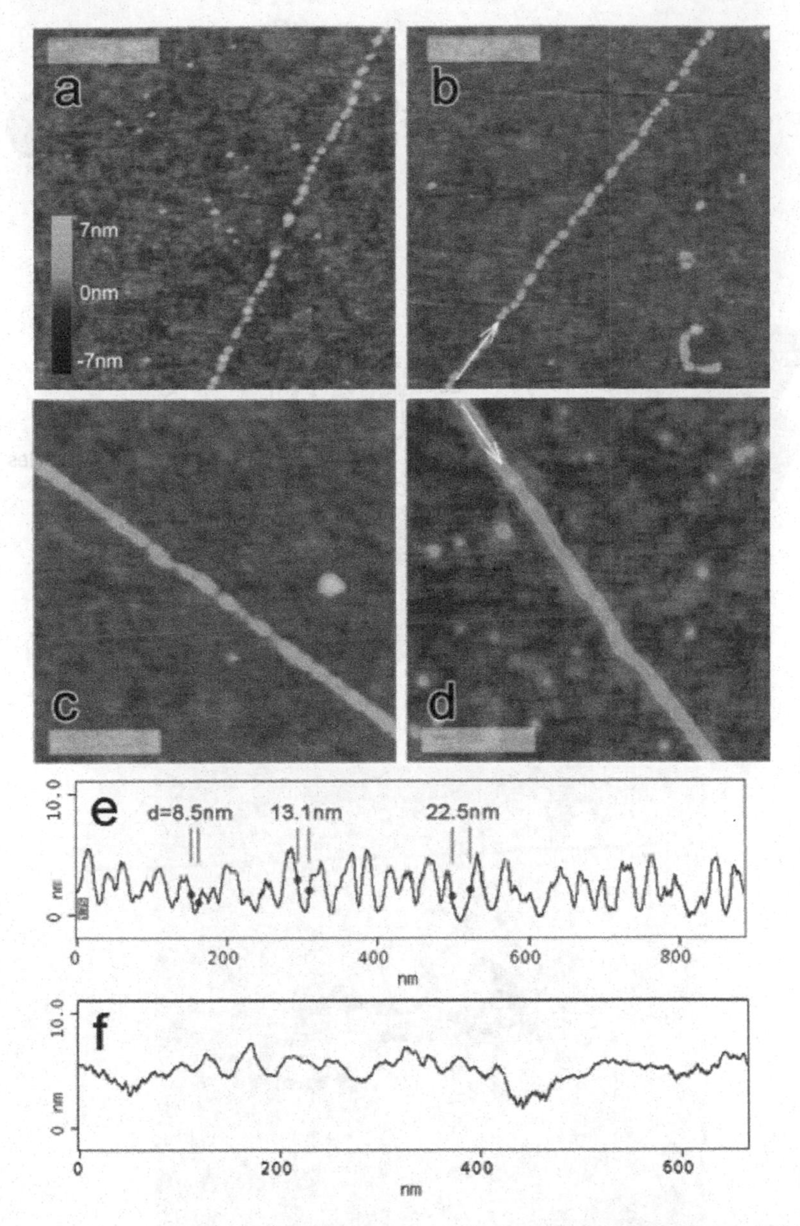

FIG. 1.14. AFM images of DNA decorated with 3.5-nm thiocholine stabilized gold nanoparticles. DNA was combed onto Si/SiO_2 surfaces that were modified with APDMES and HMDS before. A droplet containing the aqueous nanoparticle solution was deposited for variable periods of time. **A.** APDMES, 5 min deposition time. **B.** APDMES, 30 min. **C.** APDMES, 1 hour. **D.** HMDS, 30 min. The white scale bars are 200 nm for *(A)* and *(B)* and 150 nm for *(C)* and *(D)*. Height analyses along the axis of assembly in panels *(B)* and *(D)* are presented in panels *(E)* and *(F)*, respectively (Reprinted with permission from Braun et al. 2005, copyright 2005 American Chemical Society.).

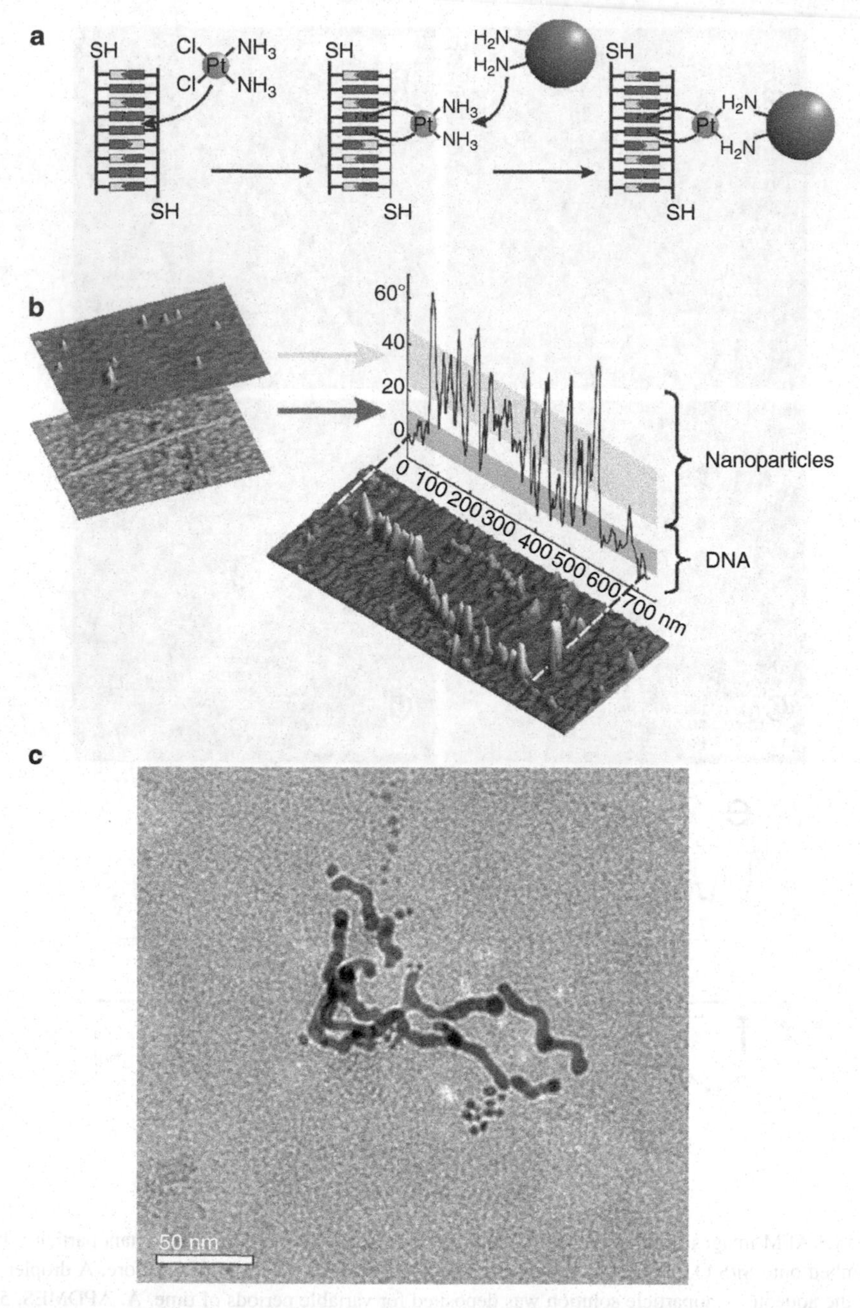

FIG. 1.15. **A.** Schematic illustration of the binding of amine-terminated nanoparticles to *cis*-Pt-modified DNA-strands. **B.** AFM image (phase contrast) of DNA aligned gold nanoparticles assembled on a silicon substrate (**A** and **B** reprinted from Noyong et al. 2003 with permission of the Materials Research Society, copyright 2003.) **C.** TEM analysis leads to continuous wire-like structures.

This conformational change is initiated by a dehydration process under electron microscopy conditions. The compression of the length of the groove from 1.43 nm in B-DNA to 0.73 nm in A-DNA corresponds with the change of the cluster diameter from 1.4 to 0.6 nm, which could be observed during the microscopy studies. Thus, it is assumed that by the change of DNA morphology the Au_{55} clusters are degraded to Au_{13} clusters, which fit into the major groove of the dehydrated A-DNA. The attachment of Au_{55} and Au_{13} clusters to the different DNA conformations is depicted in Fig. 1.16B.

An approach towards a DNA templated electronic device was introduced by Fitzmaurice and co-workers (Ongaro et al. 2004). For the formation of a DNA-based nanogap structure, a DNA template with a biotin moiety placed centrally

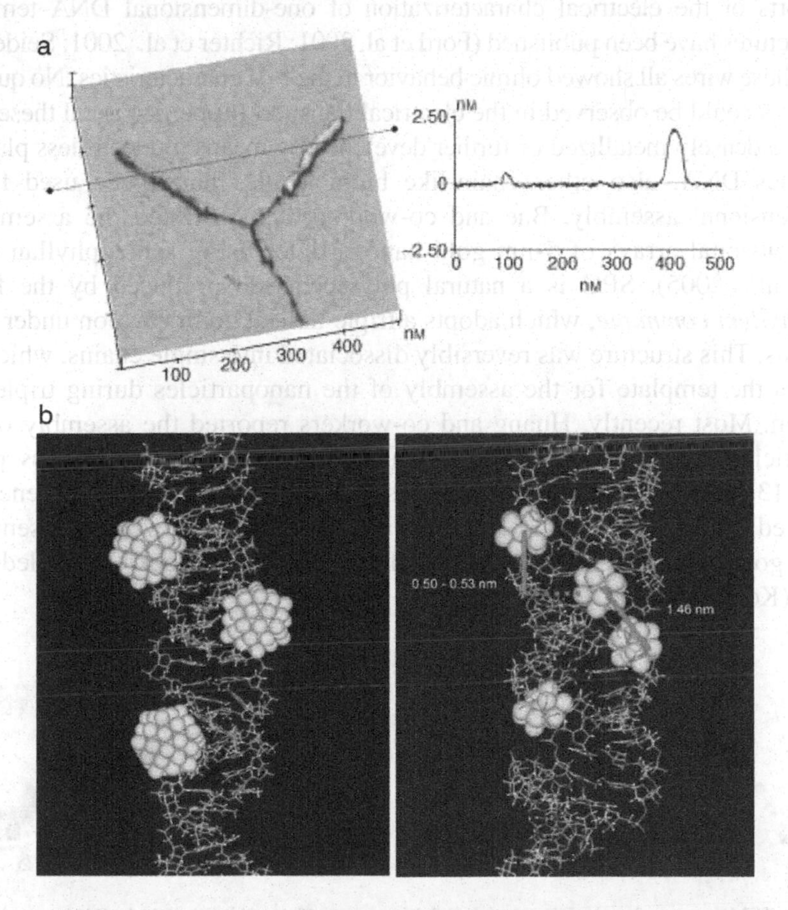

FIG. 1.16. **A.** AFM study of a DNA junction metallized with water-soluble Au_{55}-clusters. The left image shows the surface topography. The right image gives a height profile of metallized and unmetallized DNA under the applied tapping mode conditions. **B.** Model of cluster size degradation that goes along with the conformational change of DNA in the high vacuum during TEM analysis (Reprinted from Liu et al. 2003 with permission of Wiley-VCH Verlag GmbH&Co. KGA, Weinheim, copyright 2003.).

on the strand was deposited on a silicon wafer surface. The substrate was exposed to a dispersion of positively charged DMAP stabilized gold nanoparticles, as has been described by the same group for the fabrication of conductive nanowires (Ongaro et al. 2005). In a third step the surface was exposed to streptavidin, which selectively binds to the biotin function placed on the template. Thereby, the protein displaced the weakly bound nanoparticles in the vicinity of the biotin, yielding a streptavidin functionalized nanogap. By electroless gold deposition the particles were enlarged and enjoined. In a last step a biotin functionalized nanoparticle could be bound to the strepatavidin function and thus localized in the gap structure. The individual steps of the device fabrication were monitored by means of TEM and are shown in Figs. 1-17A to D.

Reports of the electrical characterization of one-dimensional DNA templated nanostructures have been published (Ford et al. 2001; Richter et al. 2001; Seidel et al. 2004). These wires all showed ohmic behavior in the I–V characteristics. No quantum size effects could be observed in the electrical transport properties as all these structures were densely metallized or further developed by means of electroless plating.

Besides DNA, also other chain-like biomolecules have been used for the one-dimensional assembly. Bae and co-workers demonstrated the assembly of one-dimensional arrays of 5-nm gold nanoparticles using schizophyllan (SPG) (Bae et al. 2005). SPG is a natural polysaccharide produced by the fungus *Schizophyllum commune*, which adopts a triple-helical conformation under native conditions. This structure was reversibly dissociated into single chains, which then served as the template for the assembly of the nanoparticles during triple-helix formation. Most recently, Huang and co-workers reported the assembly of gold nanoparticles and further development to metallic wires by electroless plating using M13 phage, a filamentous virus, as template, which had been genetically engineered (Huang et al. 2005). Koyfman and co-workers showed the assembly of cationic gold nanoparticles by a ladder of RNA-building blocks, so-called tectosquares (Koyfman et al. 2005).

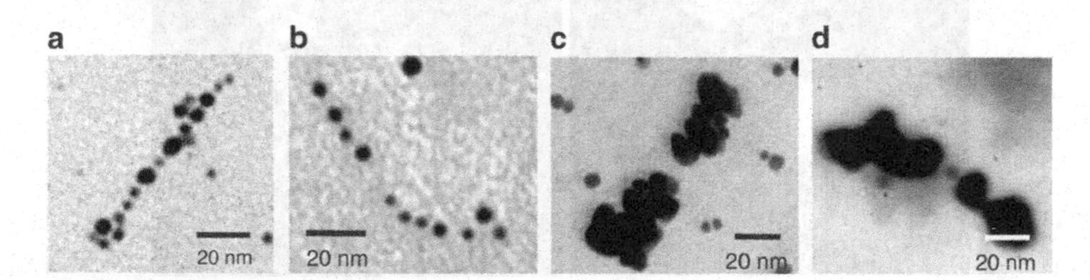

FIG. 1.17. TEM images showing the process of the nanogap formation. **A.** Biotin DNA template after incubation with DMAP stabilized nanoparticles. **B.** After exposure to Streptavidin a gap could be formed. **C.** The structure after electroless gold deposition. **D.** A biotin modified nanoparticle was localized within the gap (Reprinted with permission from Ongaro et al. with permission of Wiley-VCH Verlag GmbH&Co. KGA, Weinheim, copyright 2004.).

4 CONCLUSION

This chapter has highlighted some milestones of DNA-based assembly of metal nanoparticles in one, two, and three dimensions, which are situated at the frontier between biomolecular chemistry, inorganic chemistry, and materials science. It has been illustrated that the assembly process itself is based on the enormous binding specificity of nucleic acids and that it can be tailored by virtue of synthetic organic chemistry.

However, there is still a lot of work to be done to finally realize a reliable technology based on this approach. As a big challenge one has to face the synthesis of nanoparticle–nucleic acid conjugates that are well defined with respect to their stoichiometric composition. This is particularly important for the generation of sophisticated nanoarchitectures, what is often called "molecular engineering." Nevertheless, one may anticipate a broad scope of applications in the area of bioanalytics and advanced materials research. As a very exciting and promising direction one may judge the works of Williams and co-workers and Keren and co-workers (Williams et al. 2002; Keren et al. 2003). Both focused on the setup of higher complexity functional devices. Although Williams and co-workers used the recognition of DNA to bind carbon nanotubes to metallic contacts by self-assembly, Keren and co-workers described the self-assembly of segmented nanowires in a multi-step process to build up a molecular-based transistor element. However, there is still a great demand to develop these processes further in order to realize devices or even device architectures and to prove that these devices are robust enough for application in nanoelectronic circuitry.

REFERENCES

Alivisatos, A.P., Johnsson, K.P., Peng, X., Wilson, T.E., Loweth, C.J., Bruchez Jr., M.P. and Schultz, P.G. (1996). Organization of 'nanocrystal molecules' using DNA. Nature 382(6592), 609–611.

Bae, A.H., Numata, M., Hasegawa, T., Li, C., Kaneko, K., Sakurai, K. and Shinkai, S. (2005). 1D Arrangement of Au nanoparticles by the helical structure of schizophyllan: a unique encounter of a natural product with inorganic compounds. Angew. Chem. Int. Ed. 44(13), 2030–2033. Angew. Chem. 117(13), 2066–2069.

Braun, E., Eichen, Y., Sivan, U. and Ben-Yoseph, G. (1998). DNA-templated assembly and electrode attachment of a conducting silver wire. Nature 391(6669), 775–778.

Braun, G., Inagaki, K., Estabrook, R.A., Wood, D.K., Levy, E., Cleland, A.N., Strouse, G.F. and Reich, N.O. (2005). Gold nanoparticle decoration of DNA on silicon. Langmuir 21(23), 10699–10701.

Burley, G.A., Gierlich, J., Mofid, M.R., Nir, H., Tal, S., Eichen, Y. and Carell, T. (2006). Directed DNA metallization. J. Am. Chem. Soc. 128(5), 1398–1399.

Chen, Y., Aveyard, J. and Wilson, R. (2004). Gold and silver nanoparticles functionalized with known numbers of oligonucleotides per particle for DNA detection. Chem. Comm. (24), 2804–2805.

Cobbe, S., Connolly, S., Ryan, D., Nagle, L., Eritja, R. and Fitzmaurice, D. (2003). DNA-controlled assembly of protein-modified gold nanocrystals. J. Phys. Chem. B 107(2), 470–477.

Daniel, M.-C. and Astruc, D. (2004). Gold nanoparticles: assembly, supramolecular chemistry, quantum-size–related properties, and applications toward biology, catalysis and nanotechnology. Chem. Rev. 104(1), 293–346.

Demers, L.M., Mirkin, C.A., Mucic, R.C., Reynolds, R.A., Letsinger, R.L., Elghanian, R. and Viswanadham, G. (2000). A fluorescence-based method for determining the surface coverage and hybridization efficiency of thiol-capped oligonucleotides bound to gold thin films and nanoparticles. Anal. Chem. 72(22), 5535–5541.

Deng, Z., Tian, Y., Lee, S., Ribbe, A.E. and Mao, C. (2005). DNA-encoded self-assembly of gold nanoparticles into one-dimensional arrays. Angew. Chem. Int. Ed. 44(23), 3582-3585. Angew. Chem. 117(23), 3648–3651.

Dujardin, E., Hsin, L.B., Wang, C.R.C. and Mann, S. (2001). DNA-driven self-assembly of gold nanorods. Chem. Comm. (14), 1264–1265.

Feldheim, D.L. and Foss, C.A. (2002). Metal Nanoparticles: Synthesis, Characterization, and Applications. Marcel Dekker, New York.

Fitzmaurice, D. and Connolly, S. (1999). Programmed assembly of gold nanocrystals in aqueous solution. Adv. Mater. 11(14), 1202–1205.

Ford, W.E., Harnack, O., Yasuda, A. and Wessels, J.M. (2001). Platinated DNA as precursors to templated chains of metal nanoparticles. Adv. Mater. 13(23), 1793–1797.

Gearheart, L.A., Ploehn, H.J. and Murphy, C.J. (2001). Oligonucleotide adsorption to gold nanoparticles: a surface-enhanced Raman spectroscopy study of intrinsically bent DNA. J. Phys. Chem. B 105(50), 12609–12615.

Halperin, W.P. (1986). Quantum size effects in metal particles. Rev. Mod. Phys. 58(3), 533–606.

Hazarika, P., Giorgi, T., Reibner, M., Ceyhan, B. and Niemeyer, C.M. (2004). In: C.M. Niemeyer (Eds.), Bioconjugation Protocols: Strategies and Methods, Humana Press, Totowa New York, pp. 295–304.

Huang, Y., Chiang, C.-Y., Lee, S.K., Gao, Y., Hu, E.L., De Yoreo, J. and Belcher, A.M. (2005). Programmable assembly of nanoarchitectures using genetically engineered viruses. Nano Lett. 5(7), 1429–1434.

Jamieson, E.R. and Lippard, S.J. (1999). Structure, recognition, and processing of cisplatin-DNA adducts. Chem. Rev. 99(9), 2467–2498.

Jin, R., Wu, G., Li, Z., Mirkin, C.A. and Schatz, G.C. (2003). What controls the melting properties of DNA-linked gold nanoparticle assemblies? J. Am. Chem. Soc. 125(6), 1643–1654.

de Jongh, L.J. (1994). Physics and Chemistry of Metal Cluster Compounds, Model Systems for Small Metal Particles, Series: Physics and Chemistry of Materials with Low-Dimensional Structures, Vol. 18, Springer, Heidelberg, New York.

Keren, K., Berman, R.S., Buchstab, E., Sivan, U. and Braun, E. (2003). DNA-templated carbon nanotube field-effect transistor. Science 302(5649), 1380–1382.

Komiyama, M. and Hirai, H. (1983), Colloidal rhodium dispersions protected by cyclodextrins. Bull. Chem. Soc. Jpn. 56(9), 2833–2834.

Koplin, E., Niemeyer, C.M. and Simon, U. (2006). Formation of electrically conducting DNA-assembled gold nanoparticle monolayers. J. Mater. Chem. 16(14), 1338–1344.

Koyfman, A.Y., Braun, G., Magonov, S., Chworos, A., Reich, N.O. and Jaeger, L. (2005). Controlled spacing of cationic gold nanoparticles by nanocrown RNA. J. Am. Chem. Soc. 127(34), 11886–11887.

Kretschmer, R. and Fritzsche, W. (2004), Pearl chain formation of nanoparticles in microelectrode gaps by dielectrophoresis. Langmuir 20(26), 11797–11801.

Le, J.D., Pinto, Y., Seeman, N.C., Musier-Forsyth, K., Taton, T.A. and Kiehl, R.A. (2004). DNA-templated self-assembly of metallic nanocomponent arrays on a surface. Nano Lett. 4(12), 2343–2347.

Letsinger, R.L., Elghanian, R., Viswanadham, G. and Mirkin, C.A. (2000). Use of a steroid cyclic disulfide anchor in constructing gold nanoparticle-oligonucleotide conjugates. Bioconjug. Chem. 11(2), 289–291.

Li, H., Park, S.H., Reif, J.H., LaBean, T.H. and Yan, H. (2004). DNA-templated self-assembly of protein and nanoparticle linear arrays. J. Am. Chem. Soc. 126(2), 418–419.

Li, U. and Mirkin, C.A. (2005). G-quartet-induced nanoparticle assembly. J. Am. Chem. Soc. 127(33), 11568–11569.

Li, Z., Jin, R., Mirkin, C.A. and Letsinger, R.L. (2002). Multiple thiol-anchor capped DNA–gold nanoparticle conjugates. Nucleic Acids Res. 30(7), 1558–1562.

Liu, Y., Meyer-Zaika, W., Franzka, S., Schmid, G., Tsoli, M. and Kuhn, H. (2003). Gold-cluster degradation by the transition of B-DNA into A-DNA and the formation of nanowires. Angew. Chem. Int. Ed. 42(25), 2853–2857. Angew. Chem. 115(25), 2959–2963.

Long, H. and Schatz, G. (2003). In: U. Simon, G. Schmid, S. Hong, S.J. Stranick and S.M. Arrivo (Eds.), Bioinspired Nanoscale Hybrid Systems, Symposium Proceedings Vol. 735, Materials Research Society, Warrendale Pennsylvania, pp. 143–152.

Loweth, C.J., Caldwell, W.B., Peng, X., Alivisatos, A.P. and Schultz., P.G. (1999). DNA-based assembly of gold nanocrystals. Angew. Chem. Int. Ed. 38(12), 1808–1812. Angew. Chem. 111(12), 1925–1929.

Mbindyo, J.K.N., Reiss, B.D., Martin, B.R., Keating, C.D., Natan, M.J. and Mallouk, T.E. (2001). DNA-directed assembly of gold nanowires on complementary surfaces. Adv. Mater. 13(4), 249–254.

Mirkin, C.A. (2000). Programming the assembly of two- and three-dimensional architectures with DNA and nanoscale inorganic building blocks. Inorg. Chem. 39(11), 2258–2272.

Mirkin, C.A., Letsinger, R.L., Mucic, R.C. and Storhoff, J.J. (1996). A DNA-based method for rationally assembling nanoparticles into macroscopic materials. Nature 382(6592), 607–609.

Mitchell, G.P., Mirkin, C.A. and Letsinger, R.L. (1999). Programmed assembly of DNA-functionalized quantum dots. J. Am. Chem. Soc. 121(35), 8122–8123.

Moiseev, I., Vargaftik, M.N., Strom-nova, T.A., Gekhman, A.E., Tsirkov, G.A. and Makhlina, A.M. (1996). Catalysis with a palladium giant cluster: phenol oxidative carbonylation to diphenyl carbonate conjugated with reductive nitrobenzene conversion. J. Mol. Cat. A 108(2), 77–85.

Mucic, R.C., Storhoff, J.J., Mirkin, D.A. and Letsinger, R.L. (1998). DNA-directed synthesis of binary nanoparticle network materials. J. Am. Chem. Soc. 120(48), 12674–12675.

Niemeyer, C.M. (2001a). Bioorganic applications of semisynthetic DNA-protein conjugates. Chem. Eur. J. 7(14), 3188–3195.

Niemeyer, C.M (2001b). Nanoparticles, proteins, and nucleic acids: biotechnology meets materials science. Angew. Chem. Int. Ed. 40(22), 4128-4158. Angew. Chem. 113(22), 4254–4287.

Niemeyer, C.M., Bürger, W. and Peplies, J. (1998). Covalent DNA-streptavidin conjugates as building blocks for novel biometallic nanostructures. Angew. Chem. Int. Ed. 37(16), 2265–2268. Angew. Chem. 110(16), 2391–2395.

Niemeyer, C.M, Ceyhan, B. and Hazarika, P. (2003a). Oligofunctional DNA-gold nanoparticle conjugates. Angew. Chem. Int. Ed. 42(46), 5766–5770. Angew. Chem. 115(46), 5944–5948.

Niemeyer, C.M., Ceyhan, B., Noyong, M. and Simon, U. (2003b). Bifunctional DNA–gold nanoparticle conjugates as building blocks for the self-assembly of cross-linked particle layers. Biochem. Biophys. Res. Comm. 311(4), 995–999.

Niemeyer, C.M. and Mirkin, C.A. (2004). Nano-Biotechnology: Concepts, Methods and Applications. Wiley-VCH, Weinheim.

Noyong, M., Gloddek, K. and Simon, U. (2003). In: U. Simon, G. Schmid, S. Hong, S.J. Stranick and S.M. Arrivo (Eds.), Bioinspired Nanoscale Hybrid Systems, Symposium Proceedings Vol. 735, Materials Research Society, Warrendale Pennsylvania, pp. 153–158.

Ongaro, A., Griffin, F., Beecher, P., Nagle, L., Iacopino, D., Quinn, A., Redmond, G. and Fitzmaurice, D. (2005). DNA-templated assembly of conducting gold nanowires between gold electrodes on a silicon oxide surface. Chem. Mater. 17(8), 1959.

Ongaro, A., Griffin, F., Nagle, L., Iacopino, D., Eritja, R. and Fitzmaurice, D. (2004). DNA-templated assembly of a protein-functionalized nanogap electrode, Adv. Mater. 16(20), 1799–1803.

Parak, W.J., Pellegrino, T., Micheel, C.M., Gerion, D., Williams, S.C. and Alivisatos, A. P. (2003). Conformation of oligonucleotides attached to gold nanocrystals probed by gel electrophoresis. Nano Lett. 3(1), 33–36.

Park, S.-J., Lazarides, A.A., Mirkin, C.A., Brazis, P.W., Kannewurf, C.R. and Letsinger, R.L. (2000). The electrical properties of gold nanoparticle assemblies linked by DNA. Angew. Chem. Int. Ed. 39(21), 3845–3848. Angew. Chem. 112(21), 4003–4006.

Park, S.J., Taton, T.A. and Mirkin, C.A. (2002). Array-based electrical detection of DNA with nanoparticle probes. Science 295(5559), 1503–1506.

Reetz, M.T. and Helbig, W. (1994). Size-selective synthesis of nanostructured transition metal clusters. J. Am. Chem. Soc. 116(16), 7401–7402.

Remacle, F. and Levine, R.D. (2001). Quantum dots as chemical building blocks: elementary theoretical considerations. Chem. Phys. Chem. 2(1), 20–36.

Richards, R., Boennemann, H., Hormes, J. and Leuschner, C. (2005). In: C.S.S.R. Kumar and J. Hormes (Eds.), Nanofabrication Towards Biomedical Applications. Wiley-VCH, Weinheim, pp. 3–32.

Richter, J., Mertig, M., Pompe, W., Mönch, I. and Schackert, H.K. (2001). Construction of highly conductive nanowires on a DNA template. Appl. Phys. Lett. 78(4), 536–538.

Rosi, N.L. and Mirkin, C.A. (2005). Nanostructures in biodiagnostics. Chem. Rev. 105(4), 1547–1562.

Schmid, G. (2004). Nanoparticles: From Theory to Applications. Wiley-VCH, Weinheim.

Schmid, G., Boese, R., Pfeil, R., Bandermann, F., Meyer, S., Calis, G.H.M. and van der Velden, J.W.A. (1981). $Au_{55}[P(C_6H_5)_3]_{12}Cl_6$—Ein Goldcluster ungewöhnlicher Größe. Chem. Ber. 114(11), 3634–3642.

Schmid, G. and Lehnert, A. (1989), The complexation of gold colloids. Angew. Chem. Int. Ed. 28(6), 780–781. Angew. Chem. 101(6), 773–774.

Schmid, G., Lehnert, A., Kreibig, U., Damczyk, Z.A. and Belouschek, P. (1990). Synthese und elektonenmikroskopische Untersuchung kontrolliert gewachsener, ligandstabilisierter Goldkolloide sowie theoretische Überlegungen zur Oberflächenbelegung durch Kolloide. Z. Naturforsch. 45b, 989–994.

Schmid, G., Morun, B. and Malm, J.-O. (1989). $Pt_{309}Phen_{36}^{*}O_{30\pm10}$, a four-shell platinum cluster. Angew. Chem. Int. Ed. 28(6), 778–780. Angew. Chem. 101(6), 772–773.

Schmid, G., Pugin R., Sawitowski Th., Simon U. and Marler B. (1999). Transmission electron microscopic and small angle X-ray diffraction investigations of $Au_{55}(PPh_3)_{12}Cl_6$ microcrystals. Chem. Comm. (14), 1303–1304.

Schön, G. and Simon, U. (1995). A fascinating new field in colloid science: small ligand-stabilized metal clusters and possible application in microelectronics. Coll. Polym. Sci. 273(2), 101–117.

Seidel, R., Ciacchi, L.C., Weigel, M., Pompe, W. and Mertig, M. (2004). Synthesis of platinum cluster chains on DNA templates: conditions for a template-controlled cluster growth. J. Phys. Chem. B 108(30), 10801–10811.

Sharma, J., Chhabra, R., Liu, Y., Ke, Y. and Yan, H. (2006). DNA-templated self-assembly of two-dimensional and periodical gold nanoparticle arrays (p). Angew. Chem. Int. Ed. 45(5), 730–735. Angew. Chem. 118(5), 744–749.

Shipway, A.N., Katz, E. and Willner, I. (2000). Nanoparticle arrays on surfaces for electronic, optical, and sensor applications. Chem. Phys. Chem. 1(1), 19–52.

Simon, U. (1998). Charge transport in nanoparticle arrangements. Adv. Mater. 10(17), 1487–1492.

Spatz, J.P., Roescher, A. and Moeller, M. (1996). Gold nanoparticles in micellar poly(styrene)-b-poly(ethylene oxide) films—size and inter-particle distance control in monoparticulate films. Adv. Mater. 8(4), 337–340.

Spatz, J.P., Roescher, A., Sheiko, S., Krausch, G. and Moeller, M. (1995). Noble metal loaded block lonomers: micelle organization, adsorption of free chains and formation of thin films. Adv. Mater. 7(8), 731–735.

Stoltenberg, R.M. and Woolley, A.T. (2004), DNA-templated nanowire fabrication. Biomed. Microdevices 6(2), 105–111.

Storhoff, J.J., Elghanian, R., Mucic, R.C., Mirkin, C.A. and Letsinger, R.L. (1998). One-pot colorimetric differentiation of polynucleotides with single base imperfections using gold nanoparticle probes. J. Am. Chem. Soc. 120(9), 1959–1964.

Storhoff, J.J., Lazarides, A.A., Mucic, R.C., Mirkin, C.A., Letsinger, R.L. and Schatz, G.C. (2000). What controls the optical properties of DNA-linked gold nanoparticle assemblies? J. Am. Chem. Soc. 122(19), 4640–4650.

Storhoff, J.J. and Mirkin, C.A. (1999). Programmed materials synthesis with DNA. Chem. Rev. 99(7), 1849–1862.

Tsoli, M., Kuhn, H., Brandau, W., Esche, H. and Schmid, G. (2005). Cellular uptake and toxicity of Au_{55} clusters. Small 1(8–9), 841–844.

Turkevitch, J., Stevenson, P.C. and Hillier, J. (1951). A study of the nucleation and growth processes in the synthesis of colloidal gold. Discuss. Faraday Soc. 11, 55–75.

Vargaftik, M.N., Zagorodnikov, V.P., Stolyarov, I.P., Moiseev, I.I., Likholobov, V.A., Kochubey, D.I., Chuvilin, A.L., Zaikovsky, V.I., Zamaraev, K.I. and Timofeeva, G.I. (1985). A novel giant palladium cluster. J. Chem. Soc., Chem. Comm. (14), 937–939.

Warner, M.G. and Hutchison, J.E. (2003). Linear assemblies of nanoparticles electrostatically organized on DNA scaffolds. Nat. Mater. 2(4), 272–277.

Williams, K.A., Veenhuizen, P.T., de la Torre, B.G., Eritja, R. and Dekker, C. (2002). Nanotechnology: carbon nanotubes with DNA recognition. Nature 420(6917), 761.

Willner, I. and Katz, E. (2004). Integrated nanoparticle-biomolecule hybrid systems: synthesis, properties, and applications. Angew. Chem. Int. Ed. 43(45), 6042–6108.

Woehrle, G.H., Warner, M.G. and Hutchison, J.E. (2004). Molecular-level control of feature separation in one-dimensional nanostructure assemblies formed by biomolecular nanolithography. Langmuir 20(14), 5982–5988.

Xiao, S., Liu, F., Rosen, A.E., Hainfeld, J.F., Seeman, N.C., Musier-Forsyth, K. and Kiehl, R.A. (2002). Self-assembly of metallic nanoparticle arrays by DNA scaffolding. J. Nanoparticle Res. 4(4), 313–317.

Zanchet, D., Micheel, C.M., Parak, W.J., Gerion, D. and Alivisatos, A.P. (2001). Electrophoretic isolation of discrete Au nanocrystal/DNA conjugates. Nano Lett. 1(1), 32–35.

Zou, B., Ceyhan, B., Simon, U. and Niemeyer, C.M. (2005). Self-assembly of crosslinked DNA-gold nanoparticle layers visualized by in-situ scanning force microscopy. Adv. Mater. 17(13), 1643–1647.

2

DNA-Based Nanoelectronics

Rosa Di Felice

1 INTRODUCTION

1.1 DNA FOR MOLECULAR DEVICES

Nature bases its operation on proteins to perform actions of any sort. DNA is the material that encodes and transfers the information to fabricate proteins through a spontaneous process. Therefore, the possible exploitation of DNA in nanotechnology cannot be based on a straightforward translation of an inherent biological action.

On the one hand, various conformational transitions have been proposed to generate motion from DNA and thus realize a mechanical device (Mao, Sun, Shen, and Seeman 1999; Alberti and Mergny 2003; Brucale, Zuccheri, and Samorì 2005); alternatively, a DNA-based mechanical device can be even fueled by DNA itself by virtue of the elastic response (Yurke, Turberfield, Mills, Simmel, and Neumann 2000). In these examples, DNA is guided to perform a certain action that is not proper for it in nature, based on its recognition and structuring capabilities: In practice, the potentialities of DNA are embedded in its conformation and topology. Other notable applications envisage the two- and three-dimensional assembly of complex objects (cubes, octahedral, etc.) made with DNA (Seeman 1998, 2003) onto organized chips to recognize and position other biological materials, with applications in diagnostics and medicine.

A. Offenhäusser and R. Rinaldi (eds.), *Nanobioelectronics - for Electronics, Biology, and Medicine*,
DOI: 10.1007/978-0-387-09459-5_3,

On the other hand, scientists are fascinated by the issue of whether or not DNA can also have applications in nanoelectronics. Consequently, research initiatives were launched worldwide to explore the conductivity of DNA (Braun, Eichen, Sivan, and Ben-Yoseph 1998; Fink and Schönenberger 1999; de Pablo, Moreno-Herrero, Colchero, Gómez Herrero, Herrero, Baró, Ordejón, Soler, and Artacho 2000; Porath, Bezryadin, de Vries and Dekker 2000; Kasumov, Kociak, Guéron, Reulet, Volkov, Klinov, and Bouchiat 2001; Rakitin, Aich, Papadopoulos, Kobzar, Vedeneev, Lee, and Xu 2001; Storm, van Noort, de Vries, and Dekker 2001; Watanabe, Manabe, Shigematsu, Shimotani, and Shimizu 2001; Shigematsu, Shimotani, Manabe, Watanabe, and Shimizu 2003). Alternatively, if measurable currents cannot be sustained by DNA molecules, another interesting strategy is to realize hybrid objects (metal nanoparticles/wires, proteins/antibodies, etc.) in which electrons move and carry current flows, templated by DNA helices at selected locations (Braun, Eichen, Sivan, and Ben-Yoseph 1998; Richter, ertig, Pompe, Mönch, and Schackert 2001; Keren, Krueger, Gilad, Ben-Yoseph, Sivan, and Braun 2002; Keren, Berman, Buchstab, Sivan, and Braun 2003; Berti, Alessandrini and Facci 2005): This route also allows to embed conducting objects into the hybrid architectures, to realize, e.g., a carbon nanotube DNA-templated nanotransistor (Keren, Berman, Buchstab, Sivan, and Braun 2003). Both of these ways could lead to the development of DNA-based molecular electronics.

1.2 WHAT IS KNOWN ABOUT DNA'S ABILITY TO CONDUCT ELECTRICAL CURRENTS?

The current intense interest in the experimental/theoretical probing of the electronic structure of DNA-based polymers was stimulated by the quest for the development of molecular electronics. Good reviews of the field can be found (Di Ventra and Zwolak 2004; Endres, Cox, and Singh 2004; Porath, Cuniberti, and Di Felice 2004; Joachim and Ratner 2005). We just point out here the salient results that motivated the pursuit of optimized measurement setups on one hand, and of *DNA-derivatives* and mimics beyond *native-DNA* on the other hand. The desired "mutants" should exhibit enhanced conductivity and/or other exploitable functions, whereas maintaining the inherent recognition and structuring traits of native Watson-Crick DNA that are demanded for self-assembling.

The molecules used for electronic applications need to express three main features: (a) *Structuring*, namely, the possibility to tailor their structural properties (composition, length, etc.) "on demand"; (b) *Recognition*, namely, the ability to attach them to specific sites or to other target molecules; (c) *Electrical functionality*, namely, suitable conductivity and control of their electrical characteristics.

The use of DNA and its derivatives as building blocks for molecular electronic devices is particularly intriguing. These molecules rely on a broad base of biotechnology tools available for their synthesis and modifications through natural enzymes that provide a control on their structure. Their specific interaction with complementary strands and proteins, guided by recognition, is extremely useful for implementing self-assembly in molecular circuits. One of the main challenges with such molecules, however, is the control of their electrical conductivity. Early work in this field has yielded seemingly controversial results for *native-DNA*, showing electrical behaviors from insulating through semiconducting to conducting, with even a single report of proximity-induced superconductivity. Indeed, recent reviews of the experimental literature (Di Ventra and Zwolak 2004; Endres, Cox, and Singh 2004; Porath, Cuniberti, and Di Felice 2004) highlighted that the variety of available experiments cannot be analyzed in a unique way; for instance, electrical measurements conducted on single molecules, bundles, and networks, are not able to reveal a uniform interpretation scheme for the conductivity of DNA, because they refer to different materials or at least aggregation states. In addition, the experimental conditions were often variable, e.g., suspended molecules (either electrostatically trapped or thiol-attached between electrodes), molecules deposited onto different inorganic substrates with or without a soft organic buffer layer, vacuum vs. ambient or liquid. The situation was schematically summarized as follows: (a) charges may be transported with relatively poor conductivity in short single DNA molecules or in longer molecules organized in bundles and networks; and (b) charge flow is blocked for long molecules deposited onto "hard" inorganic substrates. An intriguing explanation was that the inability to conduct is not due to the intrinsic electronic structure, but to strong deformations induced by the substrate (Kasumov, Klinov, Roche, Guéron, and Bouchiat 2004; Porath, Cuniberti, and Di Felice 2004), that completely alter the structure and the regular π-stack. Indeed, two recent experiments devoted to optimize the molecule–electrode contacts and realize current-voltage measurements on *suspended* DNA polymers (Xu, Zhang, Li, and Tao 2004; Cohen, Nogues, Naaman, and Porath 2005; Cohen, Nogues, Ullien, Daube, Naaman, and Porath 2006), demonstrated that it is possible to solicit a current flow through short stacks. Although these two experiments cannot be interpreted evenly in terms of conduction ability and mechanisms, they both show the efficiency of avoiding the contact to a substrate along the molecule and using covalent bonding at the molecules extremities.

Given the situation illustrated in the preceding, two solutions remain to continue pursuing the route toward DNA-based electronics (Fig. 2.1): Either to reduce/avoid substrate-induced deformations (standing-molecule measurements, use of a soft organic buffer, etc.), or explore stiffer molecules. The following illustrates the background that led to this viable bifurcated strategy, and some efforts in both directions. The following also traces the origin of conductivity measurements in DNA

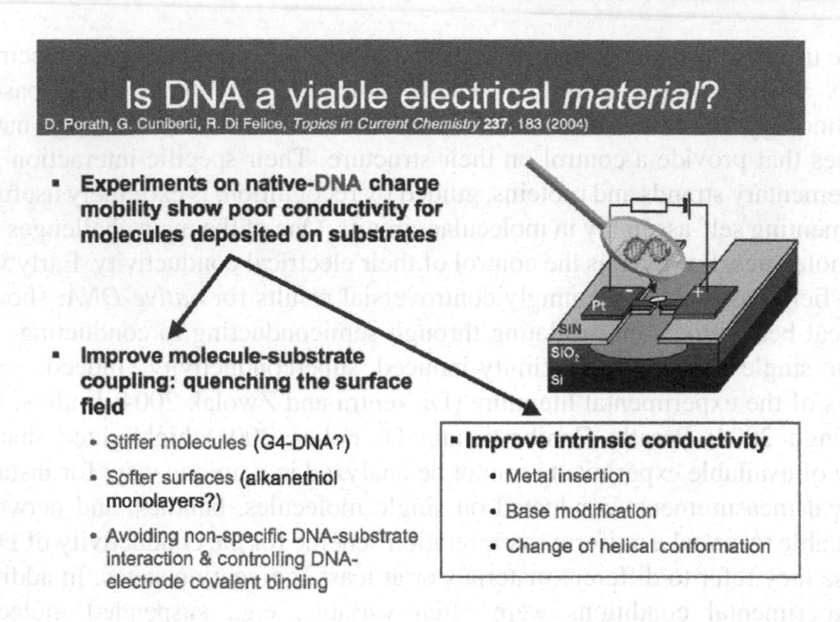

Is DNA a viable electrical material?
D. Porath, G. Cuniberti, R. Di Felice, Topics in Current Chemistry 237, 183 (2004)
Experiments on native-DNA charge mobility show poor conductivity for molecules deposited on substrates
Improve molecule-substrate coupling: quenching the surface field
Stiffer molecules (G4-DNA?)
Softer surfaces (alkanethiol monolayers?)
Avoiding non-specific DNA-substrate interaction & controlling DNA-electrode covalent binding
Improve intrinsic conductivity
Metal insertion
Base modification
Change of helical conformation
Pt
SiN
SiO2
Si

FIG. 2.1. Scheme of the problems and suggested solutions to measure the conductivity of DNA molecules. The inset illustrates a setup to measure the conductivity of a DNA molecule between two metal electrodes.

to the detection of fast long-range electron transfer in solution, divides the work between experiment and theory, and reports the main methods and selected results in both cases. Finally the current perspectives are summarized and briefly exposed.

This chapter focuses solely on the investigation of the intrinsic electronic properties of DNA, without discussing possible electronic applications that involve the construction of DNA-nanoparticle hybrids or DNA-templated nanowires. This branch, which may be named "DNA-templated nanoelectronics" (Braun, Eichen, Sivan, and Ben-Yoseph 1998; Richter, Mertig, Pompe, Mönch, and Schackert 2001; Maubach, Csáki, and Fritzsche 2003) is a very lively parallel research line and deserves a chapter by itself.

2 METHODS, MATERIALS, AND RESULTS

The investigation of the DNA conductivity was fueled by measurements of charge transfer rates of DNA molecules in solution. Therefore, here we also reserve some space for the description of the latter experiments (Section 1.1), and of the theoretical framework for their interpretation (Section 2.1). Sections 1.2 and 1.3 are devoted to conductivity measurements in different setups. Sections 2.2 and 2.3 are devoted to band structure analysis from first principles and with empirical methods.

2.1 EXPERIMENTAL INVESTIGATIONS

2.1.1 MEASUREMENT OF CHARGE TRANSFER RATES IN SOLUTION

The idea that double-stranded DNA, the carrier of genetic information in most living organisms, may function as a conduit for fast electron transport along the axis of its base pair stack, was first advanced in 1962 following the measurement of solid DNA specimens at 400 K (Eley and Spivey 1962). (The so-called π-way is naively inspired by the π-like nature of the HOMO and LUMO of the nucleobases. Such orbitals are indeed characterized by a charge density distribution that extends out of the base plane, facilitating inter-plane interactions. However, the strength of such interactions is not necessarily large enough to establish a continuous π-way!) Instead, the observation that radiation-induced conductivity at –78°C was found only with hydrated samples, led to the proposal that it was due to highly mobile charge carriers migrating within the ice-like water layer surrounding the helix, rather than through the base pair core (Warman, de Haas, and Rupprecht 1996). The long-lasting interest of the radiation community in the problem of charge migration in DNA was due to its relevance for the mechanisms of DNA oxidative damage, whose main target is the guanine (G) base (Retèl, Hoebee, Braun, Lutgerink, van den Akker, Wanamarta, Joenjie, and Lafleur 1993). Recently, the interest in the same issue has been revived and extended to other interdisciplinary research communities. In particular, the issue of electron and hole migration in DNA has become a hot topic (Turro and Barton 1998; Lewis, Wu, Liu, Letsinger, Greenfield, Miller, and Wasielewski 2000) for a number of chemistry scholars following the reports that photoinduced electron transfer occurred with very high and almost distance-independent rates between distant donor and acceptor intercalators along a DNA helix (Murphy, Arkin, Jenkins, Ghatlia, Bossman, Turro, and Barton 1993; Hall, Holmlin, and Barton 1996). This evidence led to the conclusion that double-stranded DNA may exhibit a "wire-like" behavior (Kelley, Jackson, Hill, and Barton 1999). From the large body of experimental studies on the solution chemistry that has become available and reviewed in the last years (Barbara and Olson 1999; Grinstaff 1999; Wagenknecht 2005), depending on the energetics of the base sequence and the overall structural aspects of the system under investigation, the mechanisms proposed for DNA-mediated charge migration include single-step superexchange (Murphy, Arkin, Jenkins, Ghatlia, Bossman, Turro, and Barton 1993), multistep hole hopping (Bixon, Giese, Wessely, Langenbacher, Michel-Beyerle and Jortner 1999; Giese, Amaudrut, Köhler, Spormann, and Wessely 2001; Berlin, Kurnikov, Beratan, Ratner, and Burin 2004), phonon-assisted polaron hopping (Schuster 2000), and polaron drift (Conwell and Rakhmanova 2000; Conwell 2004).

In one statement, it was the very suggestion of a "wire-like" behavior by Barton and co-workers that encouraged the idea of exploiting DNA molecules as molecular wires in devices, thus initiating a huge worldwide effort devoted to the measurement of DNA current-voltage characteristics between electrodes.

Barton and co-workers adopt a variety of spectroscopic, biochemical, and electrochemical techniques to probe the efficiency of charge transfer in DNA, employing well-characterized DNA assemblies that incorporate metallointercalators, organic intercalators, and modified bases. In the spectroscopic approach, for instance, they excite a modified base or a fluorescent intercalator by photon irradiation, and then observe by fluorescence spectroscopy the dynamics of charge transfer from femtoseconds to milliseconds, through the decay of the excited state (Kelley and Barton 1999). With this technique they investigate the distance dependence when several measurement conditions are changed (e.g., fluorescent species, inter-strand or intra-strand decay) and find β-values (the exponential factor for superexchange) ranging from 0.1 to 1.0 $Å^{-1}$, indicating several different mechanisms for different samples. For what concerns the electrochemical approach, they employ DNA films covalently bonded to a gold surface by alkane-thiol linkers (Kelley, Jackson, Hill, and Barton 1999), as shown in Fig. 2.2A. A redox-active intercalator (e.g., daunomycin) is inserted at a given controlled place along the base pair stack, namely, at a given distance from the surface. DNA-mediated transport of an electron from the gold surface to the distally bound intercalator is then monitored by cyclic voltammetry or similar techniques. The electron transfer can thus be probed as a function of distance and in the presence of stacking defects (Fig. 2.2B).

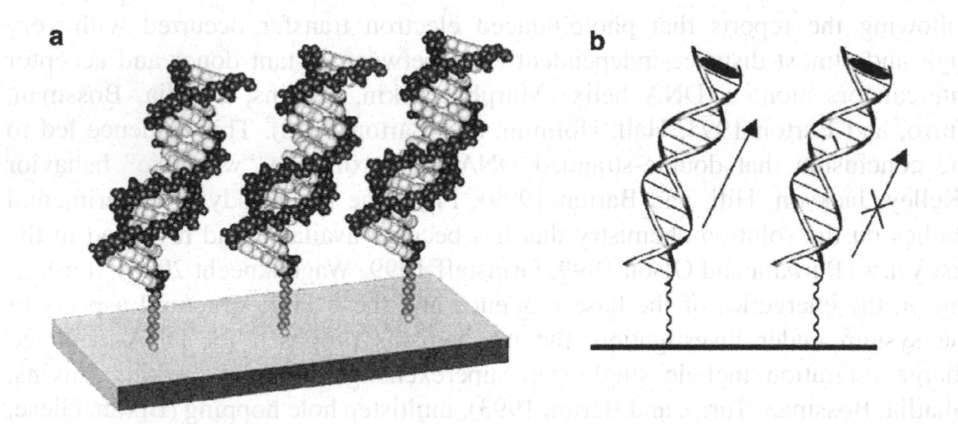

FIG. 2.2. **A.** A DNA film is connected to a gold surface by alkane-thiol chains. **B.** A redox active intercalator is inserted at a given variable point of the sequence: Electron transfer occurs efficiently in an ordered sequence, but is interrupted if a stacking defect is introduced (Adapted from Kelley, Jackson, Hill, and Barton 1999 with permission; copyright 1999 by Wiley-VCH.).

It is found that DNA-mediated charge transfer is strongly sensitive to the structure and dynamics of the base stack and to the coupling of the intercalator, but insensitive to distance: At least up to 100 Å, the rate limiting step for the charge transfer reaction is the transit through the alkane chains (Wagenknecht 2005). Some results (Kelley, Jackson, Hill, and Barton 1999) are interpreted in terms of a β value of 0.1 $Å^{-1}$, very small in comparison to values known for proteins. The electrochemical investigations of *electron transfer* are complementary to the spectroscopic investigations of the group, which mainly address *hole transfer*. In all cases the authors conclude that DNA-mediated charge transfer proceeds through the π-stack.

Giese's group at the University of Basel performed different experiments to probe charge transfer through various DNA sequences separating a hole source from a hole trap. In two topical experiments, they characterized charge hopping through guanines (Meggers, Michel-Beyerle, and Giese 1998) and adenines (Giese, Amaudrut, Köhler, Spormann, and Wessely 2001). Their measurements are based on the following multi-step chemical procedure (Giese 2004): (a) injection of a radical cation (a hole) into a G base by a chemical assay; (b) piperidine treatment and water reaction with the oxidized species, which led to strand cleavage; and (c) separation of the strands by gel electrophoresis and quantification of the product ratio, to determine the electron transfer efficiency in terms of the amount of products in which the oxidized species has moved from the place where the radical cation was initially created to another place along the same strand. In one experiment they investigated the sequences of Fig. 2.3, in which a hole source ($G_{23}^{+\bullet}$) was

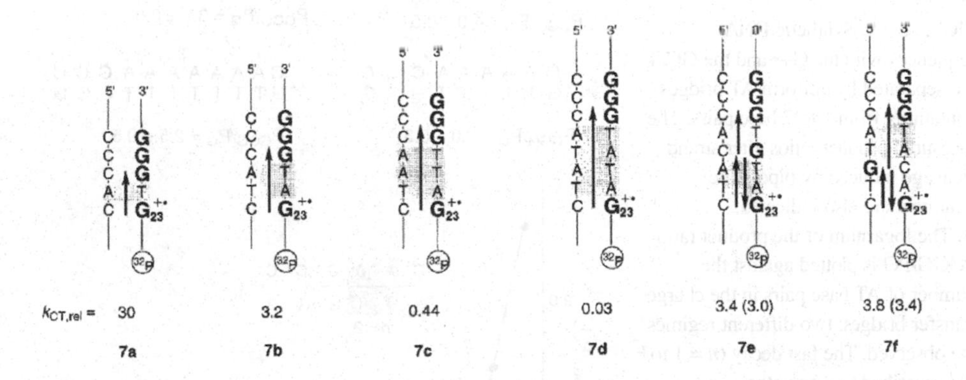

FIG. 2.3. Synthetic DNA sequences through which hole transfer was probed: The GGG triplets acted as hole traps due to the low oxidation potential. The relative charge transfer rates are reported, evaluated from the measurement of the product ratio P_{GGG}/P_{G}. In 7a-d hole transfer takes place through AT base pairs by a single-step superexchange mechanism. In double strands 7e,f a two-step mechanism is proposed in which the intervening G base is oxidized in the first step followed by an irreversible hole transfer to the GGG unit. This hopping process increases the charge transfer efficiency compared to the single-step superexchange (7d) over the same distance (Adapted from Meggers, Michel-Beyerle, and Giese 1998 with permission; copyright 1998 by the American Chemical Society.).

separated by a hole trap (GGG) by various bridge sequences that included (7e,f) or not (7a–d) other guanines. The product ratio P_{GGG}/P_G decreased exponentially with the number of bridge bases for bridges without any intervening guanines (see the values of $k_{CT,rel}$ in duplexes 7a–d in Fig. 2.3), whereas the decrease was mitigated in the case of guanine-containing bridges. These results were interpreted in terms of two different mechanisms for hole transfer: direct superexchange between G_{23} and GGG for sequences 7a–d, and guanine-mediated hopping for sequences 7e,f. In another experiment the same group interrogated hole transfer through different sequences, in which the $G^{+\bullet}$ hole source was separated by the GGG hole trap by only AT pairs, with the As all in one strand and the Ts all in the complementary strand (Fig. 2.4). It was found in this case that charge hopping can occur also through AT base pairs, and indeed it is the preferred mechanism if the bridge between the hole source and the hole trap is longer than 3 AT pairs.

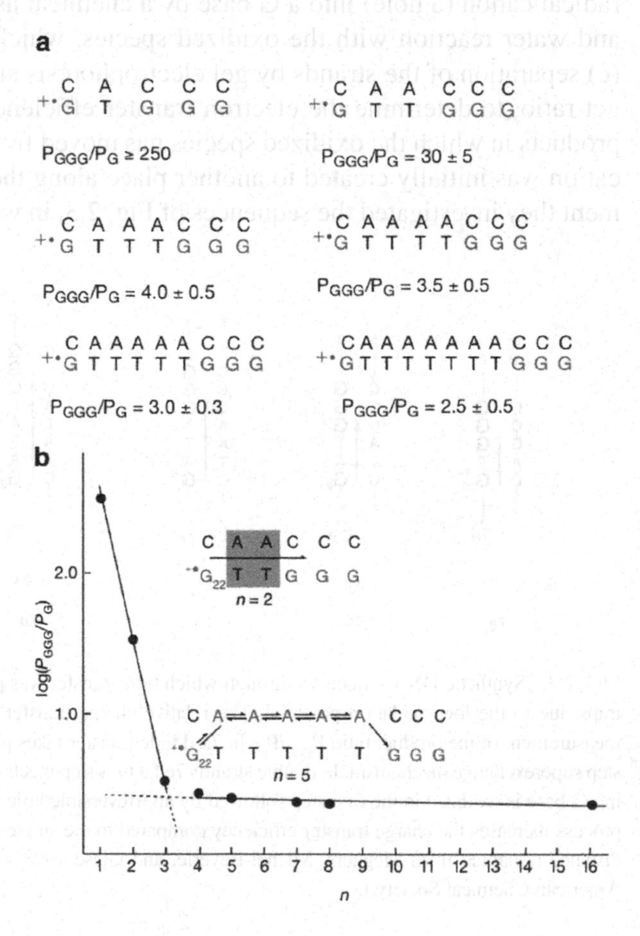

FIG. 2.4. **A.** Synthetic DNA sequences with the G+• and the GGG are separated by uniform AT bridges containing from 1 to 7 base pairs. The measured product ratios after strand cleavage induced by piperidine treatment are also indicated. **B.** The logarithm of the product ratio PGGG/PG is plotted against the number of AT base pairs in the charge transfer bridges: two different regimes are observed. The fast decay ($n = 1$ to 3) is ascribed to a one-step superexchange process. The almost distance-independent behavior ($n \geq 4$) is ascribed to multi-step hopping through adenine electron states (Adapted from Giese, Amaudrut, Köhler, Spormann, and Wessely 2001 with permission; copyright 2001 by Nature Macmillan Publishers Ltd.).

Experiments conceptually similar to those performed in Basel were carried out at Georgia Tech by Gary Schuster and co-workers by using a different hole precursor, namely anthraquinone, for charge injection. This experimental work was carried out in close collaboration with Uzi Landmann for a computational cross-analysis. By analyzing several duplex sequences containing GG traps along the same strand (Liu, Hernandez, and Schuster 2004), an exponential dependence of the charge transfer efficiency with distance was detected, with β values as low as 0.003 Å^{-1} (Fig. 2.5). The measured β values demonstrated high sensitivity to the length of the bridges between G or GG hole traps, but lower sensitivity to the sequence of such bridges. These results were interpreted in terms of a phonon-mediated polaron hopping, and were in line with several other data published previously by the same group and

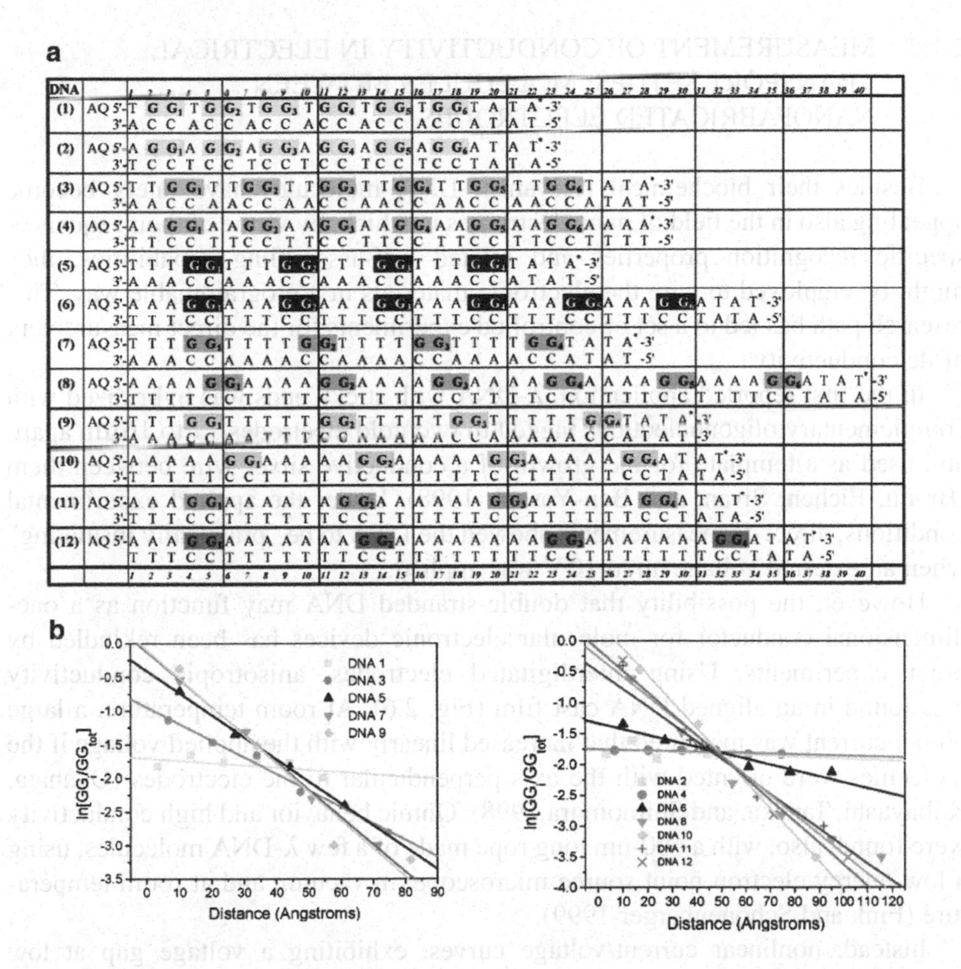

FIG. 2.5. **A.** AQ-terminated DNA sequences. **B.** Distance dependence of the various sequences (Adapted from Liu, Hernandez, and Schuster 2004 with permission; copyright 2004 by the American Chemical Society.).

recently reviewed (Schuster and Landmann 2004). Briefly, an extended electron state (the polaron) over a few stacked planes is formed and then it can move through the helix. It should be remarked that a polaron-based mechanism to explain charge transfer in DNA is also proposed on theoretical grounds by Esther Conwell (2004), although with different details.

Summarizing this section, three different mechanisms are currently proposed to explain the experiments that investigate the *kinetics of charge transfer* through DNA molecules in solution: (a) direct one-step superexchange through very short distances; (b) multistep hole hopping through G orbitals and also through A orbitals, in the latter case by thermal activation; and (c) polaron formation and migration.

2.1.2 MEASUREMENT OF CONDUCTIVITY IN ELECTRICAL TRANSPORT SETUPS: MOLECULES BETWEEN NANOFABRICATED ELECTRODES

Besides their biochemical relevance, DNA molecules have lately become appealing also in the field of nanoelectronics in which, by virtue of their sequence-specific recognition properties and related self-assembling capabilities, they might be employed to wire the electronic materials in a programmable way. This research path has led to a set of controlled experiments for the direct measurement of dc conductivity.

In the first reported application, λ-DNA with sticky ends was hybridized with complementary oligonucleotides tagged to two gold electrodes 12 to 16 μm apart, and used as a template for the growth of a conductive silver wire between them (Braun, Eichen, Sivan, and Ben-Yoseph 1998). Under the applied experimental conditions, direct dc measurements showed the DNA to be "practically insulating" when an external voltage up to 10 V was applied.

However, the possibility that double-stranded DNA may function as a one-dimensional conductor for molecular electronic devices has been rekindled by other experiments. Using interdigitated electrodes, anisotropic conductivity was found in an aligned DNA cast film (Fig. 2.6). At room temperature, a large ohmic current was measured that increased linearly with the applied voltage if the molecules were oriented with the axis perpendicular to the electrodes (Okahata, Kobayashi, Tanaka, and Shimomura 1998). Ohmic behavior and high conductivity were found, also, with a 600-nm long rope made of a few λ-DNA molecules, using a low-energy electron point source microscope, in vacuum and at room temperature (Fink and Schönenberger 1999).

Instead, nonlinear current/voltage curves, exhibiting a voltage gap at low applied voltage, were measured through a single 10 nm long poly(dG)/poly(dC)

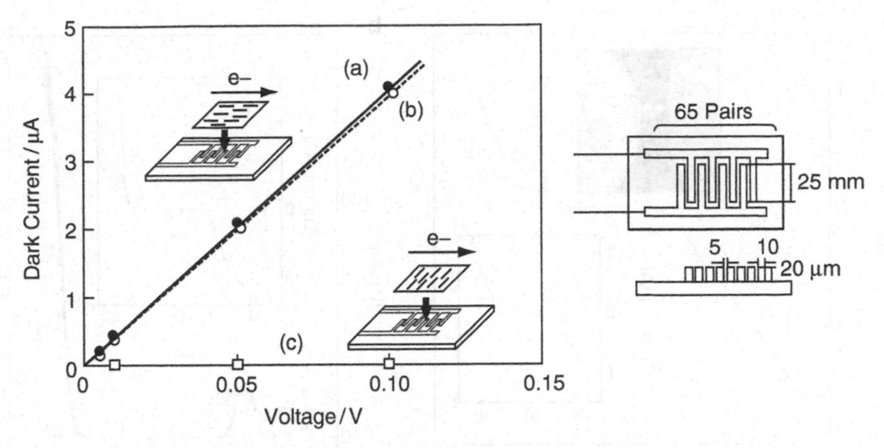

FIG. 2.6. Dark currents for aligned DNA films (20 × 10 mm, thickness 30 ± 5 μm) on comb-type electrodes at 25°C. **A.** DNA strands in the film placed perpendicular to the two electrodes and measured in atmosphere. **B.** The same film as in *(A)* measured in a vacuum at 0.1 mmHg. **C.** DNA strands in the film placed parallel to the two electrodes both in vacuum and in atmosphere. The right panel depicts top and side views of the interdigitated electrodes (From Okahata, Kobayashi, Tanaka, Shimomura 1998 with permission; copyright 1998 by the American Chemical Society.).

DNA molecule inserted between two metal nanoelectrodes by electrostatic trapping (Porath, Bezryadin, de Vries, and Dekker 2000). Large currents were observed, in air and in vacuum, both at ambient temperature and at 4 K (Fig. 2.7A, *solid curve*). The current-voltage curves in Fig. 2.7A are taken from a control experiment to demonstrate that transport was indeed measured on DNA trapped between the electrodes. In fact, the solid curve was measured after trapping a DNA molecule, whereas the dashed curve was measured after incubation of the same sample for 1 hour in a solution with 10 mg/ml DNase I enzyme. The clear suppression of the current upon DNase treatment indicates that the double-stranded DNA was cut by the enzyme. The lower inset of Fig. 2.7A shows two curves measured in a complementary experiment where the enzyme treatment was done in the absence of the Mg ions, which are necessary to activate the enzyme. This observation verifies that the DNA was indeed cut by the enzyme in the original control experiment. The authors suggested that the observed electron transport is best explained by a band model, in which the electronic states are delocalized over the entire length of the base pair stack. Additional experiments were performed in 2001 in the same laboratory (Storm, van Noort, de Vries, and Dekker 2001), in which longer DNA molecules (>40 nm) with various lengths and sequence compositions were stretched on the surface between planar electrodes in various configurations. No current was observed in these experiments, suggesting that charge transport through DNA molecules longer than 40 nm on surfaces is blocked.

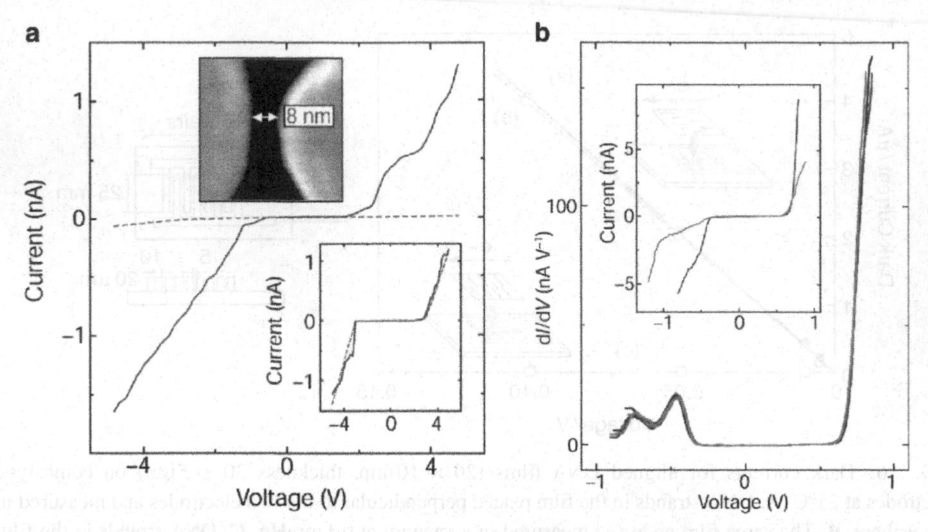

Fig. 2.7. A. Current-voltage curves measured at room temperature on a 10.4-nm-long poly(dG)-poly(dC) molecule trapped between two metal nanoelectrodes: The top inset is a SEM image of the two metal electrodes (light area) and the 8-nm gap between them (dark area). **B.** Differential conductance dI/dV versus applied voltage V at 100 K. The differential conductance manifests a clear peak structure. Good reproducibility can be seen from the six nearly overlapping curves. Peak structures are different from sample to sample. Subsequent sets of I–V measurements can show a sudden change, possibly due to conformational changes of the DNA. The inset shows an example of two typical I–V curves that were measured before and after such an abrupt change. Switching between stable and reproducible shapes can occur upon an abrupt switch of the voltage or by high current (From Porath, Bezryadin, de Vries, and Dekker 2000 with permission; copyright 2000 by Nature Macmillan Publishers Ltd.).

Transport measurements done in Paris on DNA molecules between electrodes report a proximity-induced superconductivity effect in λ-DNA molecules, when the electrodes are of a superconducting material and the resistivity of the DNA sample is determined below the critical temperature (Kasumov, Kociak, Guéron, Reulet, Volkov, Klinov, and Bouchiat 2001). Other authors report a semiconducting-to-metal transition when β-DNA is treated with Zn^{2+} ions in such a fashion that the metal ions should substitute for H-bonds in the base pairs, according to what is suggested by the authors (Rakitin, Aich, Papadopoulos, Kobzar, Vedeneev, Lee, and Xu 2001). The term M-DNA was coined for this kind of DNA derivatives. We note that these two latter experiments on single molecules have never been repeated in other laboratories so far.

Several transport measurements on DNA *films* or *bundles* between electrodes were performed by the group of Tomoji Kawai, usually reporting significant currents with a dependence on sequence and doping (e.g., Yoo, Ha, Lee, Park, Kim,

Kim, Lee, Kawai, and Choi 2001). Other nucleobase phases were also probed (Rinaldi, Branca, Cingolani, Masiero, Spada, and Gottarelli 2001). We refer the reader to the original publications and to recent reviews (Porath, Cuniberti, and Di Felice 2004; Taniguchi and Kawai 2006) for details on the available data on various aggregates.

The preceding discussion highlights a rather puzzling situation, in which the DNA is shown to have resistivities varying by orders of magnitude, from those typical of an insulator to those typical of a conductor, with evidences for ohmic or semiconducting behaviors, and with even some exotic behaviors. The following discusses other experiments done in different measurement settings, and then tries to put some order into this chaotic background.

2.1.3 MEASUREMENT OF CONDUCTIVITY BY SCANNING PROBES

The scanning tunneling microscope (STM) and the atomic force microscope (AFM) are playing an increasingly important role in imaging biomolecules, as well as in measuring their mechanical properties. The same techniques can also be exploited to probe electrical transport in DNA single molecules, in configurations in which the tip of the instrument acts as one of the electrodes in a two-terminal geometry. Scanning probes were initially employed to detect current-voltage curves for DNA molecules deposited on substrates, and only very recently for suspended molecules.

Two typical layouts based on AFM applied to DNA are shown in Fig. 2.8 and are realized in successive steps. These and similar setups in which the AFM tip, usually insulating, is covered by a metal layer, are denoted “conductive AFM” (cAFM). In the example on the left (de Pablo, Moreno-Herrero, Colchero, Gómez Herrero, Herrero, Baró, Ordejón, Soler, and Artacho 2000), one electrode is a planar gold electrode and the other is a metallic scanning tip. First DNA molecules are deposited on a mica substrate and lie horizontal on the surface. The molecules are easily evidenced in AFM images as long chains. After DNA deposition, evaporation of gold is carried out, and one electrode thus covers partially some of the DNA molecules. Finally, the second electrode is a metal-covered AFM tip that can scan the molecule along its length, thus effectively inspecting different molecular lengths. A voltage is applied between the tip and the deposited gold electrode and current-voltage curves are collected. The experiment by de Pablo and coworkers used λ-DNA molecules, with a natural total length of 16 μm, and the tip was positioned at a variable distance from the planar electrode, always greater than 70 nm. No finite current could be measured in these conditions. The authors concluded that λ-DNA molecules longer than 70 nm and deposited on a substrate are insulators. In the example on the right of Fig. 2.8 (Shigematsu, Shimotani, Manabe, Watanabe, and Shimizu 2003), one electrode is a metallic carbon nanotube (the cathode p1) and the other a metallic scanning tip appended on another carbon nanotube (the anode). First, salmon sperm commercial DNA is deposited onto a SiO_2/Si substrate. Second, the horizonal DNA molecules are contacted with two metallic nanotubes (p1 and p2). Finally, the measure is done by applying a finite voltage between the lying nanotube p1 and another nanotube functioning as a metallized AFM tip, and revealing the flowing current. Also in this case, having a scanning electrode, the electrical response can be probed as a function of the molecular length (d_{CA}): At a fixed cathode-anode voltage (V_{CA}) of 2 V, the authors found that the current dropped from 2 nA at d_{CA}~2 nm, to less than 0.1 nA in the length range d_{CA}~6 ÷ 20 nm.

These two brilliant setups are affected by a major drawback: The DNA molecules experience nonspecific interactions with the surface along their axis. It has become clear by AFM imaging of DNA molecules deposited on substrates that: (a) the images reveal an average apparent molecule height of ~0.8, nm

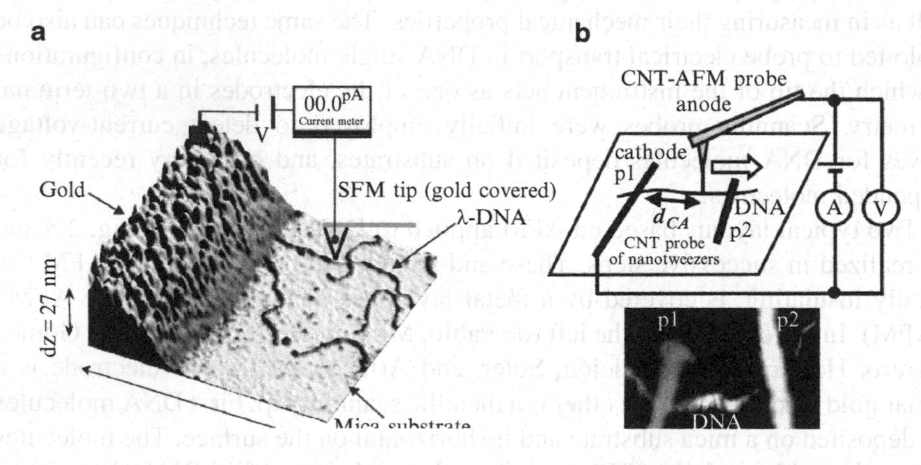

FIG. 2.8. **A.** Three-dimensional SFM image of a cAFM experimental setup, showing two DNA molecules in contact with the left gold electrode. The image size is 1.2 μm × 1.2 μm. A scheme of the electrical circuit used to measure the DNA resistivity is also shown. (From de Pablo, Moreno-Herrero, Colchero, Gómez Herrero, Herrero, Baró, Ordejón, Soler, and Artacho 2000 with permission; copyright 2000 by the American Physical Society.) **B.** Schematic of a different electric current measurement. Two CNT probes (p1 and p2) were set on a DNA. In a two-probe dc measurement, one of the CNT probes (p1) was used as the cathode. A CNT AFM probe was contacted with the DNA to act as an anode. The electric current was measured with varying a distance d_{CA} between the anode and cathode. The lower inset shows an AFM image (scale bar 10 nm) of a single DNA molecule attached with the two CNT probes p1 and p2 (From Shigematsu, Shimotani, Manabe, Watanabe, and Shimizu 2003 with permission; copyright 2003 by the American Institute of Physics.).

much smaller (Muir, Morales, Root, Kumar, Garcia, Vellandi, Jenigian, Marsh, and Vesenka 1998; Kasumov, Klinov, Roche, Guéron, and Bouchiat 2004) than the native helix diameter of ~2.1 nm in crystallized molecules (Saenger 1984); and (b) the surface field forces can induce deformations in terms of persistence length and stability, thus likely affecting the π-stacking and conductivity. These evidences may be important factors to block the current flow, or at least one can undoubtedly say that the occurrence of nonspecific molecule–surface interactions (revealed through the height deformations) prevents from measuring the intrinsic conductivity, which therefore remains elusive. As said in the Introduction, one way out is to avoid this uncontrollable longitudinal molecule–surface interaction, while at the same time strengthening the molecule–electrode coupling (possibly making it covalent). This viable solution has been pursued in two recent works that are reported in the following.

Xu and co-workers (Xu, Zhang, Li, and Tao 2004) developed an STM-based method to measure the I–V curves of molecules in a wet environment: the molecules are captured directly from the solution. They studied both uniform poly(GC)-poly(CG) sequences of length varying from 8 to 14 base pairs, and similar sequences intercalated by AT pairs in the middle, with total length changing from 8 to 12 base pairs and AT length from 0 to 4 base pairs. The molecules were terminated by $(CH_2)_3$-SH thiol groups at the 3′ ends, to form stable S-Au bonds with the gold substrate. The technique to create molecular junctions proceeds as follows: First an STM tip is brought into contact with a flat gold surface covered by the DNA solution; then the tip is retracted under the control of a feedback loop, to break the direct tip–electrode contact. During the latter step, the DNA molecules can bridge the tip and the substrate, as shown in the right inset of Fig. 2.9. For the shortest 8-bp $(GC)_4$ molecules, the results indicated the formation of different junctions with an integer number of molecules: the statistical analysis allowed to determine the value of the conductance of a single molecule. In addition, the current-voltage curves exhibited a linear behavior and currents of about 100 nA at 0.8 V were detected. The statistical analysis of longer uniform molecules and of nonuniform molecules proved that also in these cases the molecular junctions realized by retracting the STM tip contain an integer number of molecules: The conductance of a single molecule in the case when two $(GC)_2$ segments are connected by one or two (AT) segments is smaller than in the case of the uniform sequence with only GC pairs. The length dependence of the $(GC)_n$ (n = 4,5,6,7) electrical measurements indicated a linear behavior of the conductance, increasing with the inverse length, signature of a hopping charge transfer mechanism. For the $(GC)_2(AT)_m(GC)_2$ polymers, the conductance was exponentially decreasing with the number of AT pairs, revealing tunneling of the conducting electrons between the two guanine-cytosine portions.

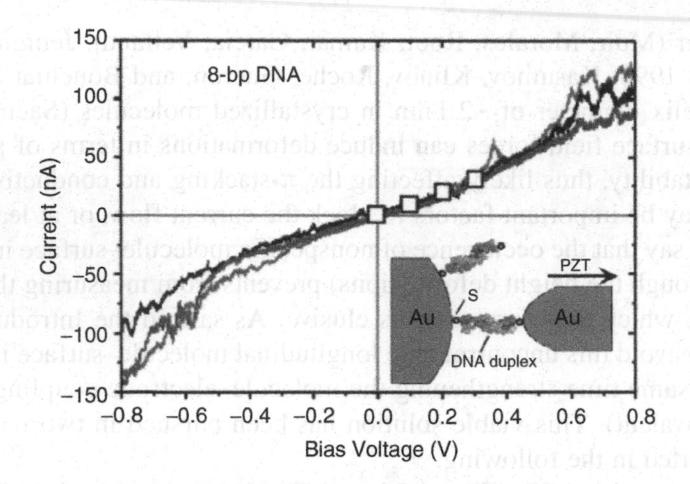

FIG. 2.9. I–V curves of three different junctions formed by the same kind of 8-base pair DNA molecules of sequence $(GC)_4$, with the STM trapping technique illustrated in the inset and described in the text (Adapted from Xu, Zhang, Li, and Tao 2004 with permission; copyright 2004 by the American Chemical Society.).

Cohen and co-workers (Cohen, Nogues, Naaman, and Porath 2005) developed a conceptually similar method using the cAFM. The experiment proceeds (Fig. 2.10A) by depositing a thiolated-ssDNA monolayer on a gold surface, and then arriving with complementary thiolated-ssDNA molecules connected to 10-nm gold nanoparticles (Nogues, Cohen, Daube, and Naaman 2004). A metallized AFM tip, covered with Cr and Au successively, then approaches the Au nanoparticles to perform the electrical measurements: Contact is established between the Au layer of the tip and the gold nanoparticle. Since the thiol-gold bond is known to be covalent (Di Felice and Selloni 2004), the covalent attachment at both electrodes is under control, with no arbitrariness. Therefore, measurements done in this way are comparable. In addition, since one end of each dsDNA polymer is attached to the surface and the other to a nanoparticle, the molecules are in a “standing” configuration—not necessarily perpendicular, but definitely not horizontal—in such a way that the helical conformation is not grossly disturbed by the measurement setup. Note that the technique does not guarantee that just one dsDNA molecule is linked to a nanoparticle, but a rough estimate based on the surface area and on the helix dimensions ensures that only a few DNA molecules may be connected in parallel between the two electrodes. By this “standing” cAFM situation, 26-bp dsDNA molecules of nonuniform sequence were measured. Currents as high as ~200 nA at 2 V are detected (Fig. 2.10B). The I–V curves generally have an S-shape; typical resistances are ~60 MΩ between −1 and +1 V, as low as 2 MΩ at 2 V. The overall shape of the current-voltage characteristics is well reproducible in consecutive sets of measurements, although the quantitative details differ. The authors also provide the results of several control experiments done in order to

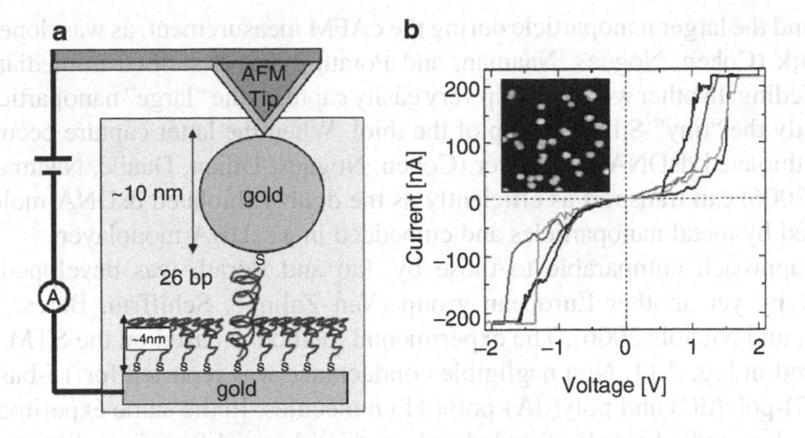

FIG. 2.10. A 26 base-pairs long dsDNA of complex sequence was connected to a metal substrate and a 10-nm metal particle on opposite ends using thiol groups. Schematic representation of the measurement configuration *(A)* and I–V curves, showing high current density *(B)*. The inset (250 × 250 nm^2) portrays an AFM image of the gold nanoparticles (Adapted from Cohen, Nogues, Naaman, and Porath 2005 with permission; copyright 2005 by the National Academy of Sciences USA.).

check that the high currents flow through the DNA molecules and are not due to any artifact.

In a more recent work (Cohen, Nogues, Ullien, Daube, Naaman, and Porath 2006), Cohen and co-workers investigated to a deeper level the role of molecule–electrode contacts within the same "standing-molecule" cAFM geometry, using the 3D mode of the instrument (Gómez-Navarro, Gil, Álvarez, de Pablo, Moreno-Herrero, Horcas, Fernández-Sánchez, Colchero, Gómez-Herrero, and Baró 2002). They compare electrical transport through ssDNA and dsDNA monolayers, with and without upper thiol end-groups. Transport through these systems is also compared with the situation illustrated in the preceding, of dsDNA molecules with gold nanoparticles on top and embedded in an ssDNA monolayer. They find that ssDNA monolayers are unable to transport current. For dsDNA monolayers covalently bonded to the substrate but without thiols on the opposite end, charge transport is generally blocked. dsDNA monolayers with thiols on both ends are in most cases able to transport currents as high as those through the dsDNA molecules with the nanoparticles on top. These results are interpreted as evidence that dsDNA molecules have a finite conductivity. However, to reveal their conductivity in an actual measurement, it is necessary that covalent contacts be established efficiently to both electrodes, otherwise charge injection is hindered. When the top thiol end-group is missing, one contact cannot be established and consequently currents are not revealed. When the top thiol end-group is present, contact may be established directly between the metallized tip and the top thiol. This, however, is less controllable than realizing the contact between the top thiol and a gold nanoparticle before double-strand hybridization, followed by establishing the direct contact between the tip and the larger nanoparticle during the cAFM measurement, as was done in the first work (Cohen, Nogues, Naaman, and Porath 2005) described immediately in the preceding. In other words, the tip very easily captures the "large" nanoparticle and less easily the "tiny" S head-group of the thiol. When the latter capture occurs, the doubly thiolated dsDNA monolayer (Cohen, Nogues, Ullien, Daube, Naaman, and Porath 2006) can transport as efficiently as the doubly thiolated dsDNA molecules decorated by metal nanoparticles and embedded in a ssDNA monolayer.

An approach comparable to those by Tao and Porath was developed most recently by yet another European group (Van Zalinge, Schiffrin, Bates, Haiss, Ulstrup, and Nichols 2006). The experimental method is based on the STM and is illustrated in Fig. 2.11. Non negligible conductance was reported for 15-base pair poly(dG)-poly(dC) and poly(dA)-poly(dT) molecules. In the same experiment the authors also studied single stranded polynucleotides and found smaller conductance for the latter.

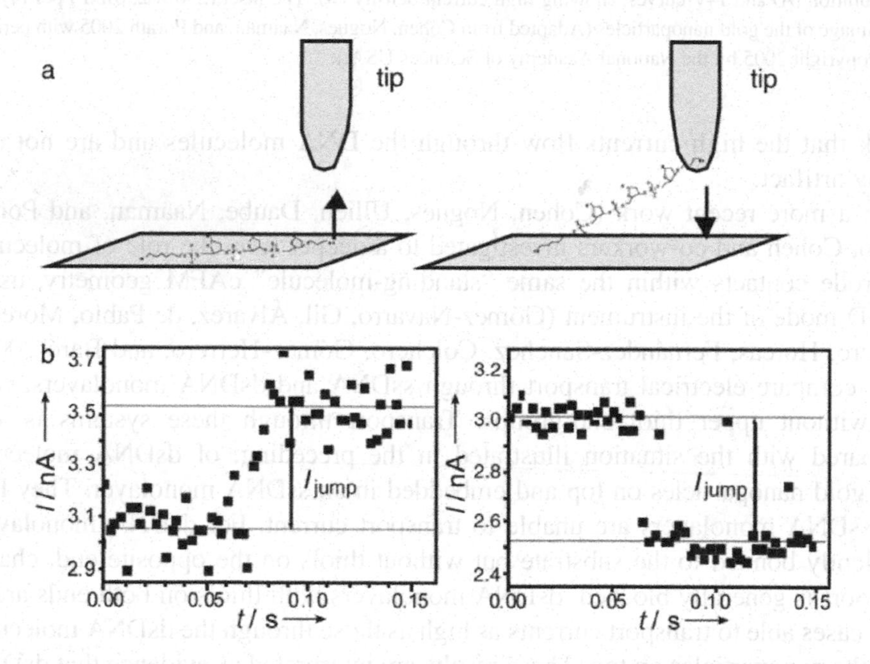

FIG. 2.11. **A.** Another variety of the "standing-molecule" setup with thiol contacts can be realized by trapping molecules with an STM tip. Single- or double-stranded oligonucleotides thiolated at both ends are initially deposited at low coverage on a gold surface. The terminating thiols ensure appropriate chemical and electrical contact between each end of the molecule and the tip and the surface. After positioning the tip, the feedback loop of the STM is switched off and the tunneling current is measured as a function of time. **B.** The current revealed with this method shows characteristic jumps, which are attributed to the attachment or detachment of a molecule to or from the tip. The amplitude of the current jump is used to estimate the conductance, which thus pertains to a single molecule (Adapted from Van Zalinge, Schiffrin, Bates, Haiss, Ulstrup, and Nichols 2006 with permission; copyright 2006 by Wiley-VCH.).

Note that an attempt to realize clear-cut current-voltage measurements with optimal molecule–electrode contacts *within the planar-electrode geometry* was also recently published (Iqbal, Balasundaram, Ghosh, Bergstrom, and Bashir 2005).

The salient results are collected in Table 2.1. From this table we see that until 2003 the reports on DNA conductivity were really disparate. However, one can also clearly see that the experiments were conducted in very different conditions: (a) single molecules or bundles or films; (b) dry or humid atmosphere; and (c) suspended or deposited molecules. Several other discrepancies can be traced as well. Therefore, one cannot say that a given material, the DNA, was probed, but rather several different materials. In order to get reproducible results, the measurements should be performed on a reproducible material. Even in inorganic crystals the conductivity can depend on the sampled phase!

Only very recently experiments aimed at controlling the measured molecular phase were carried out, in particular on synthetic short B-DNA molecules, with different sequences, which do not lie on the substrate and are covalently bonded to the electrodes. All these data indicate that significant currents can flow through the DNA in such conditions. Although there are residual discrepancies to be cleared, and although the mechanism by which the currents can flow is still debated, and although it is not clear whether the same results can be obtained on longer molecules in a controlled way for exploitation in nanoelectronic devices, a huge progress was made in 2004–2006 toward the rationalization of our knowledge and understanding of the ability of DNA molecules to sustain electrical currents (namely, the performance as nanowires).

2.2 THEORETICAL INVESTIGATIONS

2.2.1 ELECTRON TRANSFER THEORY AND TRANSFER INTEGRALS

The theoretical framework that was initially adopted to analyze the results from measurements of charge transfer (CT) rates on DNA in solution is the Marcus–Hush–Jortner electron transfer (ET) theory (Jortner and Bixon 1999). This theory describes the process in which one electron (or hole) is initially located at a given site (the donor D) and finally is detected at another site (the acceptor A), as expressed in the equation:

$$k_{CT} = \frac{2\pi}{\hbar}\left|H_{DA}^{eff}\right|^2 (DWFC) \qquad (1)$$

where H_{DA}^{eff} is the effective coupling between the initial and final states D and A, and DWFC is the density-weighted Franck–Condon factor. The latter accounts for high-frequency intramolecular vibrations and low-frequency solvent modes,

TABLE 2.1. Summary of main experimental results reviewed here*.

Reference	$\rho(\Omega cm)$	$R(\Omega)$	C(nS)	~L(nm)	"Phase" (remarks)	Configuration
Braun et al. 1998, insulator		$>10^{13}$		12000	Single molecule	On substrate
de Pablo et al. 2000, insulator	$>10^{4}$	$>10^{12}$		100	Single molecule	On substrate
Storm et al. 2001, insulator		$>10^{13}$		$40 \div 1000$	Single molecule	On substrate
Porath et al. 2000. semiconductor		$^{\S}2\times10^{8}$		10	Up to 10 molecules	Suspended
Fink and Schönenberger 1999, ohmic	10^{-3}	$<3\times10^{3}$		$600 \div 900$	Up to 10 molecules	Suspended
Kasumov et al. 2001, ohmic		$<10^{5}$		500	Single molecule	On substrate
Shigematsu et al. 2003		$^{\S}10^{9} \div 10^{11}$		$1 \div 20$	Single molecule (very short!)	On substrate
Yoo et al. 2001	$^{\dagger}0.025$	$^{\dagger}1.3\times10^{6}$ $^{\int}10^{8}$		$500 \div 3000$ (electrode gap 20nm)	Film/network	On substrate
Cai et al. 2000	$^{\dagger}1$	$^{\dagger}10^{9}$ $^{\int}10^{10}$		$^{\#}500 \div 3000$ (electrode-tip spacing >40 nm)	Bundles/network	On substrate
Xu et al. 2004, ohmic			100	2.7	Single molecule	Suspended
Cohen et al. 2005, semi conductor		$^{\S}10^{7}$		8.8	Up to 10 molecules	Suspended
Van Zalinge et al. 2006			$^{\dagger}0.5$ $^{\int}1.0$	5.1	Single molecules (relative conductance measurement)	Suspended

*This summary, as well as the whole chapter, is far from being complete. It is designed to make a concise analysis and identify few clear features.

§Estimated by assuming a linear dependence in the reported I–V curves.

† poly(dG)-poly(dC)

∫ poly(dA)-poly(dT)

#The authors measure a length dependence of the resistance. The reported values of R and ρ are estimates for the length of 50 nm.

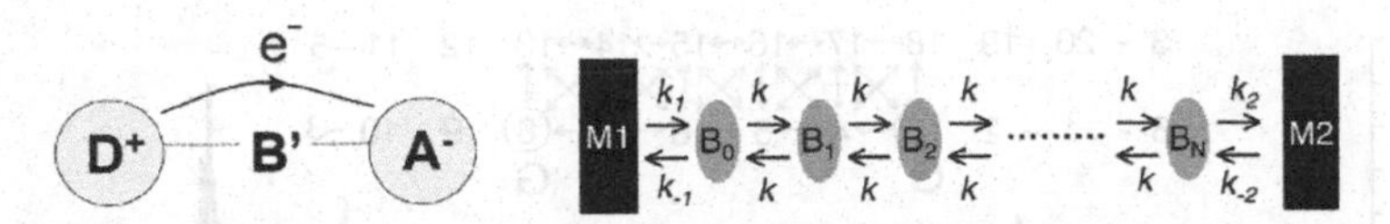

FIG. 2.12. **Left:** A schematic donor-acceptor system. Before the CT process both D and A are neutral. After an electron moves from D to A, D becomes positively charged and A becomes negatively charged. The CT step can occur directly or through a molecular bridge (indicated as B′). **Right:** Scheme of a possible geometry in which the DNA base-pairs are intermediate sites of a molecular bridge (B0, …, BN) and the molecule–electrode contacts are accounted for (M1 and M2 are the left and right electrode, respectively, in an electrode-DNA-electrode nano-junction).

often treated as a unique "solvent reorganization" mode. The scheme of a CT reaction is illustrated in Fig. 2.12. This scheme, based on residence sites for the mobile charge, is a viewpoint somehow alternative to the itinerant-charge picture that is normally employed to describe charge motion in inorganic crystals, in which the mobile charges, occupying the continuum of energy bands, are instantaneously delocalized throughout the material.

On the basis of Hartree–Fock computations of the electronic structure and of $\left|H_{DA}^{eff}\right|$ (Voityuk, Jortner, Bixon, and Rösch 2001; Rösch and Voityuk 2004), the overall emerging picture is that in an arbitrary complex DNA sequence the motion of charge is most likely to occur via successive hopping events (Jortner, Bixon, Langenbacher, and Michel-Beyerle 1998; Bixon, Giese, Wessely, Langenbacher, Michel-Beyerle, and Jortner 1999). Direct superexchange is viable only over extremely short distances (see also Fig. 2.4). Recently, Bixon and Jortner extrapolated the results of electron transfer theory to infer the conditions under which currents may flow through a DNA bridge connecting two metal pads, analyzing different schemes of voltage drops (Bixon and Jortner 2005). Their work clearly establishes some general aspects of the connection between maximal observed currents and structural/electronic features of metal–DNA junctions. The quantitative determination of currents, however, is still inhibited, because the currents depend on several parameters that are unknown at the present level of theory.

The ET equation can also be useful to inquire into the influence of structural fluctuations on the efficiency of charge transfer through DNA. Troisi and Orlandi computed at the Hartree–Fock level the evolution inter-base transfer integrals with time, by selecting several structures of duplex fragments from a classical MD simulation (Troisi and Orlandi 2002). They considered the effective coupling between two guanines in various duplexes in which the two guanines, acting as hole traps with respect to the neighboring bases with a higher oxidation potential, are separated by a bridge comprising four base pairs (Fig. 2.13). The evolution of this parameter with time (plot in Fig. 2.13), due to instantaneous changes in the atomic conformation, is remarkable, and indicates that the CT process is extremely

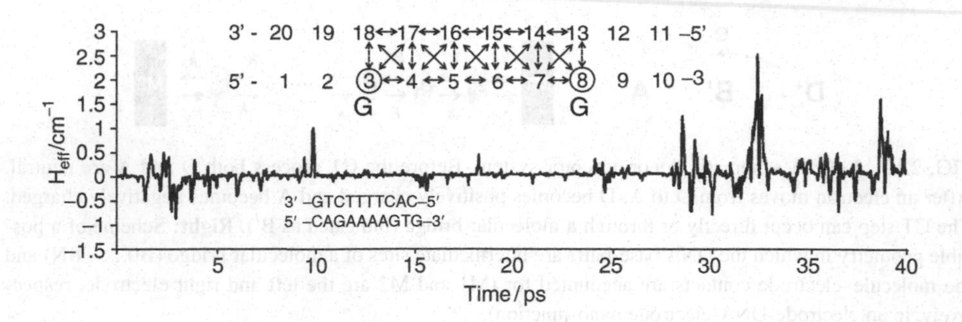

FIG. 2.13. Time evolution of the effective coupling between the two G sites in the simulated duplex fragment *(bottom inset)*. The effective coupling was computed by including the transfer integrals between the intervening bridge bases (the AAAA sequence). The transfer integrals were computed by including all the interactions pointed out in the upper inset. The electronic structure calculations were done every 10 fs for the atomic configurations obtained at the corresponding time from a molecular dynamics simulation (Adapted from Troisi and Orlandi 2002 with permission; copyright 2002 by the American Chemical Society.).

sensitive to the atomic structure. Such a sensitivity indirectly suggests also a major role of the environment, in particular through electron–photon coupling. The authors could also identify the dynamic mode that mostly facilitates electronic coupling: a parallel sliding of the purines. This combined classical-quantum approach represents so far the most advanced treatment of the fluctuation dependence of electronic coupling in DNA and proteins. A similar approach, namely, the electronic structure computed for MD snapshots, was also recently applied to visualize the localization of hole wave functions in very short DNA duplexes. This work revealed an ion-gated hole transfer mechanism (Barnett, Cleveland, Joy, Landman, and Schuster 2001). A single report of fully ab initio Car-Parrinello MD exists, in which the electronic structure is evolved in time along with the atomic positions in the same simulation. It was not used to compute transfer integrals and their time evolution (Gervasio, Laio, Parrinello, and Boero 2005).

2.2.2 ELECTRONIC STRUCTURE BY QUANTUM CHEMISTRY AND DENSITY FUNCTIONAL THEORY

The most sophisticated computational techniques to tackle the electronic structure of molecules are based on quantum chemistry methods that explicitly write the electron wave functions in terms of Slater determinants. For what specifically concerns DNA, important information was gained during the years on gas-phase isolated nucleobases and simple assemblies/complexes by applying the Møller–Plesset second order perturbation theory (MP2) (Helgaker, Jørgensen, and Olsen 2000) to the ground-state Hartree–Fock (HF) configurations (Sponer, Leszczynski and Hobza 1996; Burda, Sponer, Leszczynski, and Hobza 1997). Such studies yield accurate geometries and relative formation energies, as well as excited-state electronic properties including correlation effects. However, they are very cumbersome and thus limited to few-atom systems. Quantum chemistry methods mostly limited to the HF level were also applied to inquire on the electronic structure and transfer integrals of base and base pair stacks (Voityuk, Jortner, Bixon, and Rösch 2001; Rösch and Voityuk 2004).

The calculations performed by Šponer and co-workers to evaluate the structure and stability of hydrogen-bonded and stacked base pairs (Šponer, Leszczynski, and Hobza 1996a,b) should be retained as important milestones for the application of first-principle computational methods to interacting nucleotides. The authors demonstrated that this kind of theoretical analysis is able to reproduce many experimental features, and has a predictive power. The conclusions of their investigations may be summarized in the following information: (a) report and analysis of the structure of the most favorable hydrogen-bonded and stacked dimers; (b) description of the rotation- and distance-dependence of the relative energy of stacked pairs; (c) understanding of the relevant interactions that determine the relative stability of base pairs. Concerning the latter issue, it was found that the energetics of stacked pairs is essentially determined by correlation effects. On the other hand, the energetics of hydrogen-bonded pairs is well described already at the HF level, and also the density functional theory (DFT) treatment is reliable in this context: Van der Waals interactions may be added a-posteriori (Elstner, Hobza, Frauenheim, Suhai, and Kaxiras 2001). The investigations by Šponer and co-workers remained limited to the analysis of the structure and energetics of DNA base pairs. The electronic properties were addressed by quantum chemistry methods mainly at the HF level (Voityuk, Jortner, Bixon, and Rösch 2001; Rösch and Voityuk 2004), which completely lacks correlation terms. The notable exception to this restriction is the work performed by Ladik and coworkers (Ye, Chen, Martinez, Otto, and Ladik 1999), who evaluated the shifts of the electron levels and gaps due to correlation effects in the MP2 scheme.

With the advent of clear-cut experiments on single DNA molecules, the demand for addressing realistic helical conformations became ever-compelling and scientists turned to DFT. The computational load is drastically quenched with respect to wave function-based methods, at the expense of quantitative accuracy and predictivity: In fact, correlation effects are treated only at the mean-field level within an effective potential, and dispersion forces are missing. For the sake of clarity we wish to point out that, in the current level of theory/implementation development, purely DFT studies give access to the electronic structure but do not yield a direct interpretation of charge motion mechanisms. In order to proceed further and to get close to the simulation of experimental data in the field of molecular electronics, progress is required in at least two directions: (a) on one hand, the gap between understanding transport through extended states (as in solids) and describing charge transfer between localized states (as in molecules) should be closed, based on theoretical developments;[1] and (b) on the other hand, the coupling of the electronic structure with intrinsic scattering mechanisms, vibrations, defects, and the environment (Gutierrez, Mandal, and Cuniberti 2005), is desired.[2]

Di Felice and co-workers addressed the performance of plane-wave pseudopotential DFT on isolated bases and base assemblies (H-bonded and stacked base pairs), proving that DFT gives reliable values of bond lengths and bond angles, as well as the correct characters of the frontier electron states (Di Felice, Calzolari, Molinari, and Garbesi 2002). (The early work was based on the BLYP functional, but later on the authors proved that the same accuracy is also obtained with the PW91 and PBE functionals.) They thus applied the method to investigate "mechanical" effects in periodic infinite model guanine stacks, namely the effect of the rotation between adjacent guanines on the HOMO and LUMO bandwidths and effective masses (Fig. 2.14). For such periodic assemblies, prototypes of DNA-based wires, it is possible to define a "crystal" lattice in one dimension. This allows the extension

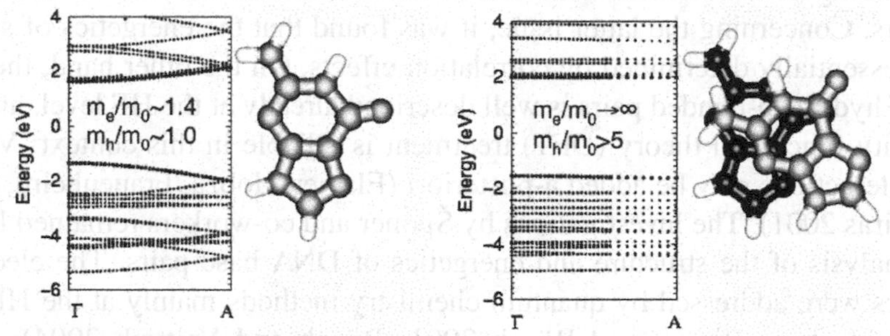
4
2
0
-2
-4
-6
Energy (eV)
m_e/m_0~1.4
m_h/m_0~1.0
Γ
A
m_e/m_0→∞
m_h/m_0>5
▪ Band dispersion and small effective masses for eclipsed guanines (left)
▪ Poor band dispersion and huge effective masses for twisted guanines by 36 degrees as in B-DNA (right)

FIG. 2.14. Band diagrams and atomic geometries of infinite guanine stacks obtained by periodically repeating along the normal to the base plane a structural building-block containing two guanines. Only the building-block is shown. In the left *(right)* configuration the building-block guanines are eclipsed (are rotated by 36°). The DFT-computed effective masses for such fictitious homo-guanine piles are indicated.

---

[1] Ab initio DFT band structure calculations give an equilibrium description of the electronic properties, whereas transport characteristics require the account of non-equilibrium effects. The opposite approach starting from molecular states has an inherent nonequilibrium nature but was traditionally based on empirical electronic parameters. A significant attempt to bridge the two opposite starting points was recently done by A. Nitzan (2001, 2002). Despite the approximations of his formulation, this is to date the most advanced endeavor to match Marcus' electron transfer theory with the theory of transport through extended electron states.

[2] Whereas methods to describe electronic correlation and electron-phonon coupling in an ab initio fashion exist, their application to realistic DNA molecules in a nanojunction setup is presently hindered by the high computational cost. Software and hardware progress will help bypass the current limitations.

of the concept of band dispersion and Bloch-like conduction to molecular wires to which one may assign a periodicity length. The advantage of defining a band structure for a DNA-based wire is the possible interpretation of experimental results in terms of conventional semiconductor-based device conductivity, using the concept of a band of allowed energy values within which a delocalized electron is mobile. Among the several relative rotation angles that were considered, the configurations most representative for the discussion about a viable band-like conductivity mechanism in guanine π-stacks are illustrated in Fig. 2.14, along with the computed bandstructure diagrams. (The authors focus on band-like conductivity because this was suggested from some experimental data [Porath, Bezryadin, de Vries, and Dekker 2000]). The conclusion that can be drawn from the band structure analysis of these model guanine strands is that dispersive bands may be induced only by a large spatial π overlap of the HOMO (LUMO) orbitals of adjacent bases in the periodic stack. Such an overlap is maximum for eclipsed guanines (Fig. 2.14, *left*), and very small for guanines rotated by 36° (Fig. 2.14, *right*) as in B-DNA. These results suggested that a band-like conductivity mechanism occurring via band dispersion and almost free-like mobile carriers (which should be injected through a suitable doping mechanism) is not viable along frozen G-rich stacks. It cannot be excluded that atomic fluctuations locally induce an amount of overlap larger than in frozen B-DNA, with partial interaction and band structure formation at least over a typical coherence length. This is possible for a short length, whereas other dynamical mechanisms should be invoked to explain long-range charge migration. The effect of structural fluctuations in realistic helices is indeed a hotter and hotter topic in the literature about charge transfer in DNA, and is tackled by a variety of methods comprising molecular dynamics (MD), empirical models, quantum chemistry, and DFT (Barnett, Cleveland, Joy, Landman and Schuster 2001; Troisi and Orlandi 2002; Rösch and Voityuk 2004; Gutierrez, Mandal, and Cuniberti 2005).

In the last 5 years, DFT studies on a variety of DNA double helices and focusing on different physical phenomena were conducted (pure electronic structure of frozen systems, effects of water and counterions, effects of vibrations, etc.). Several DFT codes are inherited from solid state applications and are commonly applied to periodic systems. In this spirit, periodic infinite double helices obtained by a finite repeat unit were described: Z-DNA with a 12-plane unit (Gervasio, Carloni, and Parrinello 2002; Gervasio, Laio, Parrinello, and Boero 2005), A-DNA with an 11-plane unit (de Pablo, Moreno-Herrero, Colchero, Gómez Herrero, Herrero, Baró, Ordejón, Soler, and Artacho 2000; Alexandre, Artacho, Soler and Chacham 2003; Artacho, Machado, Sánchez-Portal, Ordejón, and Soler 2003), and B-DNA with a 10-plane unit (Wang, Lewis, and Sankey 2004; Hübsch, Endres, Cox, and Singh 2005). DFT simulations also were performed for short finite DNA oligomers in a solution environment with water and counterions (Barnett, Cleveland, Joy, Landman, and Schuster 2001), to inquire on hole localization associated to counterion motion. A further application of DFT concerns the extraction of transfer integrals from the ab initio Hamiltonian matrix elements (Adessi, Walch and Anantram 2003; Mehrez and Anantram 2005), to construct a simplified tight-binding Hamiltonian for transport calculations. The thus computed intra- and inter-strand couplings are related to the electronic factor of Marcus' theory, whose range of values in different DNA contexts (sequences, environment, length, etc.) has been critically interpreted in terms of mechanisms for charge motion (Jortner, Bixon, Langenbacher, and Michel-Beyerle 1998; Bixon, Giese, Wessely, Langenbacher, Michel-Beyerle, and Jortner 1999). Most such contributions until 2003 were already commented, and the reader is referred to recent reviews on DNA-based molecular electronics (Di Ventra and Zwolak 2004; Endres, Cox, and Singh 2004; Porath, Cuniberti, and DiFelice 2004).

In just one critical short paragraph, the authors summarize their personal understanding of the conclusions from DFT studies of infinite DNA wires in different conformations in view of the topics developed in the rest of the chapter and according to transport paradigms usually employed to describe crystals: namely, by labeling the materials as metal or insulator or semiconductor. (We warn the reader that these notions employed for the description of conduction mechanisms of conventional materials may not be suitable for flexible one-dimensional DNA polymers. The development of the field of DNA-based electronics may reveal a new conceptual framework to properly represent the effective phenomena for charge motion.)

DNA polymers behave as "wide-band-gap semiconductors." Because of the well-known failure of DFT to evaluate band-gap values, it is impossible to know the exact entity, but it spans the range between 4 and 8 eV. However, the "bands" do not exactly have the same meaning as in perfect crystals. True coherent dispersive bands with very small bandwidths are obtained only by imposing the helical symmetry, which corresponds to setting up perfect one-dimensional wires (Hjort and Stafström 2001; Artacho, Machado, Sánchez-Portal, Ordejón, and Soler 2003). This perfection may be very weak under natural circumstances and the environmental perturbations are not obviously manageable as weak perturbations. Alternatively, by imposing only the translational symmetry to build periodic wires, the electronic structure is made of incoherent manifolds stemming from the energy levels of the single bases: each level yields a quasi-flat band; each whole manifold contains a number of such bands equal to the number of planes in the periodicity unit (Gervasio, Carloni, and Parrinello 2002; Gervasio, Laio, Parrinello, and Boero 2005), and its amplitude should be compared to the bandwidth of the corresponding unfolded band obtained under the helical constraint. For instance, in the symmetric description of poly(G)-poly(C) A-DNA given by Artacho and co-workers (Artacho, Machado, Sánchez-Portal, Ordejón, and Soler 2003), the π HOMO of isolated guanine gives origin to a band with an amplitude of 40 meV and the π LUMO of isolated cytosine to a band with an amplitude of 270 meV, unfolded into 11 k points (because of the modulo-11 periodicity) of the Brillouin Zone relative to a single inter-plane separation (Fig. 2.15). In the nonsymmetric description of poly(GC)-poly(CG) Z-DNA given by Gervasio and collaborators (Gervasio, Carloni, and Parrinello 2002), according to their description, 12 quasidegenerate states (because of the modulo-12 periodicity) are positioned at the top of the valence band. These states have a π character and are mostly localized on the guanine nucleobases and basically constitute the HOMO manifold; the "quasidegeneracy" claimed by the authors is consistent with the very small amplitude reported in the preceding for the symmetric case, although a strict comparison is hindered by the difference in base

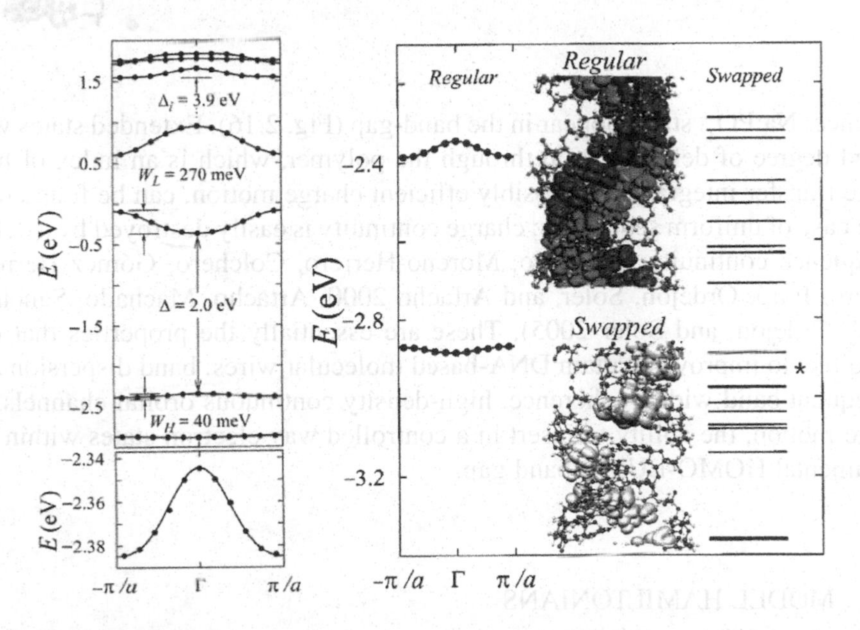

FIG. 2.15. **Left:** The bandstructure of the periodic infinite polymer with regular sequence poly(dG)-poly(dC) in the A-DNA form is reported (11 base pairs per helix turn; 11 base pairs are included in the periodic supercell in the simulation). The bottom panel shows a zoom on the computed HOMO band (dots) superimposed to a tight binding model *(solid curve)*. **Right:** The energy level diagram of a defected configuration is compared with the bandstructure of the regular sequence around the HOMO-LUMO gap: The swapped configuration is obtained by swapping the C and the G between the two strands in a single plane within the periodic unit of 11 planes. The Bloch-like band dispersion is lost with even a single defect. The insets show the orbital convolution: for the regular polymer, the 11 HOMOs *(darker)* plus 11 LUMOs *(lighter)* are combined; for the swapped polymer, only the 11 HOMOs are combined. It is evident that the defect causes a localization in the wave functions, thus acting against a *wire-like* behavior (Adapted from Artacho, Machado, Sánchez-Portal, Ordejón, and Soler 2003 with permission; copyright 2003 by Taylor & Francis Ltd.).

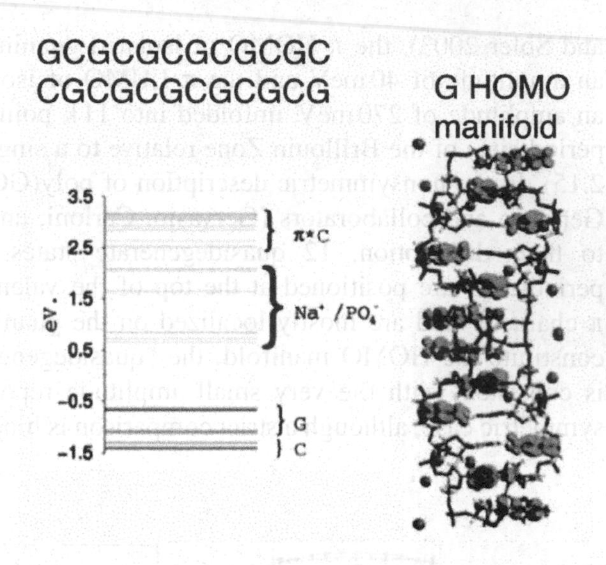

FIG. 2.16. **Left:** Energy level diagram obtained by a periodic-supercell Car-Parrinello simulation of poly(dGC)-poly(dCG) in the Z-DNA conformation (12 base-pairs per helix turn). **Right:** Isosurface plot of the convolution of 12 guanine HOMO electron states from the same simulation (Adapted from Gervasio, Carloni, and Parrinello 2002 with permission; copyright 2002 by the American Physical Society.).

sequence; $Na^{+}PO_4^{-}$ states appear in the band-gap (Fig. 2.16). Extended states with a good degree of delocalization through the polymer, which is an index of high charge transfer integrals and possibly efficient charge motion, can be found only in the case of uniform sequences; charge continuity is easily destroyed by a defect in sequence continuity (de Pablo, Moreno-Herrero, Colchero, Gómez Herrero, Herrero, Baró, Ordejón, Soler, and Artacho 2000; Artacho, Machado, Sánchez-Portal, Ordejón, and Soler 2005). These are essentially the properties that one would like to improve to attain DNA-based molecular wires: band dispersion and consequent band-width, coherence, high-density continuous orbital channels for charge motion, the ability to insert in a controlled way electron states within the fundamental HOMO-LUMO band gap.

2.2.3 MODEL HAMILTONIANS

It was pointed out several times in the previous sections that quantum calculations from first principles are very cumbersome and cannot be applied to really simulate the experiments. Two main issues are still inaccessible, due partly to computer limitations and partly to theory development: (a) the computation of the quantum conductance in the non-equilibrium regime, for a direct comparison to transport data; and (b) the role of the electron-phonon coupling, with full account of internal (DNA) and external (environment/solvent) vibrations.

Model Hamiltonians can fill these gaps. These methods rely on a simplified description of the electronic structure through tight-binding parameters that are either taken from more sophisticated calculations, when available, or adjusted to fit experimental data. Being computationally much lighter, they can give access to more complicated quantities, such as the quantum conductance and the I–V curves.

Cuniberti and co-workers fitted the experimental transport data that showed a voltage gap in the I–V curves by the so-called "fishbone model," which is the simplest scheme to represent a DNA molecule between two electrodes (Cuniberti, Craco, Porath, and Dekker 2002). The metal pads were represented by their Fermi functions (electron reservoirs), without any atomistic information. The quantum conductance that enters the current-voltage characteristics was described according to the Landauer formalism, in which the computation of Green's function needs the electronic structure of the system. For electronic structure computations the model Hamiltonian comes into play: Each base pair of the poly(dG)-poly(dC) DNA molecule was represented as a single tight-binding site with the Hamiltonian:

$$H_{\text{mol}} = \sum_{i,\sigma} \varepsilon_i b^{\dagger}_{i\sigma} b_{i\sigma} - \sum_{\langle i,j \rangle,\sigma} t_{ij} b^{\dagger}_{i\sigma} b_{j\sigma} \tag{2}$$

where ε_i are on-site energies, t_{ij} are inter-site hopping terms. Given the uniform sequence of the investigated duplex, a single ε and a single $t_{\parallel}$ were considered. Depending on the parameters, such a linear chain can be anything from an insulator to a metal. In particular, if the hopping integral $t_{\parallel}$ is large enough, then the electroniccoupling ensures a metallic behavior of the wire. The opening of the band-gap was instead explained in terms of coupling of the linear chain to transversal degrees of freedom standing for the backbone: inter-site hopping through the backbone was inhibited. The base-backbone coupling was expressed through an additional term in the tight-binding Hamiltonian. The complete model is pictorially shown in Fig. 2.17. Whereas it is remarkable that such a simple model

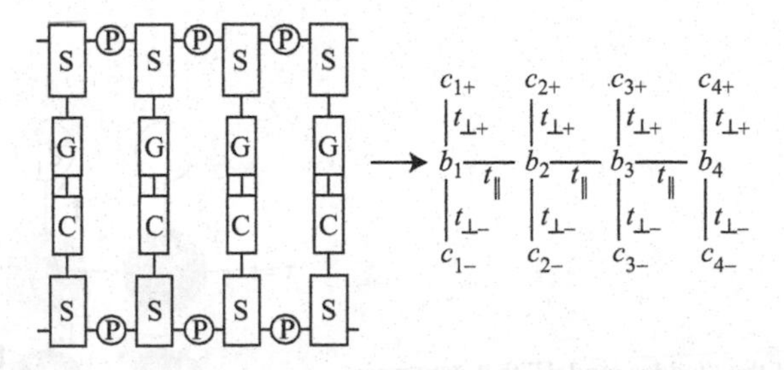

FIG. 2.17. Scheme of the simplest "fishbone model". Each base pair (GC on the left) is represented with a single tight-binding site (b_i on the right). Each chain site b_i is coupled to the side backbone (Adapted from Cuniberti, Craco, Porath, and Dekker 2002 with permission; copyright 2002 by the American Physical Society.).

can reproduce experimental data, we warn the reader that the quantitative agreement is affected by the values of the employed parameters, and that the hopping parameter that reproduces the experimental gap is indeed very large with respect to predictions from quantum chemistry and DFT.

The model in Fig. 2.17 can be complicated in a rather straightforward manner by: (a) adding the coupling with the environment through a thermal phonon bath (Gutierrez, Mandal, and Cuniberti 2005); (b) writing down a "ladder model" (Fig. 2.18), in which the two strands are left explicit instead of substituting each base pair with a single site (thus, both stacking and H-bonding parameters are needed); or (c) combining these two points.

The first way was recently followed with success in Cuniberti's group. Gutierrez and co-workers exclusively focused on low-energy transport, i.e., the charge injection energies are small compared with the π–π^* band-gap of the *isolated* molecule ($2 \div 4$ eV). Thus, equilibrium transport could be considered. At low energies, only the frontier orbitals of the molecule are expected to contribute to transport. Since the inter-strand hopping is very weak (a few meV), any inter-strand contribution to transport was neglected. Finally, the sugar-phosphate backbones were viewed as inducing a perturbation of the π-stack that leads to gap opening. So far: the "fishbone model." The environment was added in the simplest approximation as a collection of harmonic oscillators, which linearly couple to the charge density on the backbone sites. The Hamiltonian is reported in Fig. 2.19D. The bath was completely described by introducing its spectral density, and ohmic dissipation was assumed. By performing a unitary transformation, the linear coupling to the bath can be eliminated. However, the transversal coupling terms will be renormalized by exponential bosonic operators. Several coupling regimes to the bath can be analyzed. In the strong-coupling limit, the time scales of the charge–bath interaction are much shorter compared with typical electronic time scales. The impact of the bath on the electronic structure is twofold. On one hand, the strong coupling to the environment leads to the

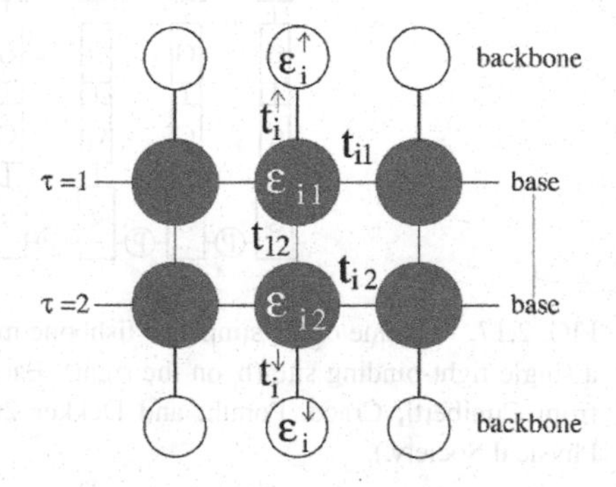

FIG. 2.18. Scheme of the "ladder model," that expresses the two strands explicitly and need both stacking and H-bonding coupling parameters (Adapted from Klotsa, Römer, and Turner 2005 with permission; copyright 2005 by the Biophysical Society.).

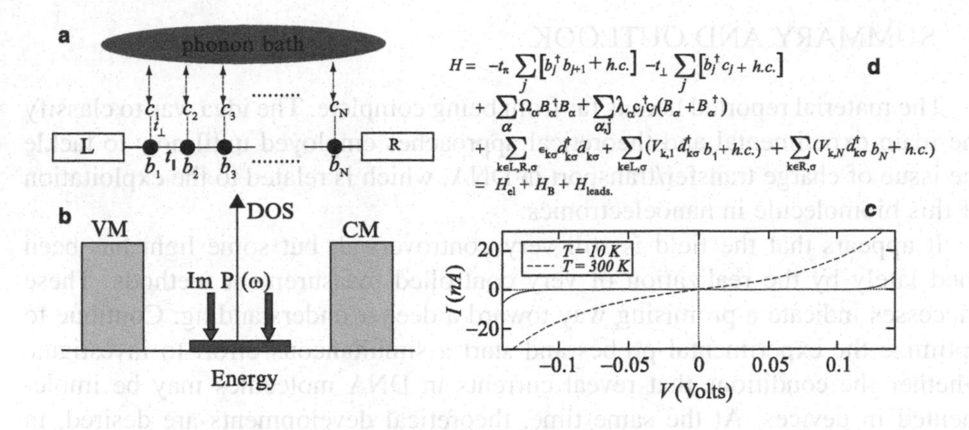

FIG. 2.19. **A.** Scheme of the "fishbone and bosonic model." **B.** Scheme of the band structure resulting from the model: A low-density of states is induced at the Fermi level by the coupling to the bosonic bath. The "fishbone model" alone gives only VM (valence band) and CM (conduction band). **C.** Current-voltage characteristics computed in the Landauer formalism at different temperatures. **D.** Model Hamiltonian. $H_{el} = H_b + H_c$ is the Hamiltonian of the electronic system *(first row)*. H_B contains both the Hamiltonian of the bath and the mutual interaction of the bath with the electronic degrees of freedom *(second row)*. H_{leads} contains the electrode Hamiltonian as well as the tunneling Hamiltonian describing the propagation of a charge from the leads onto the wire and vice versa. In the absence of coupling to the bath, the eigenstates of H_{el} yield two manifolds (VM and CM) containing N states each and separated by a band gap, whose magnitude basically depends on the size of the transversal coupling $t_\perp$. It is important to note that this gap survives the thermodynamic limit, so that the signatures of the environment inside the band gap can be addressed without any mixing with intrinsic electronic states (Adapted from Gutierrez, Mandal, and Cuniberti 2005 with permission; copyright 2005 by the American Chemical Society.).

emergence of new bath-induced electronic states inside the molecular band-gap. Although such states are strongly damped by the dissipative action of the bath and do not appear as resonances in the transmission, they induce an *incoherent temperature-dependent* background that leads to a finite density of states inside the gap (Fig. 2.19B). Charges injected at low energies will now find states supporting *transport;* thus, a finite current at low bias may exist (Fig. 2.19C).

The "ladder model" (see Fig. 2.18) without coupling to a phonon bath was employed to investigate the localization length as a function of sequence (Klotsa, Römer, and Turner 2005). Efforts devoted to combine the phonon bath with the "ladder model" are still ongoing. We underline again that the main potentiality of model Hamiltonians is the *qualitative* analysis of trends, for instance, behavior of transport features by varying environment, sequence, electronic coupling, etc. However, the *quantitative* agreement with experiment is affected by the choice of input parameters and should be interpreted with care. For a deeper understanding of the models and the various approximations the reader is referred to the original papers and recent dedicated reviews (Starikov, Lewis, and Tanaka 2006).

3 SUMMARY AND OUTLOOK

The material reported here is far from being complete. The idea was to classify the main experimental and theoretical approaches employed until now to tackle the issue of charge transfer/transport in DNA, which is related to the exploitation of this biomolecule in nanoelectronics.

It appears that the field is still very controversial, but some light has been shed lately by the realization of very controlled measurement methods. These successes indicate a promising way toward a deeper understanding: Continue to optimize the experimental probes and start a simultaneous effort to investigate whether the conditions that reveal currents in DNA molecules may be implemented in devices. At the same time, theoretical developments are desired, in particular to simulate the experiments as closely as possible. Although it was not mentioned in the previous sections, it should to pointed out here that the STM, used in the spectroscopy mode (STS), may also play a strong role in understanding the electronic properties of DNA molecules lying on surfaces. This instrument should in fact be able to probe the molecular density of electronic states, which contains the information on the ability of electrons to move longitudinally and/or transversally. Very few STM and STS investigations have been reported (Lindsay, Li, Pan, Thundat, Nagahara, Oden, DeRose, Knipping, and White 1991; Kanno, Tanaka, Nakamura, Tabata, and Kawai 1999; Iijima, Kato, Nakanishi, Watanabe, Kimura, Suzuki, and Maruyama 2005; Xu, Endres, Tsukamoto, Kitamura, Ishida, and Arakawa 2005; Shapir, Cohen, Sapir, Borovok, Kotlyar, and Porath 2006), and progress in this direction may bring important new insight, especially in comparison to theoretical analysis of the images and of the electronic structure (Shapir, Yi, Cohen, Kotlyar, Cuniberti, and Porath 2005). It is interesting to note that the scanning tunneling spectroscopy has also been suggested as a tool for DNA sequencing (Zwolak and Di Ventra 2005), even though other methods for sequencing (e.g., flowing through nanopores) appear more promising (Zwolak and Di Ventra 2006).

Last but not least, the other way to pursue DNA-based nanoelectronics is by looking at derivatives that may exhibit intrinsic conductivity better than the double helix. One of the most appealing candidates in the guanine quadruple helix G4-DNA (Fig. 2.20). Other viable candidates are DNA hybrids with metal ions and double helices in which the native bases are substituted with more aromatic bases that may improve the longitudinal π-overlap (Calzolari, Di Felice, Molinari, and Garbesi 2002; Di Felice, Calzolari, and Zhang 2004; Di Felice, Calzolari, Garbesi, Alexandre, and Soler 2005; Kotlyar, Borovok, Molotsky, Cohen, Shapir, and Porath 2005).

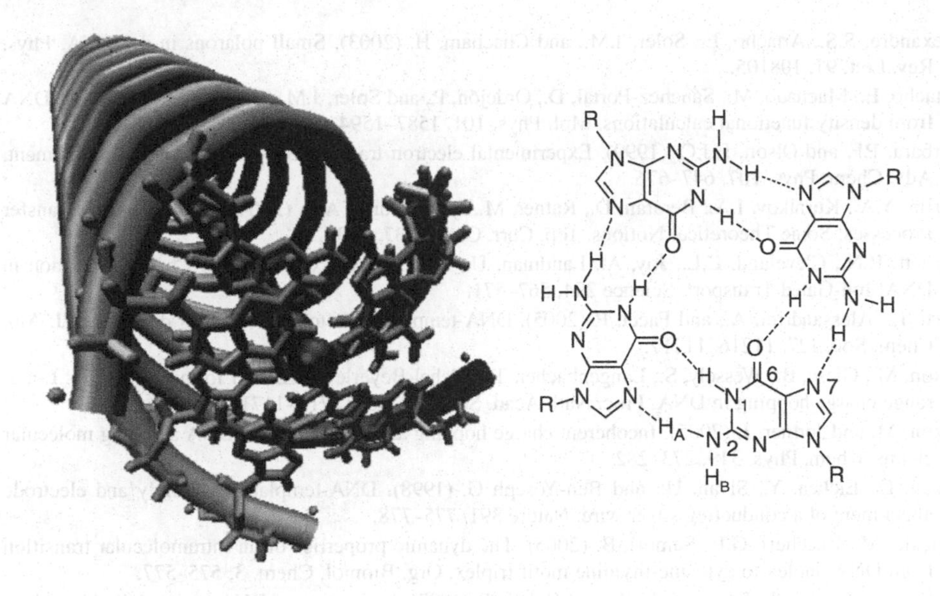

FIG. 2.20. **Left:** Perspective view of a guanine quadruplex: The tubules represent the four strands; the front plane is a stick representation of a guanine tetrad, which is the basic motif in each plane; the central spheres stand for metal cations. **Right:** Formula of the guanine tetrad, the planar unit.

ACKNOWLEDGMENTS

I am indebted first of all to my direct collaborators in Modena: Arrigo Calzolari, Anna Garbesi, Elisa Molinari, HouYu Zhang, Manuela Cavallari, Daniele Varsano, Giorgia Brancolini, Agostino Migliore, Stefano Corni. I am also grateful to several other colleagues for collaboration and fruitful discussions: Danny Porath, Joshua Jortner, Julio Gómez-Herrero, Jose Soler, Simone Alexandre, Angel Rubio, Giovanni Cuniberti, Rafael Gutierrez, Sasha Kotlyar, Shlomo Yitzchaik. The research activity is funded by: the EC through FET-Open projects "DNA-based Nanowires" (IST-2001-38951) and "DNA-based Nanodevices" (FP6-029192) and Marie Curie network "EXCITING" (contract HPRN-CT-2002-00317); the Italian Ministry for University and Research (MIUR) by contract "FIRB-NOMADE" (2003–2006); the INFM Parallel Computing Initiative and CINECA.

REFERENCES

Adessi, C., Walch, S., and Anantram, M.P. (2003). Environment and structure influence on DNA conduction. Phys. Rev. B 67, 081405.

Alberti, P. and Mergny, J.L. (2003). DNA duplex-quadruplex exchange as the basis for a nanomolecular machine. Proc. Natl. Acad. Sci. USA 100,1569–1573.

Alexandre, S.S., Artacho, E., Soler, J.M., and Chacham, H. (2003). Small polarons in dry DNA. Phys. Rev. Lett. 91, 108105.

Artacho, E., Machado, M., Sánchez-Portal, D., Ordejón, P., and Soler, J.M. (2003). Electrons in dry DNA from density functional calculations. Mol. Phys. 101, 1587–1594.

Barbara, P.F. and Olson, E.J.C. (1999). Experimental electron transfer kinetics in a DNA Environment. Adv. Chem. Phys. 107, 647–676.

Berlin, Y.A., Kurnikov, I.A., Beratan, D., Ratner, M.A., and Burin, A.L. (2004). DNA Electron Transfer Processes: Some Theoretical Notions. Top. Curr. Chem. 237, 1–36.

Barnett, R.N., Cleveland, C.L., Joy, A., Landman, U., and Schuster, G.B. (2001). Charge Migration in DNA: Ion-Gated Transport. Science 294, 567–571.

Berti, L., Alessandrini, A., and Facci, P. (2005). DNA-templated photoinduced silver deposition. J. Am. Chem. Soc. 127, 11216–11217.

Bixon, M., Giese, B., Wessely, S., Langenbacher, T., Michel-Beyerle, M.E., and Jortner, J. (1999). Long-range charge hopping in DNA. Proc. Natl. Acad. Sci. USA 96, 11713–11716.

Bixon, M. and Jortner, J. (2005). Incoherent charge hopping and conduction in DNA and long molecular chains. Chem. Phys. 319, 273–282.

Braun, E., Eichen, Y., Sivan, U., and Ben-Yoseph G. (1998). DNA-templated assembly and electrode attachment of a conducting silver wire. Nature 391, 775–778.

Brucale, M., Zuccheri, G.P., Samorì, B. (2005). The dynamic properties of an intramolecular transition from DNA duplex to cytosine-thymine motif triplex. Org. Biomol. Chem. 3, 575–577.

Burda, J.V., Sponer, J., Leszczynski, J., and Hobza, P. (1997). Interaction of DNA base pairs with various metal cations (Mg^{2+}, Ca^{2+}, Sr^{2+}, Ba^{2+}, Cu^{+}, Ag^{+}, Au^{+}, Zn^{2+}, Cd^{2+}, and Hg^{2+}): Nonempirical ab initio calculations on structures, energies, and nonadditivity of the interaction. J. Phys. Chem. B 101, 9670–9677.

Cai, L., Tabata, H., and Kawai, T. (2000). Self-assembled DNA networks and their electrical conductivity. Appl. Phys. Lett. 77, 3105–3106.

Calzolari, A., Di Felice, R., Molinari, E., and Garbesi, A. (2002). G-quartet biomolecular nanowires. Appl. Phys. Lett. 80, 3331–3333.

Cohen, H., Nogues, C., Naaman, R., and Porath, D. (2005). Direct measurement of electrical transport through single DNA molecules of complex sequence. Proc. Natl. Acad. Sci. USA 102, 11589–11593.

Cohen, H., Nogues, C., Ullien, D., Daube, S., Naaman, R., and Porath, D. (2006). Electrical characterization of self-assembled single- and double-stranded DNA monolayers using conductive AFM. Faraday Discussions 131, 367–376.

Conwell, E.M., Rakhmanova, S.V. (2000). Polarons in DNA. Proc. Natl. Acad. Sci. USA 97, 4556–4560.

Conwell, E. (2004). Polarons and Transport in DNA. Top. Curr. Chem. 237, 73–101.

Cuniberti, G., Craco, L., Porath, D., and Dekker, C. (2002). Backbone-induced semiconducting behavior in short DNA wires. Phys. Rev. B 65, 241–314.

de Pablo, P.J., Moreno-Herrero, F., Colchero, J., Gómez Herrero, J., Herrero, P., Baró, A.M., Ordejón, P., Soler, J.M., Artacho, E. (2000). Absence of conductivity in λ-DNA. Phys. Rev. Lett. 85, 4992–4995.

Di Felice, R., Calzolari, A., Garbesi, A., Alexandre, S.S., and Soler, J.M. (2005). Strain-dependence of the electronic properties in periodic quadruple helical G4-wires. J. Phys. Chem. B 109, 22301–22307

Di Felice, R., Calzolari, A., Molinari, E., and Garbesi, A. (2002). Ab initio study of model guanine assemblies: The role of π-π coupling and band transport. Phys. Rev. B 65, 045104.

Di Felice, R., Calzolari, A., and Zhang, H. (2004). Towards metalated DNA-based structures. Nanotechnology 15, 1256–1263.

Di Felice, R. and Selloni A. (2004). Adsorption modes of cysteine on Au(111): Thiolate, amino-thiolate, disulfide. J. Chem. Phys. 120, 4906–4914.

Di Ventra, M. and Zwolak, M. (2004). In: Encyclopedia of Nanoscience and Nanotechnology, vol. 2, pp. 475–493, edited by Nalwa, S.H., American Scientific Publishers, CA.

Elstner, M., Hobza, P., Frauenheim, T., Suhai, S., and Kaxiras, E. (2001). Hydrogen bonding and stacking interactions of nucleic acid base pairs: a density-functional-theory based treatment. J. Chem. Phys. 114, 5149–5155.

Eley, D.D. and Spivey, D.I. (1962). Semiconductivity of organic substances. Part 9 – Nucleic acid in the dry state. Trans Faraday Soc 58, 416–428.

Endres, R.G., Cox, D.L., and Singh, R.R.P. (2004). The quest for high-conductance DNA. Rev. Mod. Phys. 76, 195–214.

Fink, H.W. and Schönenberger, C. (1999). Electrical conduction through DNA molecules. Nature 398, 407–410.

Giese, B. (2004). Hole injection and hole transfer through DNA: the hopping mechanism. Top. Curr. Chem. 236, 27–44.

Gervasio, F.L., Carloni, P., and Parrinello, M. (2002). Electronic structure of wet DNA. Phys. Rev. Lett. 89, 108102.

Gervasio, F.L., Laio, A., Parrinello, M., and Boero, M. (2005). Charge localization in DNA fibers. Phys. Rev. Lett. 94, 158103.

Giese, B., Amaudrut, J., Köhler, A.K., Spormann, M., and Wessely, S. (2001). Direct observation of hole transfer through DNA by hopping between adenine bases and by tunnelling. Nature 412, 318–320.

Gómez-Navarro, C., Gil, A., Álvarez, M., de Pablo, P.J., Moreno-Herrero, F., Horcas, I., Fernández-Sánchez, R., Colchero, J., Gómez-Herrero, J., and Baró, A.M. (2002). Scanning force microscopy three-dimensional modes applied to the study of the dielectric response of adsorbed DNA molecules. Nanotechnology 13, 314–317.

Grinstaff, M.W. (1999). How do charges travel through DNA? An update on a current debate. Angew. Chem. Int. Ed. 38, 3629–3635.

Gutierrez, R., Mandal, S., and Cuniberti, G. (2005). Quantum transport through a DNA wire in a dissipative environment. Nano Lett. 5, 1093–1097.

Hall, D.B., Holmlin, R.E., and Barton, J.K. (1996). Oxidative DNA damage through long-range electron transfer. Nature 382, 731–735.

Helgaker, T., Jørgensen, P., and Olsen, J. (2000). Molecular Electronic-Structure Theory, Wiley, Chichester.

Hjort, M. and Stafström, S. (2001). Band resonant tunneling in DNA molecules. Phys. Rev. Lett. 87, 228101.

Hübsch, A., Endres, R.G., Cox, D.L., and Singh, R.R.P. (2005). Optical conductivity of wet DNA. Phys. Rev. Lett. 94, 178102.

Iijima, M., Kato, T., Nakanishi, S., Watanabe, H., Kimura, K., Suzuki, K., and Maruyama, Y. (2005). STM/STS study of electron density of states at the bases sites in the DNA alternating copolymers. Chem. Lett. 34, 1084–1085.

Joachim, C. and Ratner, M.A. (2005). Molecular electronics: some views on transport junctions and beyond. Proc. Natl. Acad. Sci. USA 102, 8801–8808.

Jortner, J. and Bixon, M., eds. (1999). Electron transfer: from isolated molecules to biomolecules. Part I. Wiley, New York.

Jortner, J., Bixon, M., Langenbacher, T., and Michel-Beyerle, M.E. (1998). Charge transfer and transport in DNA. Proc. Natl. Acad. Sci. USA 95: 12759–12765.

Kanno, T., Tanaka, H., Nakamura, T., Tabata, H., and Kawai, T. (1999). Real space observation of double-helix DNA structure using a low temperature scanning tunneling microscopy. Jpn. J. Appl. Phys. 38, L606–L607.

Kasumov, A.Y., Kociak, M., Guéron, S., Reulet, B., Volkov, V.T., Klinov, D.V., and Bouchiat, H. (2001). Proximity-induced superconductivity in DNA. Science 291, 280–282.

Kasumov, A.Y., Klinov, D.V., Roche, P.E., Guéron, S., and Bouchiat, H. (2004). Thickness and low-temperature conductivity of DNA molecules. Appl. Phys. Lett. 84, 1007–1009.

Kelley, S.O. and Barton, J.K. (1999). Electron transfer between bases in double helical DNA. Science 283, 375–381.

Kelley, S.O., Jackson, N.M., Hill, M.G., and Barton, J.K. (1999). Long-range electron transfer through DNA films. Angew. Chem. Int. Ed. 38, 941–945.

Keren, K., Krueger, M., Gilad, R., Ben-Yoseph, G., Sivan, U., and Braun E. (2002). Sequence-specific molecular lithography on single DNA molecules. Science 297, 72–75.

Keren, K., Berman, R.S., Buchstab, E., Sivan, U., and Braun, E. (2003). DNA-templated carbon nanotube field-effect transistor. Science 302, 1380–1382.

Klotsa, K.D., Römer, R.A., and Turner, M. (2005). Electronic transport in DNA. Biophys. J. 89, 2187–2198.

Kotlyar, A.B., Borovok, N., Molotsky, T., Cohen, H., Shapir, E., and Porath, D. (2005). Long monomolecular guanine-based nanowires. Adv. Mater. 17, 1901–1905.

Lewis, F.D., Wu, T., Liu, X., Letsinger, R.L., Greenfield, S.R., Miller, S.E., and Wasielewski, M.R. (2000). Dynamics of photoinduced charge separation and charge recombination in synthetic DNA hairpins with stilbenedicarboxamide linkers. J. Am. Chem. Soc. 122, 2889–2902.

Lindsay, S.M., Li, Y., Pan, J., Thundat, T., Nagahara, L.A., Oden, P., DeRose, J.A., Knipping, U., and White, J.W. (1991). Studies of the electrical properties of large molecular adsorbates. J. Vac. Sci. Technol. 9, 1096–1101.

Liu, C.S., Hernandez, R., and Schuster, G.B. (2004). Mechanism for radical cation transport in duplex DNA oligonucleotides. J. Am. Chem. Soc. 126, 2877–2884.

Mao, C., Sun, W., Shen, Z., and Seeman, N.C. (1999). A nanomechanical device based on the B–Z transition of DNA. Nature 397, 144–146.

Maubach, G., Csáki, A., and Fritzsche, W. (2003). Controlled positioning of a DNA molecule in an electrode setup based on self-assembly and microstructuring. Nanotechnology 14, 546–550.

Meggers, E., Michel-Beyerle, M.E., and Giese, B. (1998). Sequence dependent long range hole transport in DNA. J. Am. Chem. Soc. 120, 12950–12955.

Mehrez, H. and Anantram, M.P. (2005). Interbase electronic coupling for transport though DNA. Phys. Rev. B 71, 115405.

Muir, T., Morales, E., Root, J., Kumar, I., Garcia, B., Vellandi, B., Jenigian, D., Marsh, T., and Vesenka J. (1998). The morphology of duplex and quadruplex DNA on mica. J. Vac. Sci. Technol. A 16, 1172–1177.

Murphy, C.J., Arkin, M.A., Jenkins, Y., Ghatlia, N.D., Bossman, S., Turro, N.J., and Barton, J.K. (1993). Long-range photoinduced electron transfer through a DNA helix. Science 262, 1025–1029.

Nitzan, A. (2001). Electron transmission through molecules and molecular interfaces. Ann. Rev. Phys. Chem. 52, 681–750.

Nitzan, A. (2002). The relationship between electron transfer rate and molecular conduction. 2. The sequential hopping case. Isr. J. Chem. 42, 163–166.

Nogues, C., Cohen, S.R:, Daube, S.S., and Naaman R. (2004). Electrical properties of short DNA oligomers characterized by conducting atomic force microscopy. Phys. Chem. Chem. Phys. 6, 4459–4466.

Okahata, Y., Kobayashi, T., Tanaka, K., Shimomura, M. (1998). Anisotropic electric conductivity in an aligned DNA cast film. J. Am. Chem. Soc. 120, 6165–6166.

Porath, D., Bezryadin, A., de Vries, S., and Dekker, C. (2000). Direct measurement of electrical transport through DNA molecules. Nature 403, 635–638.

Porath, D., Cuniberti, G., and Di Felice, R. (2004). Charge transport in DNA-based devices. Topics Curr. Chem. 237, 183–227.

Rakitin, A., Aich, P., Papadopoulos, C., Kobzar, Y., Vedeneev, A.S., Lee, J.S., and Xu, J.M. (2001). Metallic conduction through engineered DNA: DNA nanoelectronic building blocks. Phys. Rev. Lett. 86, 3670–3673.

Retèl, J., Hoebee, B., Braun, J.E.F., Lutgerink, J.T., van den Akker, E., Wanamarta, A.H., Joenjie, H., and Lafleur, M.V.M. (1993). Mutational specificity of oxidative DNA damage. Mutat. Res. 299:165–182.

Richter, J., Mertig, M., Pompe, W., Mönch, I., and Schackert, H.K. (2001). Construction of highly conductive nanowires on a DNA template. Appl. Phys. Lett. 78, 536–538.

Rinaldi, R., Branca, E., Cingolani, R., Masiero, S., Spada, G.P., and Gottarelli, G. (2001). Photodetectors fabricated from a self-assembly of a deoxyguanosine derivative. Appl. Phys. Lett. 78, 3541–3543.

Rösch, N. and Voityuk, A.A. (2004). Quantum chemical calculation of donor-acceptor coupling for charge transfer in DNA. Top. Curr. Chem. 237, 337–372.

Saenger W. (1984). Principles of Nucleic Acid Structure. Springer, Berlin.

Schuster, G.B. (2000). Long-range charge transfer in DNA: transient structural distortions control the distance dependence. Acc. Chem. Res. 33, 253–260.

Schuster, G.B. and Landmann, U. (2004). The mechanism of long-distance radical cation transport in duplex DNA: ion-gated hopping of polaron-like distortions. Top. Curr. Chem. 236, 139–161.

Seeman, N.C. (1998). Nucleic acid nanostructures and topology. Angew. Chemie Int. Ed. 37, 3220–3238.

Seeman, N.C. (2003). DNA in a material world. Nature 421:427–431.

Shapir, E., Cohen, H., Sapir, T., Borovok, N., Kotlyar, A.B., and Porath, D. (2006). High-resolution STM imaging of novel poly(dG)-poly(dC) DNA. J. Phys. Chem. B 110, 4430–4433.

Shapir, E., Yi, J., Cohen, H., Kotlyar, A.B., Cuniberti, G., and Porath, D. (2005). The puzzle of contrast inversion in DNA STM imaging. J. Phys. Chem. B 109, 14270–14274.

Shigematsu, T., Shimotani, K., Manabe, C., Watanabe, H., and Shimizu, M. (2003). Transport properties of carrier-injected DNA. J. Chem. Phys. 118, 4245–4252.

Šponer, J., Leszczynski, J., and Hobza, P. (1996a). Structures and energies of hydrogen-bonded DNA base pairs. A nonempirical study with inclusion of electron correlation. J. Phys. Chem. 100, 1965–1974.

Šponer, J., Leszczynski, J., and Hobza, P. (1996b). Nature of nucleic acid-base stacking: nonempirical ab initio and empirical potential characterization of 10 stacked base dimers. Comparison of stacked and H-bonded base pairs. J. Phys. Chem. 100, 5590–5596.

Starikov, E.B., Lewis, J.P., and Tanaka, S., eds. (2006). Modern Methods for Theoretical Physical Chemistry of Biopolymers. Elsevier, St. Louis.

Storm, A.J., van Noort, J., de Vries, S., and Dekker, C. (2001). Insulating behavior for DNA molecules between nanoelectrodes at the 100 nm length scale. Appl. Phys. Lett. 79, 3881–3883.

Taniguchi, M. and Kawai, T. (2006). DNA electronics. Physica E 33, 1–12.

Troisi, A. and Orlandi, G. (2002). Hole migration in DNA: a theoretical analysis of the role of structural fluctuations. J. Phys. Chem. B 106, 2093–2101.

Turro, N.J. and Barton, J.K. (1998). Paradigms, supermolecules, electron transfer and chemistry at a distance. What's the problem? The science or the paradigm? J. Biol. Inorg. Chem. 3, 201–209.

Van Zalinge, H., Schiffrin, D.J., Bates, A.D., Haiss, W., Ulstrup, J., and Nichols, R.J. (2006). Measurement of single- and double-stranded DNA oligonucleotides. Chem. Phys. Chem. 7, 94–98.

Voityuk, A.A., Jortner, J., Bixon, M., and Rösch, N. (2001). Electronic coupling between Watson-Crick pairs for hole transfer and transport in deoxyribonucleic acid. J. Chem. Phys. 114, 5614–5620.

Wagenknecht, H.A., ed. (2005). Charge Transfer in DNA. From Mechanism to Application. Wiley-VCH, Weinheim, 2005.

Wang, H., Lewis, J.P., and Sankey, O.F. (2004). Band-gap tunneling states in DNA. Phys. Rev. Lett. 93, 016401.

Warman, J.M., de Haas, M.P., and Rupprecht, A. (1996). DNA: a molecular wire? Chem. Phys. Lett. 249:319–322.

Watanabe, H., Manabe, C., Shigematsu, T., Shimotani, K., Shimizu, M. (2001). Single molecule DNA device measured with triple-probe atomic force microscope. Appl. Phys. Lett. 79, 2462–2464.

Xu, B., Zhang, P., Li, X., and Tao, N. (2004). Direct conductance measurement of single DNA molecules in aqueous solution. Nano Lett. 4, 1105–1108.

Xu, M.S., Endres, R.G., Tsukamoto, S., Kitamura, M., Ishida, S., and Arakawa, Y. (2005). Conformation and local environment dependent conductance of DNA Molecules. Small 1, 1–4.

Ye, Y.-J., Chen, R.S., Martinez, A., Otto, P., and Ladik, J. (1999). Calculation of hopping conductivity in aperiodic nucleotide base stacks. Solid State Commun. 112, 139–144.

Yoo, K.-H., Ha, D.H., Lee, J.-O., Park, J.W., Kim, J., Kim, J.J., Lee, H.-Y., Kawai, T., Choi, H.Y. (2001). Electrical conduction through poly(dA)-poly(dT) and poly(dG)-poly(dC) DNA molecules. Phys. Rev. Lett. 87, 198102.

Yurke, B., Turberfield, A.J., Mills, Jr, A.P., Simmel, F.C., Neumann, J.L. (2000). A DNA-fuelled molecular machine made of DNA. Nature 406, 605–608.

Zwolak, M. and Di Ventra, M. (2005). Electronic signature of DNA nucleotides via transverse transport. Nano Lett. 5, 421–424.

Zwolak, M. and Di Ventra, M. (2006). Fast DNA sequencing via tranverse electronic transport. Nano Lett. 6, 779–782.

Electronics for Genomics

3

DNA Detection with Metallic Nanoparticles

Robert Möller, Grit Festag, and Wolfgang Fritzsche

1 INTRODUCTION

Through the fast increasing knowledge about biomolecules and their interaction with other biomolecules, the study of those biorecognition events has become more and more important. Especially the knowledge gained through the sequencing of the human and other organisms' genomes makes the detection of DNA a powerful tool in diagnostics. This growing information is also coupled with many important technological advances that enable us to monitor the interaction between biomolecules and use this information to identify microorganisms or other pathogens. In recent years the detection of DNA sequences has gained more and more importance in fields such as genetics, criminology, pathology, and food safety.

One of the most remarkable technologies that had a strong impact on DNA detection is probably the DNA chip (or gene chip) technology. This allows researchers to conduct thousands or even millions of different DNA sequence tests simultaneously on a single chip or array (Schena 2003). Nevertheless, the DNA chip technology also comes with some limitations. The miniaturized probe spots on a DNA chip need expensive fabrication procedures, which are also used in microfabrication (Ermantraut, Schulz, Tuchscheerer, Wölfl, Saluz, and Köhler 1998; Lipshutz, Fodor, Gingeras, and Lockhart 1999; Pirrung 2002). Also the readout of

A. Offenhäusser and R. Rinaldi (eds.), *Nanobioelectronics - for Electronics, Biology, and Medicine*,
DOI: 10.1007/978-0-387-09459-5_4, © Springer Science+Business Media, LLC 2009

the DNA arrays must be miniaturized. Finally, the detection scheme must be sensitive enough to detect just a few copies of target and selective enough to discriminate between target DNAs with slightly different compositions. DNA labeled with fluorescent dyes in combination with confocal fluorescence imaging of DNA chips has provided the high sensitivity needed (Schena 2003). Fluorescence labeling also allows multicolor labeling, making possible the multiplexed detection of differently labeled single-stranded DNA targets on one array. However, fluorescent dyes have significant drawbacks. Fluorescent labels are expensive and susceptible to photobleaching. Furthermore, they have broad emission and absorption bands, which limit the number of dyes that can be used in a multiplexed setup. Expensive equipment is needed for the readout of fluorescently labeled DNA chips and sophisticated algorithms are used to analyze the data. These disadvantages have limited the use of DNA chips mainly to specialized laboratories, whereas their full potential is not used in on-site and point-of-care applications, which require rapid, cost efficient measurement systems that are easy to use.

Recently, metal nanoparticles have approached as alternative labels in a variety of DNA detection schemes. They enable researchers to circumvent the disadvantages of fluorescent labeling as well as making new DNA detection schemes possible. These new detection schemes are based on the unique properties of metal nanoparticles, such as large optical extinction and scattering coefficients, catalytic activity, and surface electronics. Most notably, gold nanoparticles have been used for the DNA detection, because they can be easily modified with biomolecules. Therefore, this chapter mainly focuses on DNA detection with gold nanoparticles. However, other metal nanoparticles, such as Ag, Pt, and Pd, have also been used for DNA detection, but are mentioned only briefly.

2 NANOPARTICLE-BASED MOLECULAR DETECTION

2.1 NANOPARTICLE SYNTHESIS AND BIOCONJUGATION

As mentioned, gold nanoparticles are the most prominent nanoparticle labels for DNA detection. They can be synthesized in organic (Brust, Walker, Bethell, Schiffrin, and Whyman 1994; Brust, Fink, Bethell, Schiffrin, and Kiely 1995) as well as in aqueous solution (Hayat 1989; Grabar, Freeman, Hommer, and Natan 1995) through the reduction of metal salts by different reducing agents (Daniel and Astruc 2004). These methods with certain modifications can also be applied to synthesize other metal colloids. For the synthesis of gold colloids gold (III) salts are commonly used. As reducing agents organic acids, substituted ammonias, formaldehyde, and many others are described in the literature, but the citrate reduction of $HAuCl_4$ is commonly used (Turkevich, Stevenson, and Hiller 1951). Today, colloidal gold is available from a number of commercial distributors such as Ted-Pella, Sigma-Aldrich, and British Biocell. These particles are coated with a layer of surfactant molecules to stabilize the particles. In order to control the stability of the colloidal solution and the composition of surfactant molecules, one has to do one's own particle synthesis. Normally, the commercially available colloidal gold is not very stable. Higher concentrations of salts, such as NaCl, cause these particles to aggregate and change the color of the solution from red to blue to black. In order to stabilize the particles in solution they can be modified with surfactants such as phosphines or carboxyls (Parak, Pellegrino, Zanchet, Micheel, Williams, Boudreau, LeGros, Larabell, and Alivisatos 2003). These molecules are negatively charged; by modifying the surface of the particles with these molecules the particles strongly repel each other electrostatically. This leads to an increased stabilization of the nanoparticle solution.

Different approaches can be used to modify gold colloids with DNA or oligonucleotides. The simplest method is the nonspecific absorption of biomolecules to the surface of the particles (Gearheart, Ploehn, and Murphy 2001). This method does not lead to a stable and covalent binding of the oligonucleotides, as it is necessary for the use of gold nanoparticles as labels in DNA detection. It is also possible to attach biotin modified DNA to nanoparticles that are modified with avidin or streptavidin (Alexandre, Hamels, Dufour, Collet, Zammatteo, De Longueville, Gala, and Remacle 2001). The most commonly used method to functionalize gold nanoparticles with oligonucleotides is the use of thiol-modified oligonucleotides, taking advantage of the high attraction of gold surfaces and thiol groups. Colloidal gold can be modified with thiol-modified oligonucleotides just by mixing and incubation of oligonucleotides and gold nanoparticle solution (Alivisatos, Johnsson, Peng, Wilson, Loweth, Bruchez Jr., and Schultz 1996; Mirkin, Letsinger, Mucic, and Storhoff 1996). If stored under the right conditions (so that the exposure to light, high temperatures, oxidants, and microbes is minimized) these conjugates are stable for years. However, a dissociation of the thiol-bound oligonucleotides has been described when heated above temperatures of 60°C (Letsinger, Elghanian, Viswanadham, and Mirkin 2000) or the influence of a competing thiol reagent (Li, Jin, Mirkin, and Letsinger 2002). In order to improve the stability of the thiol bond, different ligands have been designed to increase the strength of the bond by using multiple thiol groups to bind one oligonucleotide to the gold surface (Fritzsche and Taton 2003). To further improve the orientation and conformation of oligonucleotides bound to gold nanoparticles, a modification with mercaptohexanol can be applied (Park, Brown, and Hamad-Schifferli 2004). This leads to better results when those nanoparticle conjugates are hybridized to their target structures. Because of the weaker bond energies between sulfur and other metals the modification of colloids made of other metals (Ag, Cu, Pd, and Pd) is less successful.

The creation of so-called core shell particles might be a solution to this problem. If a thin layer of gold (~1 atm thick) is grown around those particles one can take advantage of the high bond energies between gold and sulfur, while the particle retains the characteristics of the core (Cao, Jin, and Mirkin 2001; Taton 2002).

A true covalent coupling of biomolecules can also be achieved by using Nanogold™ (Nanoprobes Inc.). These commercially available 1.4-nm gold nanoparticles bear a functional group that makes the defined coupling of oligonucleotides possible. Through the use of a maleimide or N-hydroxysuccinimide groups the gold nanoparticles can specifically react with thiol or amino groups, by this forming 1:1 nanoparticle:biomolecule conjugates (Hainfeld, Powell, and Hacker 2004).

2.2 DETECTION METHODS FOR NANOPARTICLE-LABELED DNA

In the last 10 years, a great number of different detection schemes for nanoparticle-labeled DNA have been described in the literature. The described detection schemes can be divided into solution-based and chip or surface-based methods. Because the same properties of the nanoparticles as well as the same methods of detection are often used for the detection of biomolecular interactions both in solution-based and surface-bound detection, the detection schemes are divided by the method of detection. Two major approaches for the detection of nanoparticle-labeled DNA are used: One uses the interesting optical properties of metal nanoparticles for the optical detection of the binding events, whereas the other approach tries to simplify the readout by generating an electrical signal. This electrical readout renounces the use of any optical equipment for the detection of binding events, using electrochemical, electromechanical, or electrical detection methods, and by this often makes the detection easier, more cost efficient, and robust.

Because of their large surface-to-volume ratios metal nanoparticle solutions have different properties compared with the respective bulk materials of which they are made. The bright colors displayed by many metal nanoparticle solutions are one of the most interesting properties of metal colloids. They are caused by the collective resonance of their conductive electrons (plasmons) in the metal (Kreibig and Vollmer 1995; Yguerabide and Yguerabide 1998a). Particle size, shape, and composition mainly influence the optical properties of metal colloids (Kelly, Coronado, Zhao, and Schatz 2003). The interesting optical properties make a number of different detection schemes possible. However, other properties, such as their high specific mass or electric and electrochemical properties, have been exploited for the development of different detection schemes as well. Because those methods renounce the use of costly and sensitive equipment needed for optical detection they might be used for the construction of cost-efficient and robust devices needed for point-of-care diagnostics.

2.2.1 OPTICAL ABSORBANCE

Gold nanoparticles have been used for decades to label certain structures in histochemical samples for analysis with electron and light microscopy (Horisberger 1981). Through a specific deposition of silver onto the nanometer-sized particles it is possible to grow the particles to a size that can be imaged with a conventional light microscope (Hacker, Springall, Van Noorden, Bishop, Grimelius, and Polak 1985; Hacker, Grimelius, Danscher, Bernatzky, Muss, Adam, and Thurner 1988; Hacker, Danscher, Graf, Bernatzky, Schiechl, and Grimelius 1991) (Fig. 3.1).

A first indication how gold nanoparticles could be used for the optical detection of DNA was reported in 1996. Thirteen-nanometer gold nanoparticles, modified with two species of oligonucleotides, were bound together when a piece of single-stranded DNA (complementary with its ends to the two oligonucleotides immobilized on the particles) was added. Through hybridization of the nanoparticles, the interparticle distance decreased and the color of the solution changed from red to blue. This effect was reversible by heating the solution above melting temperature of the formed DNA double strand (Elghanian, Storhoff, Mucic, Letsinger, and Mirkin 1997) (Fig. 3.2A–D). This reversibility makes this aggregation and color change different from the one observed when the particles were incubated in high salt buffer. An interesting observed phenomenon is that the gold nanoparticle DNA conjugates displayed extremely sharp melting curves with a narrower melting temperature range than unlabeled or fluorophore-labeled DNA (Elghanian, Storhoff,

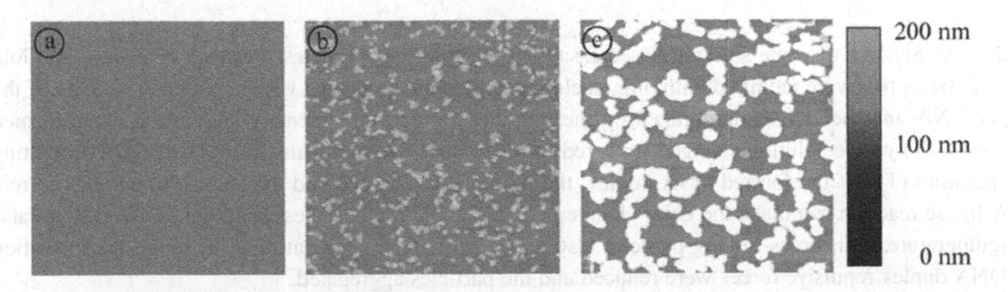

FIG. 3.1. Growth of surface bound gold nanoparticles imaged with an AFM. 15 nm gold nanoparticles were immobilized: **A.** on a chip; **B.** a 2-min silver deposition was performed; and **C.** an 8-min silver deposition was performed.

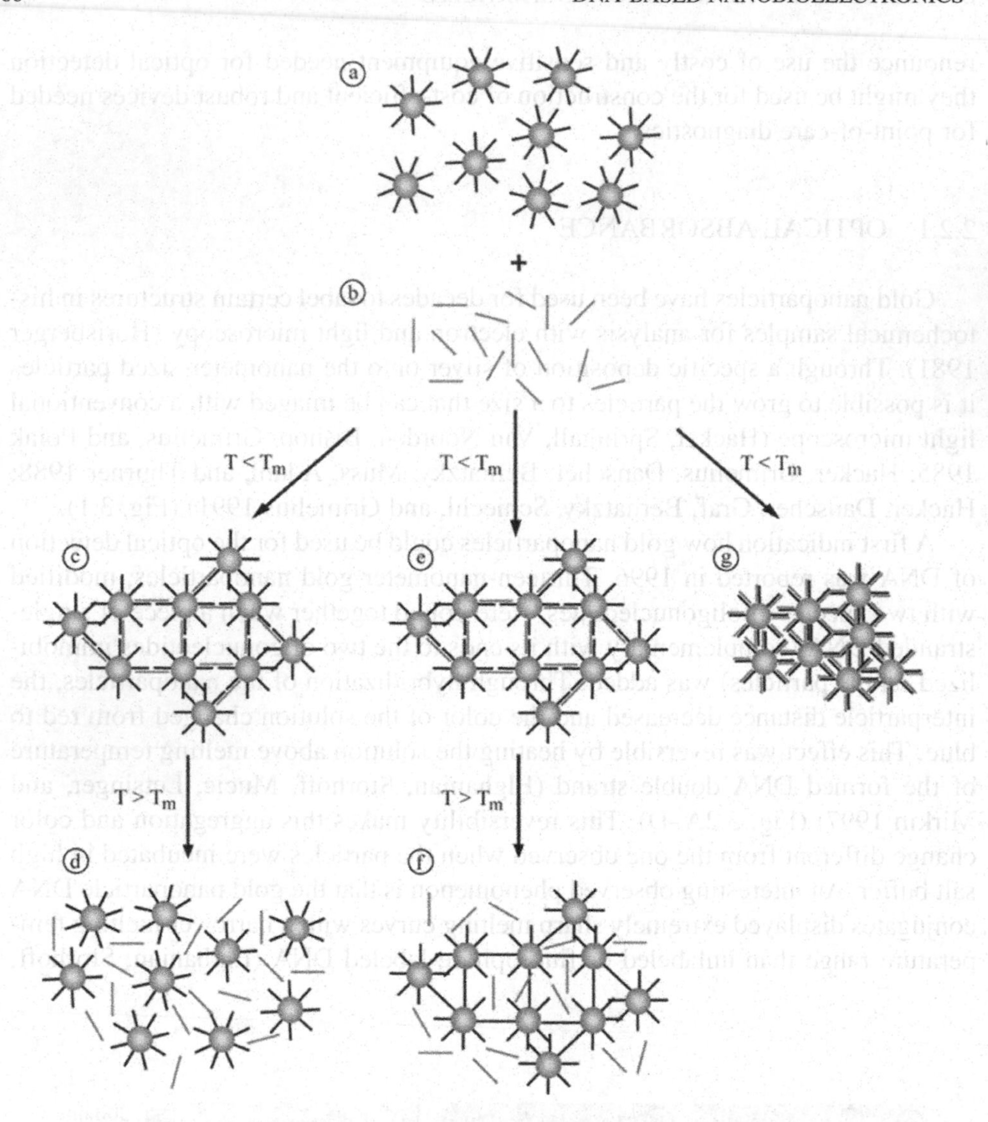

FIG. 3.2. Scheme of some solution-based assays using gold nanoparticles for DNA detection. **A.** Gold nanoparticles that were modified with oligonucleotides, **B.** were incubated with target DNA. **C,E,G.** If the target DNA and the DNA immobilized on the particles were complementary, aggregates were formed and the color of the solution changed from red to blue. **D.** If the aggregates were heated above melting temperature (T_m) of the formed DNA duplex, the particles were freed and the color change back to red. **F.** A ligase reaction can make the color change permanent and the particles could not be freed by a raise of temperature. Non–cross-linking particles also led to permanent aggregates. **G.** Through the formation of DNA duplex repulsive forces were reduced and the particles aggregated.

Mucic, Letsinger, and Mirkin 1997; Storhoff, Elghanian, Mucic, Mirkin, and Letsinger 1998; Reynolds, Mirkin and Letsinger 2000; Jin, Wu, Li, Mirkin, and

Schatz 2003). This simple test for the detection of DNA with nanoparticle labels showed how easily gold nanoparticles could be used in the detection of biomolecular interaction. The effect of the color change could be further enhanced by spotting the formed conjugates onto a white solid support and thereby providing a permanent record of the test (Elghanian, Storhoff, Mucic, Letsinger, and Mirkin 1997; Storhoff, Elghanian, Mucic, Mirkin, and Letsinger 1998). This approach allows the use of at least two color pairs (red to blue for gold nanoparticles; yellow to black for gold-coated silver nanoparticles). The described spectral shift is caused by a change of the interparticle spacing, and thus influences the plasmon resonance (Quinten 2001). If the spacing between the particles is larger than the average particle diameter the solution appears red, but if the spacing is reduced by biomolecular recognition, the color changes to blue. The same principle of the spectral shift detection was also used in a surface-bound test. The chip substrate was modified with nanoscale metal islands. If nanoparticles were bound to the surface through a biorecognition event a shift in the absorbance spectra could be detected (Hütter and Pileni 2003), enabling the detection of DNA sequence mismatches.

Modifications of the described system made the detection of a single mismatch even more feasible; therefore, a ligase reaction was added to the system. Only if the two nanoparticle species and the target DNA formed a perfect double strand, the ligase could bind the two particles together, making the color change irreversible (Fig. 3.2E,F). In the case of a mismatch the particles were not bound together by the ligase, making the color change reversible by heating the sample (Li, Chu, Liu, Jiang, He, Zhang, Shen, and Yu 2005). An irreversible color change is also achieved in the approach with so-called non–cross-linking nanoparticles. In this case the particles are not bound together by the binding of target DNA. Rather, the target DNA binds to individual particles. If a perfect duplex is formed, the repulsive forces between the particles are reduced, causing the particles to aggregate in a high salt buffer (Fig. 3.2G). The formation of not perfectly matched duplexes caused a smaller reduction of the repulsive forces and therefore no aggregation (Sato, Hosokawa, and Maeda 2003; Sato, Hosokawa, and Maeda 2005). In another approach, the gold nanoparticles were not even specifically modified with thiol-functionalized oligonucleotides. Short oligonucleotides adsorb unspecifically to the nanoparticles. In the presence of complementary single-stranded target DNA the oligonucleotides would disorb, thus causing the aggregation of the nanoparticles due to reduced electrostatic repulsion (Li and Rothberg 2004a,b). If the oligonucleotides were additionally modified with a fluorophore, this system could also be used for fluorescence detection. As long as the oligonucleotides were bound to the gold surface, the fluorescence of the dye was quenched. However, after hybridization with the target DNA the fluorescence signal could be detected (Li and Rothberg 2004c).

The specific interaction between the oligonucleotides immobilized on the gold nanoparticles and their complementary target sequence can also be used on DNA chips. DNA spots on a microstructured surface can be detected in reflected or transmitted light. The detected absorbance is directly related to the amount of bound particles or target DNA (Reichert, Csaki, Kohler, and Fritzsche 2000; Köhler, Csaki, Reichert, Möller, Straube, and Fritzsche 2001; Csaki, Kaplanek, Möller, and Fritzsche 2003). A disadvantage of this method is the high density of surface-bound nanoparticles that is needed for such a simple optical detection, meaning that lower concentrations of target DNA can not be detected. However, the signal can be significantly enhanced by the deposition of silver on the surface-bound gold nanoparticles. This process makes it possible to optically detect a single nanoparticle (Csaki, Kaplanek, Möller, and Fritzsche 2003) and low target DNA concentrations even in the femtomolar region (Taton, Mirkin, and Letsinger 2000) (Fig. 3.3). The combination with chip-integrated photodiodes shows how a miniaturized point-of-care device could be realized based on the described detection method (Li, Xu, Zhang, Wang, Peng, Lu, and Chan 2005).

2.2.2 OPTICAL SCATTERING

For the detection via optical absorbance the large extinction coefficients of 10^6 to 10^{12} M^{-1} cm^{-1}, which exceed those of organic dyes by many order of magnitudes, are exploited. Furthermore, metal nanoparticles have the ability to scatter light very efficiently. This property has also been used for a number of detection schemes. The flux of light from a single 80-nm particle has been estimated to be equal to the fluorescence light flux from 6×10^6 colocalized, individual fluorescein molecules, demonstrating the great potential that light scattering might have in the DNA detection with metal nanoparticles as label (Yguerabide and

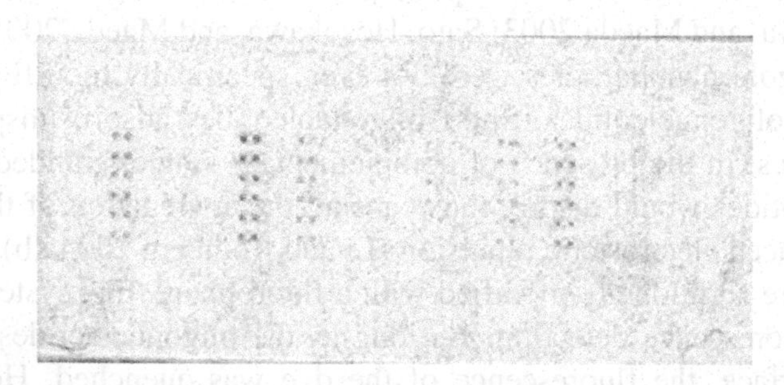

FIG. 3.3. DNA chip after labeling with gold nanoparticles and subsequent silver enhancement. The spots show gold nanoparticles specifically bound by DNA–DNA hybridization and clearly visible through their dark color after silver deposition onto the particles.

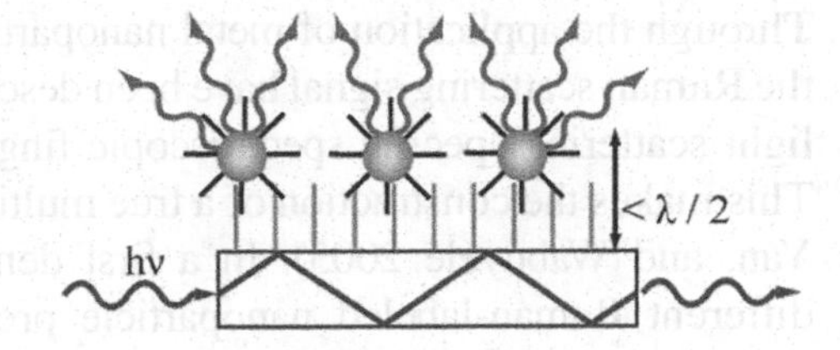

FIG. 3.4. Scheme of using light scattering on surface-bound metal nanoparticles for the readout of a DNA chip based on a wave guide. Only the particles bound to the surface are within the evanescent field of the wave guide and can be detected.

Yguerabide 1998a,b; Schultz, Smith, Mock, and Schultz 2000; Schultz 2003). The spectra of the scattered light depend on the size, shape, and composition of the nanoparticles.

Through DNA–DNA interaction surface bound metal nanoparticles can be easily analyzed when using an array slide as an internal wave guide (Fig. 3.4). This detection method makes even an online monitoring of DNA hybridization and denaturation possible. Only surface-bound nanoparticles will scatter light, because unbound particles are not within the evanescent field of the wave guide, and hence are not illuminated (and so not imaged) (Stimpson, Hoijer, Hsieh, Jou, Gordon, Theriault, Gamble, and Baldeschwieler 1995). By choosing different particle sizes and compositions the realization of a multilabel system is possible. This has been demonstrated with two sizes of gold nanoparticles (50 and 100nm). The bound 50-nm particles glow green, whereas the 100-nm particles appear orange upon illumination (Taton, Lu, and Mirkin 2001). The use of shape effects for the development of a multilabel system seems problematic, because the spectra of the scattered light strongly depend on the orientation of the nonspherical particles on the surface. Moreover, light scattering can also be used to analyze the formation of nanoparticle aggregates as described. When the aggregates were spotted on a wave guide the conjugates appeared yellow to orange because of a plasmon band red shift caused by a change in interparticle spacing. Individual particles, which where not conjugated by hybridization, appeared green because of the larger interparticle spacing (Storhoff, Lucas, Garimella, Bao, and Muller 2004). When compared with standard fluorescence labeling techniques the detection of light scattering proved to be 60 times more sensitive (Oldenburg, Genick, Clark, and Schultz 2002). As for the detection of RNA on a gene array, there could also be more genes detected with the nanoparticle labels than with the conventional fluorescence detection (Bao, Frutos, Greef, Lahiri, Muller, Peterson, Warden, and Xie 2002).

2.2.3 RAMAN SCATTERING

In addition, metal nanoparticles can also be used to enhance the signal of specific labels rather than to detect the binding of the nanoparticles directly. This approach is used when using metal nanoparticles in a detection scheme with Raman scattering.

Through the application of metal nanoparticles enhancement factors of up to 10^{14} of the Raman scattering signal have been described (Nie and Emory 1997). With Raman light scattering specific spectroscopic fingerprints of organic dyes can be detected. This makes the construction of a true multilabel detection system possible (Vo-Dinh, Yan, and Wabuyele 2005). In a first demonstration of this detection scheme six different Raman-labeled nanoparticle probes could be specifically detected in a chip-based setup. For this approach, gold nanoparticles were modified with thiol oligonucleotides and a specific Raman active dye. After binding to the surface no surface-enhanced Raman scattering (SERS) was detected, and it was assumed that the particles were not densely enough packed for an electromagnetic field enhancement. However, after a deposition of silver onto the bound gold nanoparticles the expected SERS signal could be measured (Cao, Jin, and Mirkin 2002) (Fig. 3.5). The SERS effect can also be used for a solution-based assay. For this approach a DNA stem-loop structure was bound with its one end to the metal nanoparticle, while the other end was modified with a Raman active dye. If the stem-loop structure was intact the dye remained in close proximity to the metal surface and a SERS signal was detected (Fig. 3.6A). Upon the binding of target DNA (Fig. 3.6B), the hairpin structure was opened and the dye removed from the metal surface. This led to the vanishing of the previously detected SERS signal (Wabuyele and Vo-Dinh 2005) (Fig. 3.6C).

2.2.4 SURFACE PLASMON RESONANCE IMAGING

For the detection of biomolecular interactions with surface plasmon resonance (SPR), nanoparticle labels are not needed. However, the addition of metal nanoparticles as labels greatly enhances the sensitivity of an SPR-based detection system.

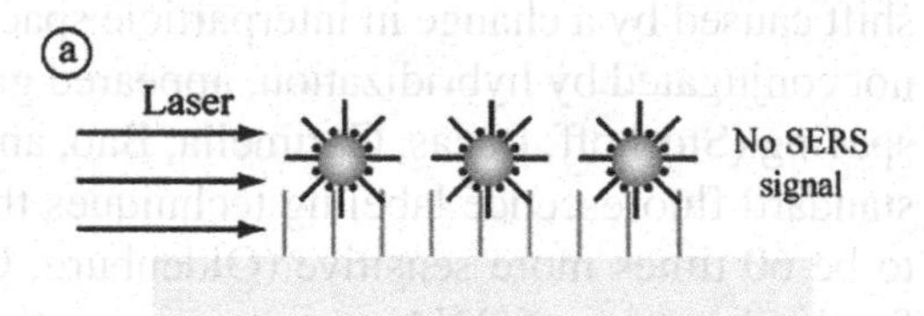

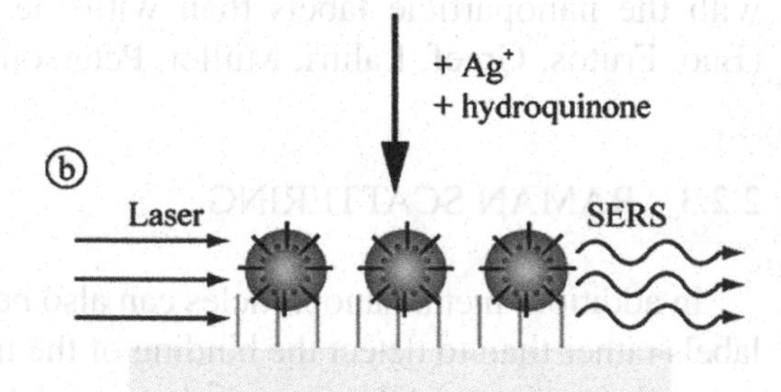

FIG. 3.5. Principle of the detection of surface-bound metal nanoparticles with surface enhanced raman spectroscopy (SERS).The nanoparticles are additionally modified with a Raman active dye. **A.** No signal can be measured when the particles are bound to the surface. **B.** The deposition of silver leads to an enhancement of the signal, which can then be detected.

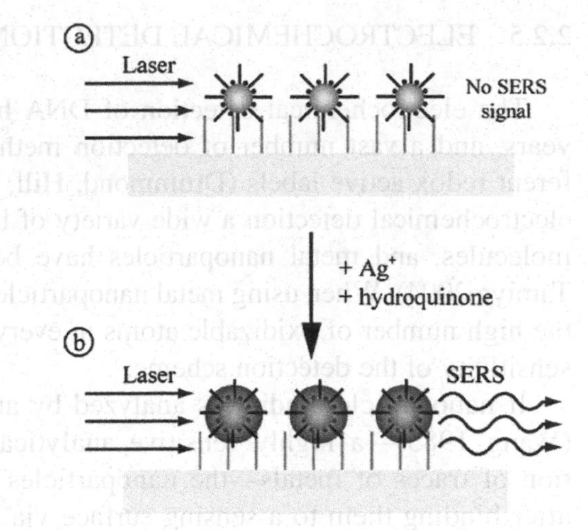

FIG. 3.6. Scheme of a solution-based DNA detection using SERS. **A.** Gold nanoparticles were modified with a hairpin DNA structure, bringing a dye in close proximity to the metal surface and enabling the detection of a SERS signal. **B.** Upon binding the target DNA, **C.** the hairpin structure is solved and no SERS signal detected.

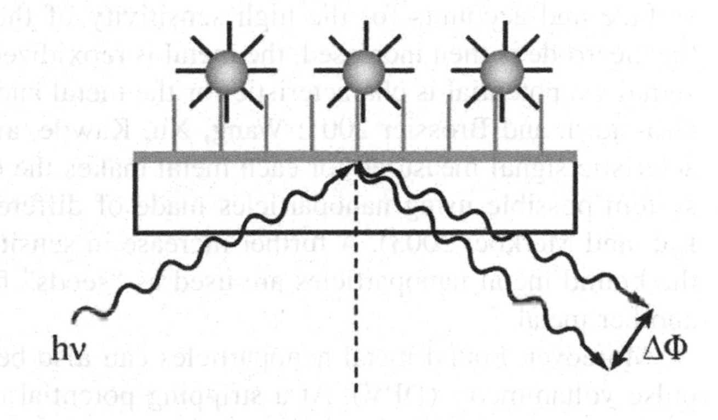

FIG. 3.7. Principle of enhanced SPR spectroscopy by nanoparticle binding. The shift of the resonant reflection angle Φ is greatly enhanced through the binding of metal nanoparticles to the sensing surface, leading to an increase of the system's sensitivity.

SPR detects the binding of biomolecules to the sensing surface through the change of the deflection angle of a laser beam that is aimed at the back side of the chip substrate (Fig. 3.7). The sensing surface of the substrate is normally covered with a thin (±50-nm) gold film. The binding of molecules to the gold film causes a modification of the evanescent field, which is induced by the laser light, and results in the change of the deflection angle, which can be measured. SPR enables the researcher to monitor binding events in real time and it has been shown that hybridization and denaturation of DNA can be studied with SPR (Peterlinz and Georgiadis 1997; Thiel, Frutos, Jordan, Corn, and Smith 1997). However, gold nanoparticles as labels greatly enhance the sensitivity of the system, because the bound particles have a great impact on the refractive index (Lyon, Musick, and Natan 1998; He, Musick, Nicewarner, Salinas, Nekovic, Natan, and Keating 2000).

2.2.5 ELECTROCHEMICAL DETECTION

The electrochemical detection of DNA has drawn a lot of interest in recent years, and a vast number of detection methods have been described using different redox active labels (Drummond, Hill, and Barton 2003). As labels for the electrochemical detection a wide variety of labels such as enzymes, redox active molecules, and metal nanoparticles have been used (Kerman, Kobayashi, and Tamiya 2004). When using metal nanoparticles as labels one can take advantage of the high number of oxidizable atoms in every particle. This accounts for the high sensitivity of the detection scheme.

If nanoparticle binding is analyzed by anodic stripping voltammetry (ASV) (Wang 1985)—a highly sensitive analytical method also used for the detection of traces of metals—the nanoparticles are dissolved in an acidic solution after binding them to a sensing surface via biomolecular interactions. In a first detection step, the metal ions are preconcentrated at an electrode with a negative potential and reduced, which leads to an accumulation of metal at the electrode surface and accounts for the high sensitivity of the method. The potential of the electrode is then increased, the metal is reoxidized, and the measured current signal vs. potential is characteristic for the metal and its concentration (Authier, Grossiord, and Brossier 2001; Wang, Xu, Kawde, and Polsky 2001). The characteristic signal measured for each metal makes the construction of a multilabel system possible using nanoparticles made of different metals as labels (Wang, Liu, and Merkoci 2003). A further increase in sensitivity can be achieved when the bound metal nanoparticles are used as "seeds" for a deposition of silver or another metal.

Moreover, bound metal nanoparticles can also be detected using differential pulse voltammetry (DPV). At a stripping potential of ~ +1.2 V the direct electrochemical oxidation of gold nanoparticles can be observed (Ozsoz, Erdem, Kerman, Ozkan, Tugrul, Topcuoglu, Ekren, and Taylan 2003). Through the deposition of metals on the bound nanoparticles an increase of the system's sensitivity can be achieved as well. The electrochemical detection methods allow a simple DNA detection with metal nanoparticles as labels. Through the incorporation of low-cost electrode structures, such as glassy carbon electrodes or carbon-based screen printed electrodes, it is possible to create inexpensive DNA tests with an electrochemical detection scheme. However, the integration into a biochip format seems problematic, because the measurements are carried out in solution and each measurement spot has to be separated from the others. Such a discrete volume handling should be possible in microtiter or nanotiter plates with prestructured microelectrodes (which could enable parallel test systems).

2.2.6 ELECTROMECHANICAL DETECTION

Although the weight of metal nanoparticles is only in the range of attograms, there are devices that can detect such small changes in mass. With quartz-crystal microbalances (QCM) and oscillating microcantilevers it is even possible to detect the hybridization of DNA to the sensing surfaces of these nanomechanical resonators (Okahata, Matsunobu, Ijiro, Mukae, Murakami, and Makino 1992; Fritz, Lang, Rothuizen, Vettiger, Meyer, Güntheroth, Gerber, and Gimzewski 2000).

In QCM a characteristic change in the frequency of the oscillating surface is detected after the binding of biomolecules, and this frequency change can be directly correlated to the mass of the molecules that are bound. The detection of binding events is even possible if the entire sensing surface is immersed in water, thus enabling kinetic studies of binding reactions. The application of nanoparticles as labels increases the sensitivity of such a detection system by many orders of magnitude because of the high specific mass of metal nanoparticles (Zhou, O'Shea, and Li 2000). Creating multiple layers of bound nanoparticles through specific biomolecular interactions can further increase sensitivity (Patolsky, Ranjit, Lichtenstein, and Willner 2000). The same effect may also be achieved if metals are deposited on the bound nanoparticles (Weizmann, Patolsky, and Willner 2001). Further miniaturization and integration of QCM is necessary for the detection of DNA in a microarray format based on the QCM detection principle.

A high parallelization has already been achieved with microcantilevers, making the construction of an array with 1000 measurement spots with simultaneous monitoring of each microcantilever possible (Vettiger, Brugger, Despont, Drechsler, Durig, Haberle, Lutwyche, Rothhuizen, Stutz, Widmer, and Binnig 1999; Lutwyche, Despont, Drechsler, Durig, Haberle, Rothhuizen, Stutz, Widmer, Binnig, and Vettiger 2000). If each cantilever can be modified with a specific DNA sequence, this approach could be used for a highly parallel detection of DNA. On a smaller scale it has already been shown that individual cantilevers can be specifically modified, resulting in specific signals when incubated with target DNA (Fritz, Lang, Rothuizen, Vettiger, Meyer, Güntheroth, Gerber, and Gimzewski 2000; Hansen, Ji, Wu, Datar, Cote, Majumdar, and Thundat 2001). Furthermore microcantilevers allow for the real-time monitoring of binding events. Again an increase in sensitivity was shown through the incorporation of nanoparticle labels, a deposition of silver onto the bound particles, or the construction of multiple layers of nanoparticles in a dendritic fashion (Su, Li, and Dravid 2003).

2.2.7 RESISTIVE AND CAPACITVE DETECTION

The electrical methods for the DNA detection using metal nanoparticles as labels basically can be grouped into two main categories: AC or capacitance measurements and DC or resistance (conductivity) measurements. Both approaches exploit the conductivity of metal nanoparticles.

If metal nanoparticles are bound via specific biomolecular interactions in a gap between two microstructured electrodes, the resistance measured over the gap should significantly drop after the nanoparticle binding. However, the detection of nanoparticle binding in an electrode gap has not yet been achieved by a simple DC resistance measurement. The reasons are that the nanoparticles could not be bound densely enough in the gap, and/or that the layer of oligonucleotides around each particle might insulate the particles from completing the circuit. These problems can be overcome by an additional deposition of metal or another conductive material on the bound nanoparticles, thus bridging the gap (Fig. 3.8). Using gold nanoparticles as labels and a subsequent silver deposition it could be demonstrated that the detection of single nucleotide polymorphisms (SNPs) was possible with such an approach (Park, Taton, and Mirkin 2002). The resistance-based detection showed very high selectivity, and because DNA-modified

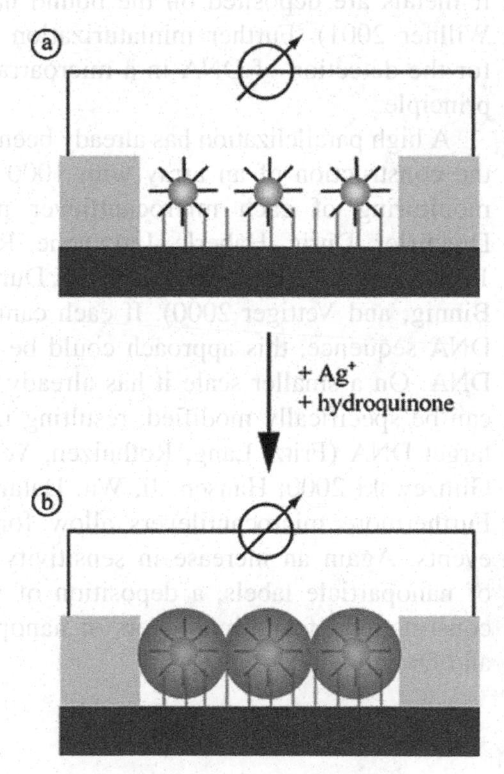

FIG. 3.8. Principle of resistive DNA detection using metal nanoparticles as labels. **A.** The nanoparticles are bound within a gap between two micro electrodes via biomolecular interactions. However, the bound nanoparticles do not lead to a detectable signal. A silver enhancement of the particles bridges the gap and causes a massive drop of the measured resistance.

gold nanoparticles were used this selectivity was achieved at room temperature using controlled ionic strength washing steps. Moreover, a quantification of the detected target DNA is possible because of the correlation among the measured resistance, the time of silver deposition, and the concentration of bound nanoparticles, which is directly related to the target DNA concentration (Möller, Csaki, Köhler, and Fritzsche 2001). The integration of this detection method into a microarray format has been demonstrated as well. Therefore, a microstructured chip with 42 individually addressable electrode gaps was constructed. Each electrode gap could be modified with a specific DNA sequence creating a low-density DNA chip. The read out of the chip was realized by a portable and robust readout device. This DNA chip reader consists of an embedded PC that controls an ohmmeter, a display, and a socket for holding and contacting the chip during the measurement (Fig. 3.9). Additionally, the chip can be modified with a layer of silicon nitride to avoid cross-talk between the measurement spots, making the chip more robust (Li, Xue, Lu, Zhang, Feng, and Chan 2003). However, the resistance measurement also has some drawbacks. No online monitoring of the silver deposition process is possible bearing the risk of under- or over-enhancing the chip. Under-enhancing is not a real problem because low signals can be enhanced with an additional step of silver deposition, although the opposite can be said for over-enhancing. If too much silver is deposited on the chip's surface unspecific signals appear and the chip can not be recovered.

No silver deposition onto the bound nanoparticles is needed when detection is achieved by AC capacitance measurement. However, this technique needs more sophisticated equipment for the measurement and different, more complex chip architecture. Interdigitated electrodes rather than a simple gap between two electrodes serve as measurement points in this approach. Gold nanoparticles, bound

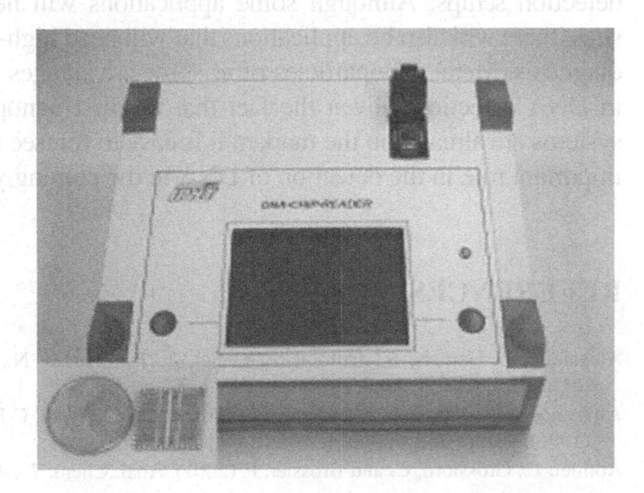

FIG. 3.9. DNA chip reader for fast and automated readout of DNA chips *(inset)* with electrical detection.

to these measurement spots, can be detected without any further modification. An increase in the sensitivity can be achieved if silver is deposited on the bound particles (Moreno-Hagelsieb, Pampin, Bourgeois, Remacle, and Flandre 2004).

3 CONCLUSION AND OUTLOOK

Metal nanoparticles have drawn a lot of interest in the last years, especially as labels for the detection of DNA. As labels they enable new and easy-to-handle detection schemes or enhance the sensitivity of already existing ones. The bottleneck for further development and establishment of DNA detection based on nanoparticle labeling will be the development of stable and robust modifications for the coupling of DNA and metal nanoparticles. Different detection systems are already commercially available using gold and silver nanoparticles as labels. These systems take advantage of the described unique optical properties using either optical absorbance (Nanosphere Inc.; Clondiag Chip Technologies; Eppendorf Array Technologies) or light scattering (Seashell Technologies; Genicon Science; Nanosphere, Inc.) for DNA detection. Besides the optical detection schemes new methods have been developed for the DNA detection using metal nanoparticle as labels. The advantage of those systems is that there is no sophisticated optical equipment needed for detection, enabling the developers to create robust and easy-to-handle devices as they are needed in point-of-care diagnostics.

As DNA detection becomes more and more important in routine diagnostics the demand for robust systems that can be operated outside of specialized laboratories grows. Parameters such as cost, portability, and robustness will become important, whereas the sensitivity, selectivity, and speed of the diagnosis should also be increased. In the future, selected applications will demand different diagnostic methods and detection setups. Although some applications will need only a few measurement sites, there will also be applications that will need high-density DNA chips for a clear diagnosis. Metal nanoparticles offer some advantages over the currently used labels in DNA detection. Given the fact that the first nanoparticle-based DNA detection systems are already on the market, it is easy to foresee that nanoparticles will play an important role in the detection of DNA in the coming years.

REFERENCES

Alexandre, I., Hamels, S., Dufour, S., Collet, J., Zammatteo, N., De Longueville, F., Gala, J.L. and Remacle, J. (2001). Anal. Biochem. 295, 1–8.

Alivisatos, A.P., Johnsson, K.P., Peng, X., Wilson, T.E., Loweth, C.J., Bruchez Jr., M.P. and Schultz, P.G. (1996). Nature 382, 609–611.

Authier, L., Grossiord, C. and Brossier, P. (2001). Anal. Chem. 73, 4450–4456.

Bao, P., Frutos, A.G., Greef, C., Lahiri, J., Muller, U., Peterson, T.C., Warden, L. and Xie, X. (2002). Anal. Chem. 74, 1792–1797.

Brust, M., Walker, M., Bethell, D., Schiffrin, D.J. and Whyman, R. (1994). J. Chem. Soc. Chem. Commun. 801–802.

Brust, M., Fink, J., Bethell, D., Schiffrin, D.J. and Kiely, C.J. (1995). J. Chem. Soc. Chem. Commun. 1655–1656.

Cao, Y., Jin, R. and Mirkin, C.A. (2001). J. Am. Chem. Soc. 123, 7961–7962.

Cao, Y.C., Jin, R. and Mirkin, C.A. (2002). Science 297, 1536–1540.

Csaki, A., Kaplanek, P., Möller, R. and Fritzsche W. (2003). Nanotechnology 14, 1262–1268.

Daniel, M.C. and Astruc, D. (2004). Chem. Rev. 104, 293–346.

Drummond, T.G., Hill, M.G. and Barton, J.K. (2003). Nat. Biotechnol. 21, 1192–1199.

Elghanian, R., Storhoff, J.J., Mucic, R.C., Letsinger, R.L. and Mirkin, C.A. (1997). Science 277, 1078–1081.

Ermantraut, E., Schulz, T., Tuchscheerer, J., Wölfl, S., Saluz, H.P. and Köhler, J.M. (1998). In: A. van den Berg (Ed.), Proceedings of Micro Total Analysis Systems. Kluwer Scientific Publishing, Dordrecht.

Fritz, J.B., Lang, H.P., Rothuizen, H., Vettiger, P., Meyer, E., Güntheroth, H.J., Gerber, Ch. and Gimzewski, J.K. (2000). Science 288, 316–318.

Fritzsche, W. and Taton, T. A. (2003) Nanotechnology 14, R63–R73.

Gearheart, L.A., Ploehn, H.J. and Murphy, C.J. (2001). J. Phys. Chem. B 105, 12609–12615.

Grabar, K.C., Freeman, R.G., Hommer, M.B. and Natan, M.J. (1995). Anal. Chem. 67, 735–743.

Hacker, G.W., Springall, D.R., Van Noorden, S., Bishop, A.E., Grimelius, L. and Polak, J.M. (1985). Virchows Arch. A Pathol. Anat. Histopathol. 406, 449–461.

Hacker, G.W., Grimelius, L., Danscher, G., Bernatzky, G., Muss, W., Adam, H. and Thurner, J. (1988). J. Histotechnol. 11, 213–221.

Hacker, G.W., Danscher, G., Graf, A.H., Bernatzky, G., Schiechl, A. and Grimelius, L. (1991). Prog Histochem Cytochem 23, 286–290.

Hainfeld, J.F., Powell, R.D., and Hacker, G.W. (2004). In: C.A. Mirkin (Ed.), Nanobiotechnology. Wiley-VCH, Weinheim, Germany, pp. 353–386.

Hansen, K.M., Ji, H.-F., Wu, G., Datar, R., Cote, R., Majumdar, A. and Thundat, T. (2001). Anal. Chem. 73, 1567–1571.

Hayat, M.H. (1989). Colloidal Gold: Principles, Methods, and Applications. Elsevier Science & Technology Books, St. Louis.

He, L., Musick, M.D., Nicewarner, S.R., Salinas, F.G., Nekovic S.J., Natan, M.J. and Keating, C.D. (2000). J. Am. Chem. Soc. 122, 9071–9077.

Horisberger, M. (1981). Scan. Elect. Microsc. II, 9–31.

Hütter, E. and Pileni, M.-P. (2003). Phys. Chem. B 107, 6497–6499.

Jin, R., Wu, G., Li, Z., Mirkin, C.A. and Schatz, G.C. (2003) J. Am. Chem. Soc. 125, 1643–1654.

Kelly, K.L., Coronado, E., Zhao, L.L. and Schatz, G.C. (2003). J. Phys. Chem. B 107, 668–677.

Kerman, K., Kobayashi, M. and Tamiya, E. (2004). Meas. Sci. Technol. 15, R1–R11.

Köhler, J.M., Csaki, A., Reichert, J., Möller, R., Straube, W. and Fritzsche, W. (2001). Sensors Actuators 76, 166–172.

Kreibig, U. and Vollmer, M. (1995). Optical Properties of Metal Clusters, Berlin.

Letsinger, R.L., Elghanian, R., Viswanadham, G. and Mirkin, C.A. (2000). Bioconjug. Chem. 11, 289–291.

Li, H. and Rothberg, L. (2004a). Proc. Natl. Acad. Sci. U S A 101, 14036–14039.

Li, H. and Rothberg, L.J. (2004b). J. Am. Chem. Soc. 126, 10958–109561.

Li, H. and Rothberg, L.J. (2004c). Anal. Chem. 76, 5414–5417.

Li, J., Xue, M., Lu, Z., Zhang, Z., Feng, L. and Chan, M. (2003). IEEE Trans. Electron Dev. 50, 2165–2170.

Li, J., Chu, X., Liu, Y., Jiang, J.H., He, Z., Zhang, Z., Shen, G. and Yu, R.Q. (2005). Nucleic Acids Res. 33, e168.

Li, J., Xu, C., Zhang, Z., Wang, Y., Peng, H., Lu, Z., and Chan, M. (2005). Sensors Actuators B 106, 378–382.

Li, Z., Jin, R., Mirkin, C.A. and Letsinger, R.L. (2002). Nucleic Acids Res. 30, 1558–1562.

Lipshutz, R.J., Fodor, S.P., Gingeras, T.R. and Lockhart, D.J. (1999). Nat. Genet. 21, 20–24.

Lutwyche, M.I., Despont, M., Drechsler, U., Durig, U., Haberle, W., Rothhuizen, H., Stutz, R., Widmer, R., Binnig, G.K. and Vettiger, P. (2000). Appl. Phys. Letts. 77, 3299–3301.

Lyon, L.A., Musick, M.D. and Natan, M.J. (1998). Anal. Chem. 70, 5177–5183.

Mirkin, C.A., Letsinger, R.L., Mucic, R.C. and Storhoff, J.J. (1996). Nature 382, 607–609.

Möller, R., Csaki, A., Köhler, J.M. and Fritzsche, W. (2001). Langmuir 17, 5426–5430.

Moreno-Hagelsieb, L.L., Pampin, R., Bourgeois, D., Remacle, J. and Flandre, D. (2004). Sensors Actuators B 98, 269–274.

Nie, S. and Emory, S.R. (1997). Science 275, 1102–1106.

Okahata, Y., Matsunobu, Y., Ijiro, K., Mukae, M., Murakami, A. and Makino, K. (1992). J. Am. Chem. Soc. 114, 8299–8300.

Oldenburg, S.J., Genick, C.C., Clark, K.A. and Schultz, D.A. (2002). Anal. Biochem. 309, 109–116.

Ozsoz, M., Erdem, A., Kerman, K., Ozkan, D., Tugrul, B., Topcuoglu, N., Ekren, H. and Taylan, M. (2003). Anal. Chem. 75, 2181–2187.

Parak, W.J.G., Pellegrino, T., Zanchet, D., Micheel, C., Williams, S.C., Boudreau, R., LeGros, M.A., Larabell, C.A. and Alivisatos, A.P. (2003). Nanotechnology 14, R15–R27.

Park, S.J., Taton, T.A. and Mirkin, C.A. (2002). Science 295, 1503–1506.

Park, S., Brown, K.A. and Hamad-Schifferli, K. (2004). Nano. Letts. 4, 1925–1929.

Patolsky, F., Ranjit, K.T., Lichtenstein, A. and Willner, I. (2000). Chem. Commun. 1025–1026.

Peterlinz, K.A. and Georgiadis, R.M. (1997). J. Am. Chem. Soc. 119, 3401–3402.

Pirrung, M.C. (2002). Angew. Chem. 114, 1326–1341.

Quinten, M. (2001). Appl. Phys. B: Lasers Optics 73, 245–255.

Reichert, J., Csaki, A., Kohler, J.M. and Fritzsche, W. (2000). Anal. Chem. 72, 6025–6029.

Reynolds, R.A., Mirkin, C.A. and Letsinger, R.L. (2000). J. Am. Chem. Soc. 122, 3795–3796.

Sato, K., Hosokawa, K. and Maeda, M. (2003). J. Am. Chem. Soc. 125, 8102–8103.

Sato, K., Hosokawa, K. and Maeda, M. (2005). Nucl. Acids Res. 33, e4.

Schena, M. (2003). Microarray Analysis. Wiley-Liss Verlag, Hoboken, NJ.

Schultz, D.A. (2003). Curr. Opin. Biotechnol. 14, 13–22.

Schultz, S., Smith, D.R., Mock, J.J. and Schultz, D.A. (2000). Proc. Natl. Acad. Sci. U S A 97, 996–1001.

Stimpson, D.I., Hoijer, J.V., Hsieh, W.T., Jou, C., Gordon, J., Theriault, T., Gamble, R., and Baldeschwieler, J.D. (1995). Proc. Natl. Acad. Sci. U S A 92, 6379–6383.

Storhoff, J.J., Elghanian, R., Mucic, R.C., Mirkin, C.A. and Letsinger, R. L. (1998). J. Am. Chem. Soc. 120, 1959–1964.

Storhoff, J.J., Lucas, A.D., Garimella, V., Bao, Y.P. and Muller, U.R. (2004). Nat. Biotechnol 22, 883–887.

Su, M., Li, S. and Dravid, V.P. (2003). Appl. Phys. Lett. 82, 3562–3564.

Taton, T.A., Mirkin, C.A. and Letsinger, R.L. (2000). Science 289, 1757–1760.

Taton, T.A., Lu, G. and Mirkin, C.A. (2001). J. Am. Chem. Soc. 123, 5164–5165.

Taton, T.A. (2002). Trends Biotechnol. 20, 277–279.

Thiel, A.J., Frutos, A.G., Jordan, C.E., Corn, R.M. and Smith, L.M. (1997). Anal. Chem. 69, 4948–4956.

Turkevich, J., Stevenson, P.L. and Hiller, J. (1951). Discuss. Faraday Soc. 11, 55.

Vettiger, P., Brugger, J., Despont, M., Drechsler, U., Durig, U., Haberle, W., Lutwyche, M., Rothhuizen, H., Stutz, R., Widmer, R. and Binnig, G. (1999). Microelectr. Eng. 46, 11–17.

Vo-Dinh, T., Yan, F. and Wabuyele, M.B. (2005). J. Raman Spectrosc. 36, 640–647.

Wabuyele, M.B. and Vo-Dinh, T. (2005). Anal. Chem. 77, 7810–7815.

Wang, J. (1985). Stripping Analysis: Principles, Instrumentation, and Aplications. VCH, Deerfield Beach, FL.

Wang, J., Xu, D., Kawde, A.N. and Polsky, R. (2001). Anal. Chem. 73, 5576–5581.
Wang, J., Liu, G. and Merkoci, A. (2003). J. Am. Chem. Soc. 125, 3214–3215.
Weizmann, Y., Patolsky, F. and Willner, I. (2001). Analyst 126, 1502–1504.
Yguerabide, J. and Yguerabide, E.E. (1998a). Anal. Biochem. 262, 137–156.
Yguerabide, J. and Yguerabide, E.E. (1998b). Anal. Biochem. 262, 157–176.
Zhou, X.C., O'Shea, S.J. and Li, S.F.Y. (2000). Chem. Commun. 953–954.

4

Label-Free, Fully Electronic Detection of DNA with a Field-Effect Transistor Array

Sven Ingebrandt and Andreas Offenhäusser

1 INTRODUCTION

Massive parallel sequence analysis of nucleic acids and their use in genetics, medicine, and drug discovery has led to a broad interest in deoxyribonucleic acid (DNA) microarrays or DNA chips. For instance, nowadays many hundreds of diseases are diagnosable by the molecular analysis of DNA. Mainly the DNA hybridization reaction is used for the detection of unknown DNA, where the target (unknown single-stranded DNA; ssDNA) is identified, when it forms a double-stranded (dsDNA) helix structure with its complementary probe (known ssDNA). This biorecognition process is based on the affinity binding reaction between the base pairs, i.e., adenine-thymine (A-T) and cytosine-guanine (C-G). The hybridization reaction is known to be highly specific and is working in a sample solution with high background concentration of non-complementary probes. By labeling of either the target DNA or the probe DNA, the hybridization reaction can be detected by radiochemical, fluorescence, electrochemical, microgravimetric, enzymatic, and electroluminescence methods (Kricka 2002). However, the labeling approach is time consuming, cost intensive, and introduces

A. Offenhäusser and R. Rinaldi (eds.), *Nanobioelectronics - for Electronics, Biology, and Medicine*,
DOI: 10.1007/978-0-387-09459-5_5,

an additional element of uncertainty into the detection process. Hence, there is a growing interest for the development of fast, simple, inexpensive, and disposable DNA chips based on a direct, fully electronic and label-free detection method.

Some of these techniques, which avoid a labeling step, are based on mass change and direct optical detection by surface plasmon resonance and imaging ellipsommetry, for example. Other approaches are based on the direct detection of the intrinsic electrical charge of the DNA molecule using electrochemical methods together with redox-active indicators (Wang 2002). Field-effect sensors, especially FETs, offer an alternative approach for the label-free detection of DNA with a direct electrical readout (Souteyrand, Cloarec, Martin, Wilson, Lawrence, Mikkelsen, and Lawrence 1997). Recently, the detection limit of potentiometric field-effect sensors was enhanced such that single nucleotide polymorphisms (SNPs) were successfully detected (Fritz, Cooper, Gaudet, Sorger, and Manalis 2002; Sakata and Miyahara 2005). The sensors used the field-effect at the electrolyte-oxide-semiconductor (EOS) interface, which was firstly described for ion-selective field-effect transistors (ISFET) (Bergveld 1970; Bergveld, Wiersma, and Meertens 1976). Typically, the response of such devices is interpreted as shift of the flat-band voltage of the field-effect structure (Tsuruta, Matsui, Hatanaka, Namba, Miyamoto, and Nakamura 1994; Kim, Jeong, Park, Shin, Choi, Lee, and Lim 2004; Pouthas, Gentil, Cote, and Bockelmann 2004; Pouthas, Gentil, Cote, Zeck, Straub, and Bockelmann 2004; Uslu, Ingebrandt, Mayer, Böcker-Meffert, Odenthal, and Offenhäusser 2004; Sakata and Miyahara 2005; Barbaro, Bonfiglio, and Raffo 2006).

Most of these sensor chips were operated as potentiometric ISFETs and used the ion-selectivity of the solid–liquid interface or artificial molecular membranes, which were attached to the FET gate structure. The detection of biomolecules such as proteins, however, was reported to be very difficult and most of the time not reliable (Bergveld 1991). In this context, the dc as well as the ac readout of the FETs has been reported and the influence of a biomembrane attached to the transistor gate structure has been described (Schasfoort, Streekstra, Bergveld, Kooyman, and Greve 1989). The ac readout for characterization of the input impedance of the FET devices has either been used to test the reliability of the encapsulation material (Chovelon, Jaffrezicrenault, Fombon, and Pedone 1991) or to detect binding of biomolecules and/or changes in membrane properties, when membranes were attached to the gate structure of the FETs (Kruise, Rispens, Bergveld, Kremer, Starmans, Haak, Feijen, and Reinhoudt 1992; Lugtenberg, Egberink, van den Berg, Engbersen, and Reinhoudt 1998; Antonisse, Snellink-Ruel, Lugtenberg, Engbersen, van den Berg, and Reinhoudt 2000; Kharitonov, Zayats, Lichtenstein, Katz, and Willner 2000; Kharitonov, Wasserman, Katz, and Willner 2001; Lahav, Kharitonov, and Willner 2001; Pogorelova, Bourenko, Kharitonov, and Willner 2002; Sallacan, Zayats, Bourenko, Kharitonov, and Willner 2002; Zayats, Raitman, Chegel, Kharitonov, and Willner 2002; Katz and Willner 2003;

Pogorelova, Kharitonov, Willner, Sukenik, Pizem, and Bayer 2004). In all previous studies employing the ac readout, only one transistor was used for detection and never a differential readout in a microarray approach was applied.

The ISFET structure can be highly integrated to multichannel sensors by using standard industrial processes (Yeow, Haskard, Mulcahy, Seo, and Kwon 1997). The inherent miniaturization of such devices, their compatibility with advanced micro fabrication technology, and hence the possibility for a massive parallel readout format is making these approaches very attractive for DNA diagnostics. A miniaturized, low-cost, fast readout, highly integrated, and addressable multichannel sensor with sensitivity high enough to detect SNPs, would be the ideal device for genetic testing and medical diagnostics.

A summary of our results for the potentiometric readout of DNA hybridization with FET arrays was already provided in a recent feature article (Ingebrandt and Offenhäusser 2006b). Here we compare the previously presented potentiometric results with a recently described method for impedimetric readout (Ingebrandt, Han, Nakamura, Schöning, Poghossian, and Offenhäusser 2006a) in terms of reliability and selectivity of the signals.

2 MATERIALS AND METHODS

2.1 FIELD-EFFECT TRANSISTORS AND AMPLIFIER SYSTEMS FOR DNA DETECTION

The p-channel FET microarrays used in this study were developed for extracellular recordings from neuronal and cardiac cells in previous works (Sprössler, Richter, Denyer, and Offenhäusser 1998; Sprössler, Denyer, Britland, Knoll, and Offenhäusser 1999; Ingebrandt, Yeung, Krause, and Offenhäusser 2001).

The transistor arrays consisted of either 16 or 8 transistor gates, which were arranged with spacing of 200 μm in a 4×4 or a 2×4 matrix in the center of a 5×5 or 2.5×5 mm^2 silicon chip, respectively.

Gate sizes of either 1×16 or 8 μm^2 or 1.5×16 or 8 μm^2 (channel length times width) were predominantly used, which results in sensitive areas of 8 to 24 μm^2 for each sensor spot. The gate oxide consisted of thermally grown SiO_2 of a thickness of 10 nm. The transconductances g_m of the FET chips were ranging from 0.2 to 0.6 mS, depending on gate size and oxide thickness. The chips were encapsulated as previously described (Offenhäusser, Sprössler, Matsuzawa and Knoll 1997) (Fig. 4.1). A sandwich layer stack of oxide-nitride-oxide for passivation of the FET contact lines was found to be stable for repeated use of the devices.

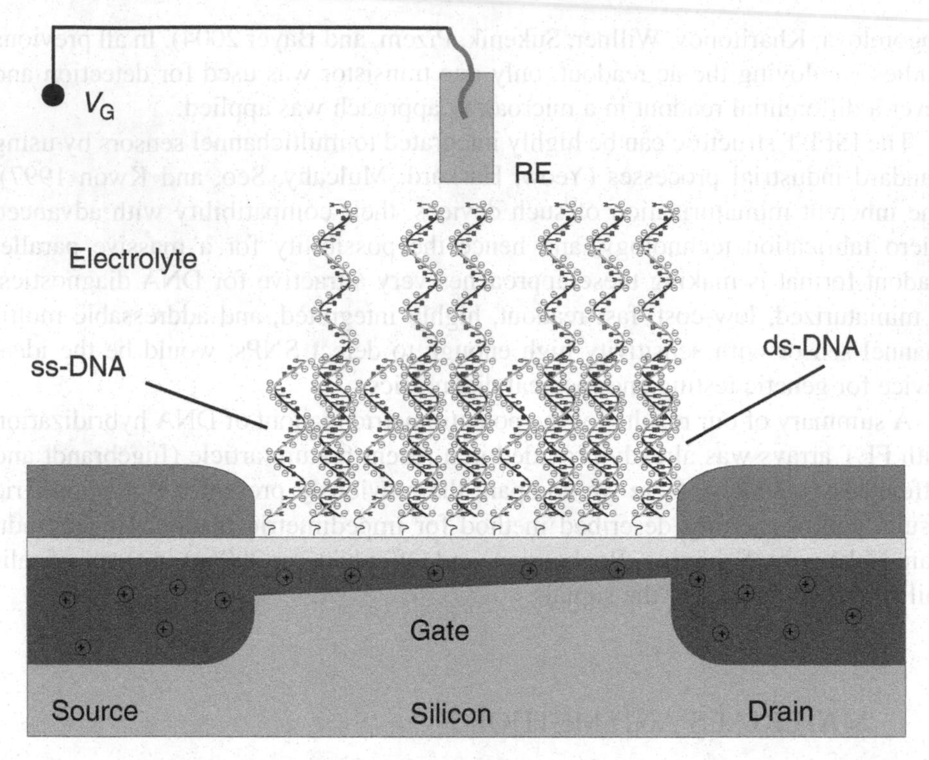

FIG. 4.1. Layout and encapsulation of the FET chips used for DNA detection. **A,B**. The 16-channel FET chip is shown. The encapsulation is done by a flip-chip process enabling the attachment of a fluidic cell onto the device. **C,D.** The 8-channel FET chip is shown. The chip is encapsulated such, that it can be dipped into an analyte solution (Ingebrandt and Offenhäusser 2006b).

2.1.1 DC READOUT

For potentiometric readout, a 16-channel amplifier system was developed, which contained a preamplifier, a main amplifier and a microprocessor (Ingebrandt, Han, Sakkari, Stockmann, Belinskyy and Offenhäusser 2005). The FET chips were operated in the constant-voltage mode. By applying a constant drain-source voltage V_{DS} and a constant gate-source voltage V_{GS}, the working point of the device was defined. At the first operational amplifier (OP), changes in the drain-source current ΔI_{DS} were converted according to $\Delta V_{out} = -10\,\text{kOhm} \cdot \Delta I_{DS}$ (Fig. 4.2B). The signals were amplified by a second stage, which can be operated in 1×, 10×, 30×, and 100× amplification. At the beginning of the time-dependent potentiometric readout, the constant drain-source currents I_{DS} of the 16 FETs were compensated by applying individual compensation voltages for each channel, separately.

2.1.2 AC READOUT

For impedimetric readout, two independent, frequency-selective amplifiers were included in the system. In total it contained 16 readout channels providing parallel operation of two 8-channel chips. The scan of the transfer function was done by feeding a sinodal test signal V_{mod} with 10 mV amplitude and varying frequency from 1 Hz – 100 kHz to the reference electrode (Fig. 4.2B). The FETs were operated in the same working point as for the dc-readout. Test signals of exact amplitude, frequency and phase were provided by a direct-digital-synthesis device. Frequency-selective amplification was confirmed via a frequency multiplier device, which was built-in into a capacitive de-coupled amplification cascade.

The whole amplifier unit was operated by a microprocessor and the data were transferred via an USB connection to the operating PC. The readout software for characterization, potentiometric time-dependent readout, differential transfer function (DTF) scan, and impedimetric, time-dependent readout was implemented in Delphi® 5.0 (Borland Software Corp., Austin, TX, USA). This software offered the possibility to set one out of the 16 channels as reference sensor while operating the remaining channels in a differential mode. This function can either be used to work with two chips containing 8 channels or to set one channel of the FET array as reference. For the differential experiments presented here, we used both options. The potentiometric, in situ detection was either done in a small container made of poly-

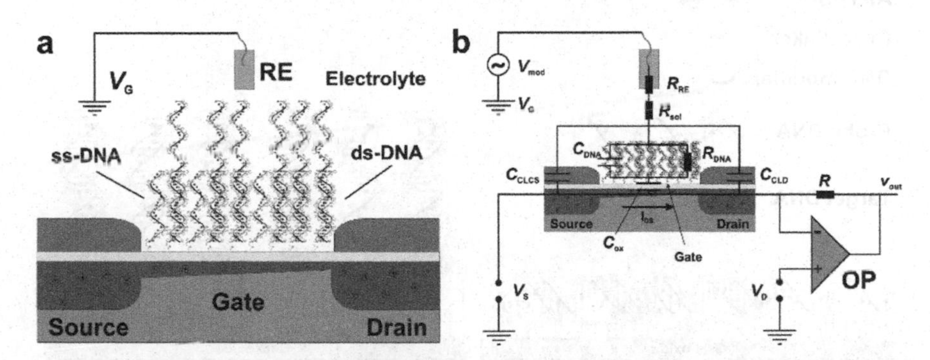

FIG. 4.2. Schematics of an open-gate FET for the label-free detection of DNA. **A.** The electrical gate contact of the transistor is maintained via a reference electrode (RE), which applies the gate voltage V_G. On top of the thin gate insulator, ssDNA probe molecules are immobilized. The FET sensor detects the hybridization reaction, when the target ssDNA forms a dsDNA helix with the immobilized probes. **B.** Electronic readout circuit: The working point of the transistor is defined by the three voltages at the gate (V_G), source (V_S), and drain (V_D). Setting constant voltages V_{GS} and V_{DS} drives constant current I_{DS} through the transistor channel. At the first operational amplifier (OP), the current I_{DS} is converted to a voltage v_{out}. In the potentiometric readout mode, changes in drain-source current ΔI_{DS}, which are caused by any potential change at the gate input, are recorded. In the impedimetric readout mode, a sinodal test signal V_{mod} of exact amplitude and frequency is additionally applied to RE. The signal transfer through the circuit is then monitored in a frequency-selective way.

tetrafluoroethylene (PTFE), which was thermally controlled by the amplifier system, or with a small fluidic cell, which was attached to the front side of the 16-channel chip. The transfer function detection was also done in the PTFE container, although heating control was not necessary for these experiments. For all dc as well as ac experiments, we used dilute buffer concentrations (~1 mM NaCl). A small liquid junction Ag/AgCl electrode (Dri-Ref™, SDR2, World Precision Instruments, Sarasota, FL, USA) was used as reference electrode in all measurements.

2.2 IMMOBILIZATION OF PROBE DNA ONTO FET SURFACES

A very defined and uniform supramolecular architecture of the immobilized probe ssDNA on the sensor spots is of high interest for all kinds of DNA chips. For FET-based sensor chips, it is in particular of interest to have the immobilized probe ssDNA end-functionalized attached to the sensor surface to allow for an efficient hybridization reaction. In Fig. 4.3 two different strategies for the immobilization of the probe ssDNA are depicted. The thin gate oxide of the FET is cleaned and activated in an acid step to expose a high density of –OH groups

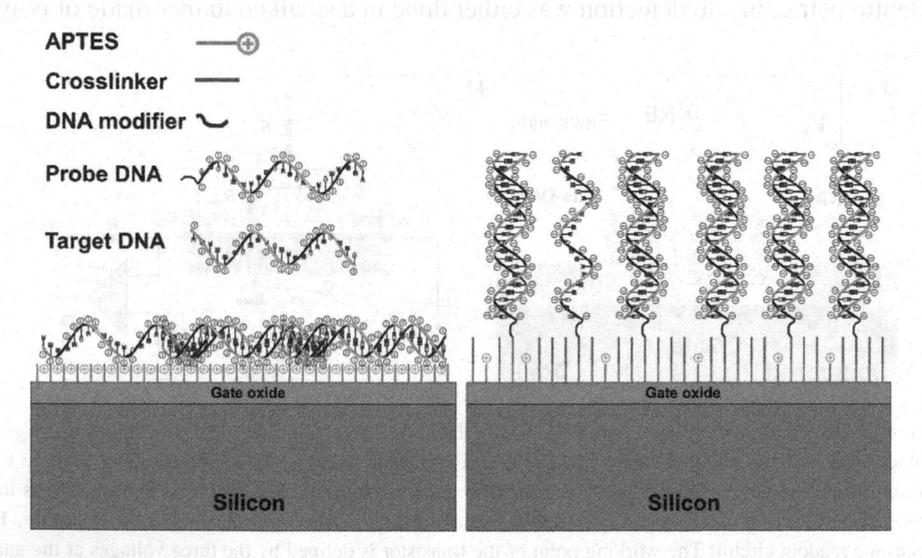

FIG. 4.3. Different strategies for the immobilization of probe DNA. On the left side, the DNA molecules are attached to the aminosilane layer by electrostatic immobilization. This usually leads to high FET signal amplitudes for the immobilization process, because the oligonucleotides are very closely attached to the gate oxide. The hybridization efficiency, however, was not very high in these experiments. On the right side, the probe ssDNA is covalently attached. This was confirmed by a cross-linker in combination with end-functionalized ssDNA.

to the electrolyte solution. Then the silicon oxide is silanized with 3-aminopropyltriethoxysilane (APTES, Sigma-Aldrich, Munich, Germany). If not stated otherwise, all chemicals used in this work were purchased from Sigma. The silane is introducing positive charges to the surface, which allows for a fast and stable electrostatic immobilization of probe ssDNA onto the surface.

In electrolyte buffers with low ionic strength, the intermolecular electrostatic interaction is strong and the molecules are very closely attached to the surface. Consequently, immobilization signals of ssDNA at amino-silanized FET structures are quite large in amplitude and the reaction times are fast (few minutes). However, in this case the hybridization efficiency is not very high, because of steric hindrance of the molecules at the surface (Uslu, Ingebrandt, Mayer, Böcker-Meffert, Odenthal, and Offenhäusser 2004). The covalent attachment of the probe ssDNA molecules (Fig. 4.3, right side) leads to a higher hybridization efficiency and reliable signals of the FET (Han, Offenhäusser, and Ingebrandt 2006b). The amino end-functionalized probe ssDNA molecules are attached via cross-linker molecules leading to more defined interface architecture at the surface.

2.2.1 ELECTROSTATIC IMMOBILIZATION

The FET devices were cleaned using cotton buds and ultrasonification for 10 min in 2% detergent (Hellmanex II, Helma GmbH & Co., Müllheim, Germany). The SiO_2 surfaces were then further cleaned and activated by 50% H_2SO_4 acid for 30 min at 80°C. After rinsing with ultrapure water (Millipore water, Milli-Q Gradient A10 18.2 MΩ, Millipore Inc., Billerica, MA, USA) the FET surfaces were silanized with APTES from Sigma, Germany. APTES was dissolved in ethanol at a concentration of 5% (v/v). The cleaned FET sensors were covered with a drop (20–30 μl) of APTES solution for 20 min. After this period, the chips were rinsed three times with ethanol and blown dry with argon.

For all measurements, the DNA samples were dissolved in a TRIS-EDTA buffer solution (TE buffer): 10 mM Tris-(hydroxymethyl) amino methane (TRIS), 1 mM ethylene-diamine-tetraacetic acid (EDTA) at pH 8 and varying concentrations of sodium chloride. Much attention was paid to the precise adjustment of the temperature and the pH value for all samples to exclude side effects.

2.2.2 COVALENT IMMOBILIZATION

In Fig. 4.4, the covalent surface modification process is depicted (Han, Sakkari, Stockmann, Belinskyy, and Offenhäusser 2006b). First, the SiO_2 surface was cleaned with MeOH/HCl (1/1) for 30 min at room temperature, rinsed with

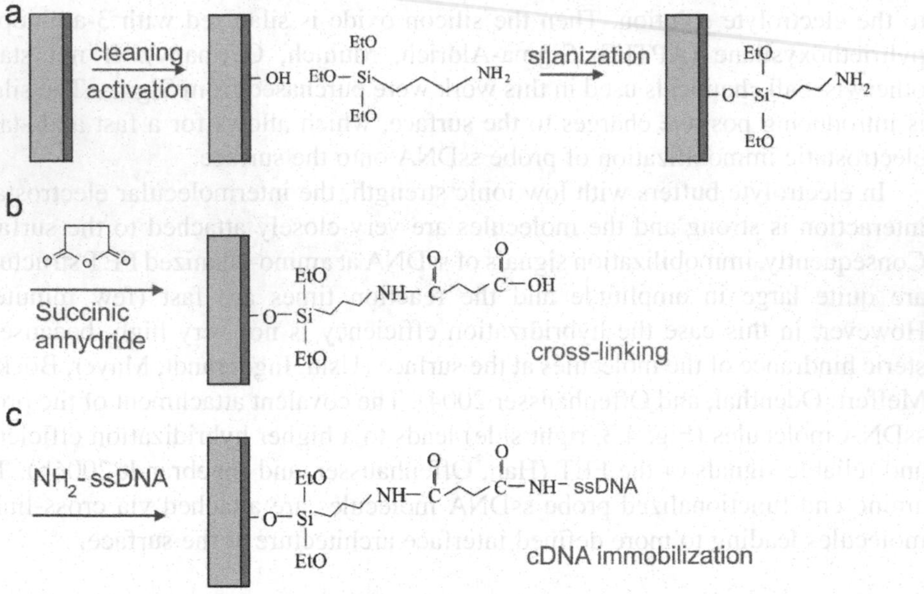

FIG. 4.4. Schematics of the surface modification protocol for the covalent immobilization of probe ssDNA. **A.** The chip surface is cleaned, activated, and silanized with APTES. **B.** In the next step, succinic anhydride is used as heterobifunctional cross-linker molecule. **C.** This finally confirms the covalent attachment of the functionalized ssDNA via an amino end group.

ultrapure water and dried with argon. In the next step, the surface was modified with amino groups by a silanization step with APTES either in gas or liquid phase (Fig. 4.4A) (Han, Mayer, Offenhäusser, and Ingebrandt 2006a). For gas phase silanization, the chips were placed in a desiccator containing a few drops of organosilane. The desiccator was sealed and heated above 100°C, and the chips were left to react for 1–2 h under a low pressure (~1 mbar) with the silane vapor. For liquid silanization, 1% APTES (v/v) in toluene was applied overnight, rinsed with 1 mM acetic acid, and dried with argon. The amino-silanized SiO_2 surface was further modified in a cross-linker reaction with a succinic anhydride solution (143 mM succinic anhydride, 87% (v/v) 1-methyl-2-pyrrolidone and 13% (v/v) sodium borate buffer) for 2 hours at room temperature, rinsed with ultra pure water and dried with argon. MES buffer solution was used as DNA immobilization buffer solution, which was composed of 0.1 M 4-morpholineethanesulfonic acid monohydrate (MES), 0.5 M NaCl and 10 mg/ml condensing agent 1-ethyl-3-(3-dimethylaminopropyl) carbodiimide hydrochloride (EDC) (pH 7). The concentration of the probe ssDNA for immobilization was 3 µM in MES buffer solution and the immobilization process was performed by overnight incubation at 37°C. After immobilization, the surface was blocked with 1% bovine serum albumin (BSA) in phosphate buffered saline buffer (PBS).

For all FET measurements, the DNA samples were dissolved in a TRIS-EDTA (TE) buffer solution: 10 mM TRIS-(hydroxymethyl) amino methane (TRIS), 1 mM ethylenediaminetetraacetic acid (EDTA) (both from Sigma) at pH 8 and with a typical sodium chloride concentration of 1 mM. Much attention was paid to the precise adjustment of the temperature and the pH value for all samples to exclude side effects. This was especially important when a potentiometric in situ readout of the hybridization reaction was used. For the impedimetric measurements presented here, we exclusively used a differential, ex situ assay: After immobilization of the probes with different sequences to the FET array, first transfer function scans we performed and the data were saved. Then the hybridization reaction was allowed for 2 h in an incubator. After this, the transfer functions were again recorded in the same low ionic strength TE buffer.

2.3 ALIGNED MICROSPOTTING

For a reliable sensor operation, the DNA detection system should use a differential readout approach between a measurement sensor and a reference sensor. A better approach, however, is the use of an aligned microspotting procedure, enabling an on-chip reference with a FMM sequence. In our DNA BioFET project an aligned microspotting system was developed, which employed a single spotting nozzle from a commercial available spotting system (Microdrop GmbH, Norderstedt, Germany) (Ingebrandt, Han, Sakkari, Stockmann, Belinskyy and Offenhäusser 2005). The nozzle was mounted on an x-y positioning table with a high lateral precision (<5 μm). In addition a small CCD camera was attached next to the nozzle.

The operation of the spotting system was confirmed via a PC and a software implemented in LabVIEW® 7.0 (National Instruments Corp., Austin, TX, USA). In this software, the image of the chip surface and the positions of the FET gates can be stored. Then the nozzle can approach the gate positions and apply individual spots of different probe ssDNA sequences. In Fig. 4.5B, the precise spotting onto 8 channels out of the 16-channel array can be seen. After incubation, the individual spots had a diameter of 100 to 120 μm without overlapping areas, which confirmed a negligible cross-contamination between the channels.

2.4 DNA SEQUENCES FOR HYBRIDIZATION DETECTION

Table 4.1 summarizes the different DNA sequences, which were used for the immobilization and hybridization experiments. All ssDNA sequences were obtained as synthesized and purified samples from MWG-Biotech AG (Ebersberg, Germany). In the first experiment (direct recording) we used a 19-bp sequence, which is specific for

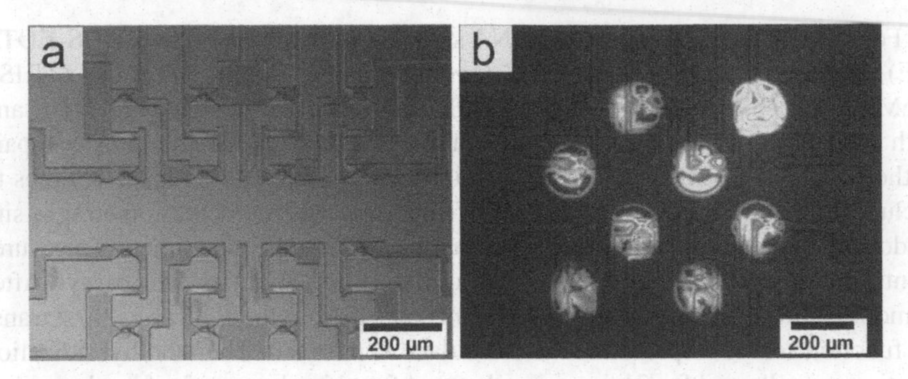

FIG. 4.5. Aligned microspotting of probe ssDNA onto the individual sensor spots. **A.** The original FET chip surface is shown. **B.** Eight channels from the 16-channel array were covered with fluorescently labeled ssDNA (see Table 4.1).

TABLE 4.1. DNA sequences, which were used for the DNA detection in this study.

Name	Sequence	Experiment
Fluoresence probe DNA	5'MMT-ATG AAC ACT GCA TGT AGT CA-3'-Cy3	*paragraph 2.3*
EpoR probe	5'-ACT CAT TCT CTG GGC TTG G-3'	direct recording
EpoR target	5'-C CAA GCC CAG AGA ATG AGT-3'	*paragraph 3.1.1*
Target DNA	5'-AC TGA TGT ACA TCA CAA GTA-3'	chip-to-chip reference
PM – probe DNA	5'MMT-TAC TTG TGA TGT ACA TCA GT-3'	*paragraph 3.1.2*
FMM – probe DNA	5'MMT-***ATG AAC ACT GCA TGT AGT CA***-3	
Target DNA	5'-TG ACT ACA TGC AGT GTT CAT-3'	on-chip reference
PM – probe DNA	5'MMT-ATG AAC ACT GCA TGT AGT CA-3'	*paragraph 3.1.3 and*
1 MM – probe DNA	5'MMT-ATG AAC ACT ***A***CA TGT AGT CA-3'	*paragraph 3.2*
2 MM – probe DNA	5'MMT-ATG AA***T*** ACT GCA TGT ***T***GT CA-3'	
3 MM – probe DNA	5'MMT-ATG AA***T*** ACT ***A***CA T***C***T AGT CA-3'	
FMM – probe DNA	5'MMT-***TAC TTG TGA TGT ACA TCA GT***-3'	

the human erythropoietin receptor (EpoR). In the second experiment, we used 20-bp sequences. The perfect match (PM) and full mismatch (FMM) probe sequences were covalently immobilized via an amino modifier (MMT) attached 5' end of the probe ssDNA samples. For the chip-to-chip reference experiments, we used a complete coating of the two dip-chips with either a PM or a FMM sequence. The reactions with a target ssDNA sequence were compared. For the on-chip reference experiment and for the impedimetric detections, we used different 20-bp sequences, which were spotted to the individual gates of the chip. The probe ssDNA sequences had either PM, a 2- or 3 MM, or a FMM to the target ssDNA sequence. The fluorescence probe DNA was used to visualize the precise microspotting with our setup.

3 RESULTS AND DISCUSSION

3.1 FET-BASED POTENTIOMETRIC DETECTION OF DNA HYBRIDIZATION

In general, the time-resolved potentiometric readout of the DNA immobilization and hybridization signals suffers from the well-known drift of ISFET devices in electrolyte solutions. This is especially true if silicon oxide was used as gate dielectrics. Therefore, in all the time-resolved, potentiometric experiments presented here, the gate-source and the drain-source voltages V_{GS} and V_{DS} were applied for 2-5 h to the FET chip. The experiments were started, when the slow sensor drift of the FETs was less than 2 mV per 20 min.

3.1.1 DIRECT RECORDING

The direct recordings were performed on 16-channel devices with electrostatically immobilized probe ssDNA. By the use of a fluidic cell, the solutions containing different DNA samples were applied. In Fig. 4.6, an exemplary recording is depicted. Here we used a natural 19-bp sequence specific for the human erythropoietin receptor at a concentration of 1 μM (for the sequences refer to Table 4.1) for both immobilization and hybridization. As buffer electrolyte we used standard, low ionic strength TE buffer with 1 mM NaCl (ionic strength at pH 8: 6 mM). The immobilization of the probe ssDNA was done by applying 1 ml of the probe ssDNA (1 μM) for three times. The immobilization signal was quite fast (90% signal amplitude after ~9 min) and after the second addition the chip surface was saturated (no reaction at the third addition of probe ssDNA).

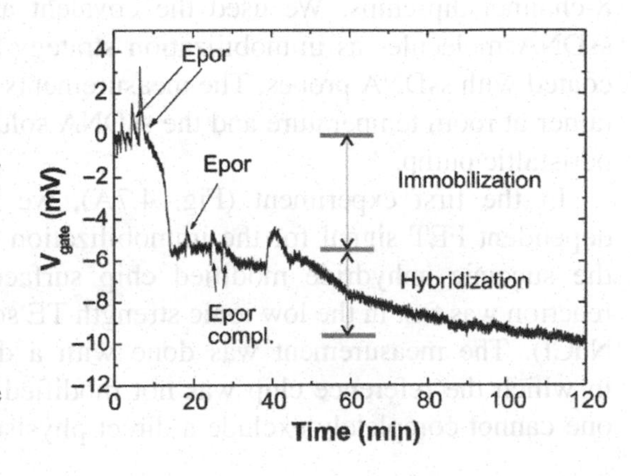

FIG. 4.6. Direct detection of immobilization and hybridization of a natural sequence specific for the human erythropoietin receptor (see Table 4.1). The electrostatically supported immobilization reaction resulted in a fast signal of 6 mV amplitude, whereas the diffusion-limited hybridization signal was much slower, with an amplitude of about 4 mV. In general, the hybridization signals in these experiments were barely distinguishable from the intrinsic long-term drift of the FETs.

The hybridization signal in this exemplary recording (Uslu, Ingebrandt, Mayer, Böcker-Meffert, Odenthal and Offenhäusser 2004) was much slower (90% signal amplitude after ~78 min) with a smaller signal amplitude compared to the immobilization (4 mV). The reason for the different binding constants of the two reactions is that the immobilization is accelerated by electrostatic attraction of the negatively charged DNA molecules to the positively charged APTES surface. Once the chip surface is saturated by DNA molecules, the surface is then electrically neutralized. The hybridization reaction is in this case diffusion-limited and reaction times in the low ionic strength buffer become very long, although a quite high target ssDNA concentration was used for these recordings.

The diffusion limitation in this time-resolved, in situ recording could possibly be overcome by the use of micro fluidic channels to restrict the diffusion volume on top of the recording devices.

A second possibility would be to use an ex situ assay in which the hybridization reaction is done in an incubator with a high ionic strength buffer solution. The field-effect detection itself would then be done in a low ionic strength buffer. In general, the reproducibility of the time-resolved, in situ measurements such as the exemplary recording shown in Fig. 4.6 was not high enough to allow for a comparison of signal amplitudes from experiment to experiments. In addition, a possible desorption of the electrostatically immobilized probe ssDNA during the hybridization reaction cannot be totally excluded in these experiments. As a result of these findings we decided for our recent studies to use the covalent attachment protocol for immobilization of the probes and to further develop a differential readout option for the FET system like it is described in the next paragraphs.

3.1.2 DIFFERENTIAL SENSOR READOUT: CHIP-TO-CHIP REFERENCE

A direct differential readout can be done with our system by the use of two 8-channel dip-chips. We used the covalent attachment protocol for the probe ssDNA molecules as immobilization strategy and the whole chip surfaces were coated with ssDNA probes. The measurements were performed in a separate container at room temperature and the ssDNA solutions were applied by the use of a peristaltic pump.

In the first experiment (Fig. 4.7A), we monitored the differential, time-dependent FET signal for the immobilization of amino-terminated ssDNA onto the succinic anhydride modified chip surface. The electrostatically-supported reaction was fast in the low ionic strength TE solution (3 μM probe ssDNA, 1 mM NaCl). The measurement was done with a differential chip-to-chip reference, in which the reference chip was not modified by aminosilane before. However, one cannot completely exclude a direct physisorption of ssDNA onto the silicon

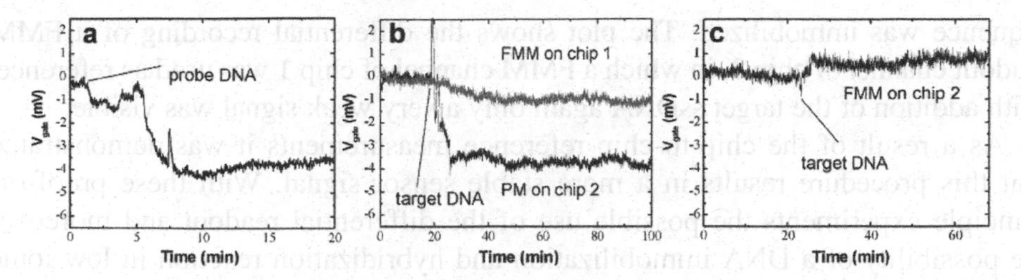

FIG. 4.7. **A.** Differential immobilization signal with chip-to-chip reference in a low ionic strength solution. The reference chip was not coated with silane. The signal corresponds to the difference between chemical covalent binding of the probes to the cross-linker on chip 1 and direct electrostatic adsorption to silicon oxide on chip 2. **B.** Differential hybridization signal with chip-to-chip reference in a low ionic strength solution. Chip 1 was completely coated with FMM probes and one channel from this chip was used as reference channel. The graph displays a second channel from chip 1, where almost no signal is visible demonstrating the analogy of the readout channels. On chip 2, which was coated by PM probes, a fast hybridization signal of 4 mV amplitude was detected. **C.** Differential hybridization assay, where both chips were coated by FMM probes. This demonstrated the possible use of the chip-to-chip reference.

oxide surface of the reference chip. However, because of the different treatments of the two chips this signal change can be attributed to the difference between direct physical adsorption onto silicon oxide and chemical covalent binding via the cross-linker molecules. In corresponding fluorescence studies, the possible immobilization reaction in the low ionic strength TE buffer was confirmed with fluorescently labelled probe ssDNA (data not shown here). The recorded signal amplitude (4 mV) and the reaction time was comparable (90% signal amplitude after ~6 min) to the previous result shown in Fig. 4.6.

For the experiment shown in Fig. 4.7B, we also used a differential chip-to-chip readout. The two electronically similar chips were treated identically except for the immobilized 20-bp probe sequences. One chip was covered with a FMM probe (chip 1) and the second chip was covered with a PM probe (chip 2). The complementary target ssDNA used for hybridization was diluted in standard TE buffer at a concentration of 3 µM as well. Exact sequences of the 20-bp ssDNA are provided in Table 4.1. For the differential recording shown in Fig. 4.7B, we used one channel of chip 1 as reference. In this graph, another channel of chip one is shown, where only a weak signal is visible, which is demonstrating the analogy of the readout channels from the 8-channel array. The PM channel, which was located on chip 2, records a signal amplitude of 4 mV for the hybridization reaction with a quite fast reaction time (90% signal amplitude after ~ 7 min) even in the low ionic strength TE solution (3 µM target ssDNA, 1 mM NaCl). For this binding assay we also confirmed in a fluorescence experiment that the hybridization reaction was successful in the same solution (data not shown here) .

In Fig. 4.7C, we confirmed the stability of the readout by an identical experiment similar to the experiment in Fig. 4.7B, except that here on both chips the FMM sequence was immobilized. The plot shows the differential recording of a FMM readout channel of chip 2, in which a FMM channel of chip 1 was used as reference. With addition of the target ssDNA again only a very weak signal was visible.

As a result of the chip-to-chip reference measurements it was demonstrated that this procedure results in a more stable sensor signal. With these proof-of-principle experiments the possible use of the differential readout and moreover the possibility of a DNA immobilization and hybridization reaction in low ionic strength buffers was demonstrated. This was the basis for the further use of the 16-channel FET chip in a microarray assay with on-chip reference like it is described in the next paragraph.

3.1.3 DIFFERENTIAL SENSOR READOUT: ON-CHIP REFERENCE

The differential readout can also be done by using an on-chip reference channel. For this purpose the aligned microspotting procedure needs to be used. The probe ssDNA was covalently immobilized to the different gates of a single 8-channel FET chip. We used 20 bp ssDNA samples with different sequences, which are provided in Table 4.1. The FMM channel was used as reference channel, which mostly canceled out the sensor drift and other side effects (e.g., temperature or pH variations). The selectivity of this recording, however, was not high enough to detect SNPs. In Fig. 4.8, the recordings from the 3 MM, 2 MM and PM channels are compared.

In summary it can be said, that the in situ detection of SNPs in the low ionic strength buffer electrolytes is at the moment not possible with our FET system. The selectivity of the hybridization reaction in this solution (at room temperature), however, was high enough to allow for a discrimination of two mismatches in

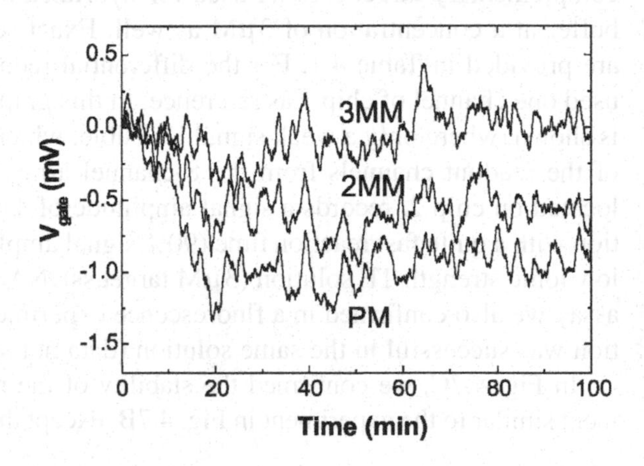

FIG. 4.8. Detection DNA hybridization using a channel, where a FMM probe ssDNA was immobilized as on-chip reference. The signals from three other channels are shown, where 3 MM, 2 MM, and PM probe ssDNA samples were immobilized. The signal amplitudes are barely distinguishable. To aid a discrimination of the different traces we reduced the plotted data by smoothing. The target DNA was added at $t=0$ in this recording.

20-bp sequences. A further improvement of the protocol (adjusting the hybridization temperature or the buffer composition for the recordings) might possibly result in a SNP selectivity in future experiments.

3.2 FET-BASED IMPEDIMETRIC DETECTION OF DNA HYBRIDIZATION

Before the impedimetric recordings can be discussed, a short survey about the general treatment of the ISFET transfer function is provided.

The detection of biomolecules using the transfer function of ISFET devices has already been described (Kruise, Rispens, Bergveld, Kremer, Starmans, Haak, Feijen and Reinhoudt 1992; Lugtenberg, Egberink, van den Berg, Engbersen and Reinhoudt 1998; Antonisse, Snellink-Ruel, Lugtenberg, Engbersen, van den Berg, and Reinhoudt 2000; Kharitonov, Zayats, Lichtenstein, Katz and Willner 2000; Kharitonov, Wasserman, Katz and Willner 2001; Lahav, Kharitonov and Willner 2001; Pogorelova, Bourenko, Kharitonov and Willner 2002; Sallacan, Zayats, Bourenko, Kharitonov and Willner 2002; Zayats, Raitman, Chegel, Kharitonov, and Willner 2002; Katz and Willner 2003; Poghossian and Schoning 2004; Pogorelova, Kharitonov, Willner, Sukenik, Pizem and Bayer 2004). The observed effects have been explained by impedance changes of the biomolecules in close contact to the transistor gate forming a membrane-like layer with a certain impedance value. The impedimetric detection of DNA hybridization with large Si/SiO_2 heterostructures in contact with electrolyte solutions has been reported as well (Souteyrand, Cloarec, Martin, Wilson, Lawrence, Mikkelsen and Lawrence 1997; Marquette, Lawrence, Polychronakos and Lawrence 2002; Cai, Peck, van der Weide and Hamers 2004; Macanovic, Marquette, Polychronakos and Lawrence 2004). A possible explanation for the observed impedance changes governed by the attachment of DNA to the Si/SiO_2/electrolyte interface has been provided (Landheer, Aers, McKinnon, Deen and Ranuarez 2005). It has been described that such effects should be more pronounced in dilute buffer electrolytes. In a simplified model, if such a membrane-like layer was attached to the gate structure of an ISFET, the transconductance g_m becomes frequency dependent (Bergveld, Vandenberg, Vanderwal, Skowronskaptasinska, Sudholter and Reinhoudt 1989; Antonisse, Snellink-Ruel, Lugtenberg, Engbersen, van den Berg, and Reinhoudt 2000; Kharitonov, Wasserman, Katz and Willner 2001).

In general, the transconductance g_m of a transistor is described by:

$$g_m = \frac{\partial I_{DS}}{\partial V_{GS}}\Big|_{V_{DS=const}} = \frac{i_{DS}}{V_{GS}} \tag{1}$$

where i_{DS} is the small-signal drain-source current and v_{GS} is the small-signal gate-source voltage. The small-signal output voltage v_{out} after the first amplification stage (Fig. 4.2b) can be calculated by:

$$v_{out} = -Ri_{DS} = -Rg_m v_{GS} \tag{2}$$

where R being the feedback resistor of the first OP, which is contacted at the inverted input. When, as in this case, a DNA layer is attached to the gate electrode, the voltage v_{GS} is divided into the voltage drop over the DNA layer and the gate oxide, respectively. This effectively reduces the potential difference between gate and source. In this case the transfer function $H(j\omega)$ needs to be introduced to account for the frequency dependency of v_{GS}:

$$H(j\omega) = \frac{1 + j\omega R_{DNA} C_{DNA}}{1 + j\omega R_{DNA}(C_{DNA} + C_{ox})} \tag{3}$$

The impedance, which is introduced by the DNA layer, can be described as RC element composed of R_{DNA} and C_{DNA} with ω being the angular frequency. The input capacitance of the gate oxide is C_{ox} (Fig. 2b). The transfer function factor (Eq. 3) can then be used to describe the frequency-dependency of v_{out} with:

$$v_{out} = -Rg_m H(j\omega) v_{GS} \tag{4}$$

It has been shown that two characteristic time constants can be extracted out of the theoretical transfer function graph (Fig. 4.9C), which describe the properties of the attached DNA layer:

$$\tau_1 = R_{DNA}(C_{DNA} + C_{ox}) = \tau_2 + R_{DNA} C_{ox} \tag{5}$$

$$\tau_2 = R_{DNA} C_{DNA} \tag{6}$$

The second time constant τ_2 provides the relaxation time of the DNA layer. Equations 5 and 6 can be analytically solved by extracting the time constants τ_1 and τ_2 from a transfer function graph with known C_{ox} of the FET device.

In Fig. 4.10 a typical experiment for impedimetric detection of the DNA hybridization using the transistor transfer function is shown (Ingebrandt, Han, Nakamura, Schöning, Poghossian and Offenhäusser 2007). The DNA immobilization protocol and microspotting procedure were identical to the potentiometric experiments. The probe ssDNA sequences used for the impedimetric experiments are provided in Table 4.1. However, in contrast to the potentiometric detection, for the impedimetric measurements an ex situ assay (2h hybridization in an incubator followed by a three-step rinse) was used. The transfer functions, $H(j\omega) = V_{out}/V_{in}$, of three of the transistor channels from the 8-channel chip after probe ssDNA immobilization

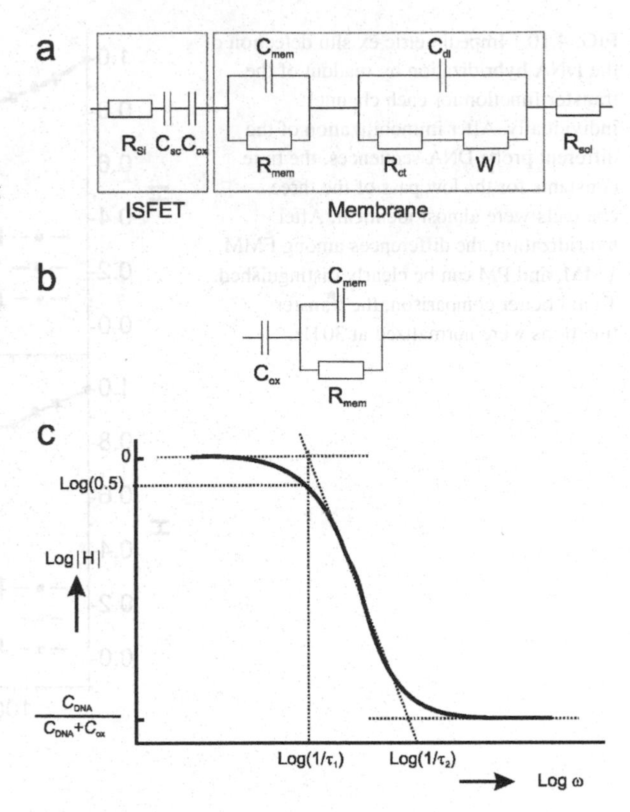

FIG. 4.9. A. Equivalent electrical circuit of a membrane attached to the gate input of an ISFET device. **B.** Simplified electrical circuit of a membrane attached to the ISFET gate with the main components influencing the ISFET transfer function. **C.** Theoretical transfer function characteristics $H(j\omega)$ for a FET with attached DNA layer (according to Antonisse, Snellink-Ruel, Lugtenberg, Engbersen, van den Berg, and Reinhoudt 2000). Two characteristic time constants τ_1 and τ_2 can be extracted out of the graph.

are shown in the range of 30 Hz to 30 kHz (Fig. 4.10). It should be noted, that the FET readout system has an overall time constant of $\tau_{co} = 1/f_{co}$, which was in the kHz frequency range for our system. Mainly because of differences in layout of the drain and source contact lines or variations in surface coverage of the probe molecules, the cutoff frequency f_{co} might vary slightly from channel to channel. For this reason, we stored the transfer function scans before hybridization as reference data.

After hybridization, the time constants of the low pass for the individual channels changed depending on the binding probability of the target DNA to the respective transistor gates (Fig. 4.10). The transistor gate, where the most target DNA molecules were reacting (PM channel), experienced the largest change. The change in impedance of the DNA layer can be recorded by the transfer function method. A higher membrane resistance R_{DNA} leads to a higher time constant τ_1. It is well accepted that dsDNA is more difficult to be protonated and deprotonated, which leads to a higher charge transfer resistance (Sponer, Leszczynski and Hobza 2001). This is mainly correlated with a change in τ_1 (Antonisse, Snellink-Ruel, Lugtenberg, Engbersen, van den Berg, and Reinhoudt 2000). In our experiments the differences between FMM, 1 MM and PM were clearly recorded, hence the

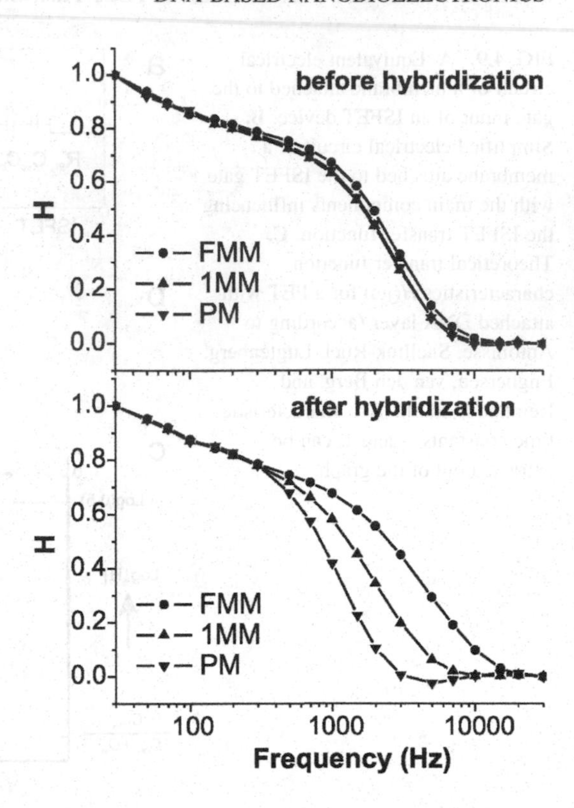

FIG. 4.10. Impedimetric ex situ detection of the DNA hybridization by readout of the transfer function for each channel individually. After immobilization of the different probe DNA sequences, the time constants for the low pass of the three channels were almost identical. After hybridization, the differences among FMM, 1 MM, and PM can be clearly distinguished. To aid better comparison, the transfer functions were normalized at 30 Hz.

selectivity of the transfer function method was high enough to enable the detection of SNPs. We additionally tested the influence of different electrolyte concentrations (0.001–100 mM). In accordance to the theoretical predicted effects (Landheer, Aers, McKinnon, Deen and Ranuarez 2005), the differences between hybridization of target ssDNA to spots of FMM, 1 MM, and PM were higher at lower electrolyte concentrations (<1 mM) in our experiments.

A possible technical realization of the Differential Transfer Function (DTF) method for large microarrays is illustrated in Fig. 4.11. After immobilization of the probe DNA to the microarray, a first frequency scan of all channels is performed. The resulting transfer function curves are stored and later used as reference for the hybridization data. In Fig. 4.11 the differences of transfer functions before and after hybridization (ex situ recording in Fig. 4-10) are plotted. In order to enable a more reliable detection of SNPs, the FMM channel is used as a chip reference. By this differential method, potential signal variations resulting from unspecific binding of DNA can mostly be canceled out. Using this differential data acquisition, the difference between PM and 1 MM can be clearly distinguished in full DTF scans (Fig. 4.11). A much quicker comparison of the channels of a large array, however, ideally would be done by a time-dependent operation of the frequency-selective amplifier at a fixed frequency of 2 to 3 kHz.

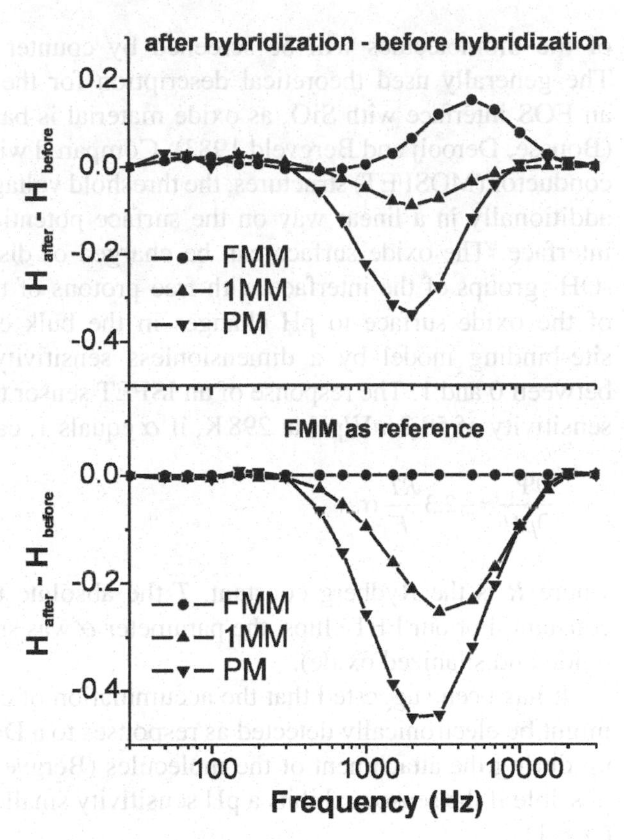

FIG. 4.11. Differential transfer function graphs. In the upper graph, the difference between the recordings before and after hybridization is shown. A differential, ex situ detection can be realized by setting the FMM channel as reference, as is shown in the lower graph. A fast ex situ readout of a large array ideally would be done at 3 kHz for this particular chip design. An in situ detection during the hybridization reaction would be feasible by time-dependent operation of the frequency-selective amplifier at 2 to 3 kHz, as well.

3.3 UNDERLYING DETECTION PRINCIPLE

3.3.1 POTENTIOMETRIC SIGNAL GENERATION

The underlying detection principle for the potentiometric, field-effect based detection of DNA is still under discussion (Fritz, Cooper, Gaudet, Sorger and Manalis 2002; Uslu, Ingebrandt, Mayer, Böcker-Meffert, Odenthal and Offenhäusser 2004; Landheer, Aers, McKinnon, Deen and Ranuarez 2005; Poghossian, Cherstvy, Ingebrandt, Offenhäusser and Schöning 2005). At the moment it is clear, that the obtained recordings are more pronounced if low ionic strength buffers are used during detection. A first attempt to describe the detection mechanism was provided by Fritz and co-workers (2002). Here mainly the Graham equation, which results from the Debye–Hückel theory, has been used.

For the field-effect based detection of charged molecules in an electrolyte solution it has been mentioned that an accumulation of the charged biomolecules at an electrolyte-oxide interface should not generate any signal, because the charge of the biomolecules will be screened by counter ions (Bergveld 1991, 1996). The generally used theoretical description for the charging and discharging of an EOS interface with SiO_2 as oxide material is based on the site-binding model (Bousse, Derooij and Bergveld 1983). Compared with common metal-oxide semiconductor (MOSFET) structures, the threshold voltage of an ISFET device depends additionally in a linear way on the surface potential ψ_0 of the oxide–electrolyte interface. The oxide surface can be charged or discharged by interaction of the -OH- groups of the interface with free protons of the electrolyte. The sensitivity of the oxide surface to pH changes in the bulk electrolyte is described in the site-binding model by a dimensionless sensitivity parameter α, which varies between 0 and 1. The response of an ISFET sensor to pH changes has a maximum sensitivity of 58.2 mV/pH at 298 K, if α equals 1, called the nernstrian response:

$$\frac{\partial \Psi_0}{\partial pH} = -2.3\frac{RT}{F}\alpha \tag{7}$$

where R is the Rydberg constant, T the absolute temperature, and F Faraday's constant. For our FET chips, the parameter α was smaller than 1 in all cases (bare oxide and silanized oxide).

It has been suggested that the accumulation of charged molecules at a surface might be electronically detected as responses to a Donnan potential, which is built up during the attachment of the molecules (Bergveld 1996). This should only be possible if the sensor exhibits a pH sensitivity smaller than the nernstrian response ($\alpha < 1$).

In a liquid environment at neutral pH values, the DNA molecule carries 0.5 – 1 negative charge per base at the phosphate groups of the molecule's backbone. The accumulation of these negative charges at the gate oxide of the FET structure shifts the flat band voltage, which results in a voltage drop of the output signal in a time-dependent measurement (Fritz, Cooper, Gaudet, Sorger and Manalis 2002; Uslu, Ingebrandt, Mayer, Böcker-Meffert, Odenthal and Offenhäusser 2004). It is well known that the electrostatic potential in an electrolyte drops according to the Debye length:

$$\lambda_D = \sqrt{\frac{kT\varepsilon_{elect}\varepsilon_0}{2z^2e^2n_0}} \tag{8}$$

where k is the Boltzmann constant, T is absolute temperature, ε_{elect} is the permittivity of the electrolyte of 78, ε_0 is vacuum permittivity, e is elementary charge, z is the valency of the ions in the electrolyte, and n_0 is the ionic strength of the electrolyte. Attachment of charged molecules to the surface results in a change of the surface potential ψ, and the resulting flat band voltage shift will be charge sensitive. The ionic strength of the buffer electrolytes used in our experiments was 6 mM (at RT), which results in a Debye length of 4 nm (Eq. 8).

According to the Debye–Hückel theory, the attachment of charges to the oxide surface should result in a corresponding surface potential change, which is described by the Graham equation:

$$\sigma_0 = \sqrt{8\varepsilon_{elect}\varepsilon_0 kTn_0}\,\sinh\left(\frac{ze\Psi_0}{2kT}\right) \tag{9}$$

The maximum surface charge density σ_0 of silicon oxide is 0.8 C/m^2 (Dong, Pappu and Xu 1998), which corresponds to a maximum surface potential of 267 mV for our experiments. An attachment of charged molecules to the surface reduces the surface potential. The resulting voltage difference at the gate surface ΔV_{gate} monitored by the FET can be calculated by:

$$\Delta V_{Gate} = \frac{2kT}{ze}\cdot\left(ar\sinh\left(\frac{\sigma_0}{\sqrt{8\varepsilon_{elect}\varepsilon_0 kTn_0}}\right) - ar\sinh\left(\frac{\sigma_0 - \sigma_{DNA}}{\sqrt{8\varepsilon_{elect+DNA}\varepsilon_0 kTn_{0+DNA}}}\right)\right) \tag{10}$$

According to Eq. 10, an additional signal might be generated if the ionic strength, electrical permittivity, or temperature of the probe solution was different from the bare buffer solution. These effects can be neglected in experiments where after attachment of the charged molecules to the surface the fluidic cell is flushed with bare buffer solution again, which results in the initial temperature, permittivity, ionic strength, and pH. For the electrostatic immobilization of the probes, the surface density of the attached 20 bp DNA after hybridization at the surface was measured to 4×10^{15} 1/m^2 (Uslu, Ingebrandt, Mayer, Böcker-Meffert, Odenthal and Offenhäusser 2004). By assuming that all oligonucleotides attach flatly and elongated to the surface and by counting one negative charge per base pair, a maximum change in surface charge density σ_0 of 0.013 C/m^2 can be reached with this DNA density. The attachment of the assumed amount of DNA strands would lead to a change in surface potential of about 0.8 mV, which is almost the detection limit of the FET system.

A more detailed description of the interface potential changes, when charged molecules attach the surface of FET-based sensors, has been provided (Landheer, Aers, McKinnon, Deen and Ranuarez 2005). This description also includes Donnan-based potential changes of biomolecular membranes attached to the surface. It is predicted that the potential changes are more pronounced when low ionic strength buffers are used for detection. This is very consistent with the experimental results in our studies. In addition, the potential changes should be more pronounced, if initially uncharged surfaces (e.g., a gold surface) would be used at the transistor gates. In another article, the potential changes are assumed to be generated by redistribution of ions at the surface of an ion-sensitive interface

(Poghossian, Cherstvy, Ingebrandt, Offenhäusser and Schöning 2005). Hence, the potential changes at the surface should be more pronounced if gate materials with a higher ion sensitivity are used.

3.3.2 IMPEDIMETRIC SIGNAL GENERATION

For explanation of the impedimetric signals, first the overall time constant of the FET readout system needs to be considered.

In test experiments we confirmed that the amplitude of the input signal V_{mod} can be treated as constant over the whole recording range (1 Hz to 100 kHz). In addition, the frequency-selective amplifier cascade was realized with fast OP devices in our readout system. Hence, the overall low pass for the system was primarily dependent on the capacitance of the contact line area C_{CL} of the chip. In order to calculate C_{CL}, the capacitance of the individual drain contact lines $C_{CLD}{}^{i}$ of all channels plus the capacitance of the common source contact line exposed to the electrolyte solution C_{CLCS} must be summed up:

$$C_{CL} = C_{CLCS} + \sum_{i} C_{CLD}^{i} \tag{11}$$

The capacitance C_{CL} in combination with the series resistor formed by the electrolyte solution R_{sol} plus *dc* resistance of the reference electrode R_{RE} resulted in the overall time constant of the system:

$$\tau_{co} = \frac{1}{f_{co}} = C_{CL} \cdot (R_{RE} + R_{sol}) \tag{12}$$

When using the commercial liquid-junction reference electrodes in combination with the lowest ionic strength buffer (0.01 mM), the resistance of the reference electrode can be neglected in Equation 12 ($R_{RE} << R_{sol}$). In addition it should be noted, that in this case the distance between the reference electrode and the transistor gate is important and should be kept constant during measurements (Kruise, Rispens, Bergveld, Kremer, Starmans, Haak, Feijen and Reinhoudt 1992). Typical values for the recordings presented here were: $C_{CL} = 1$ nF, $R_{RE} = 5$ kΩ, and $R_{sol} = 50$–400 kΩ. Therefore, the typical cutoff frequencies for the transfer functions were in the range of $f_{co} = 2.5$–20 kHz. Additionally to this overall cutoff, the time constant of the DNA layer attached to the individual gates of the chip is responsible for the individual transfer function characteristics (Fig. 4.10). The influence of a DNA layer to the transfer function characteristics of the FET has been explained in Section 3.2. In the recording it can be seen that upon hybridization the value of R_{DNA} increased in all channels, with the PM channel showing the highest increase. This can be caused by the fact that dsDNA molecules in general are less favorable to be protonated and deprotonated compared with ssDNA molecules. In addition the capacitance C_{DNA} also increased upon hybridization, which might be a result of a more dense packing of the molecules inside the layer or an effect of the higher density of fixed charges at the surfaces as described by Landheer and co-workers (Landheer, Aers, McKinnon, Deen and Ranuarez 2005). To fully explain the effects observed by the impedimetric readout of the FETs, a more detailed study will be done in future experiments.

4 CONCLUSION AND OUTLOOK

We developed 16-channel FET devices and amplifier systems for the label-free, fully electronic detection of DNA. For the sequence-specific DNA detection, we used the hybridization reaction between known sequences of probe ssDNA with a target ssDNA. The probe ssDNA sequences were immobilized to thc different sensor spots of the FET arrays, and the hybridization reaction was recorded by a differential readout between thc electronically similar FET channels. In our experiments, we used short nucleotide samples (~20 bp) for detection. The overall goal for all DNA sensor system is to reach selectivity high enough to detect SNPs. In general, FET-based, potentiometric, in situ detection of DNA mainly suffers from small signal amplitudes, long reaction times because of the need for low ionic strength buffer electrolytes, and the intrinsic sensor drift of the FET devices when operated in a liquid environment.

In a first approach, we used electrostatically immobilized probe ssDNA on amino-functionalized sensor surfaces. This results in high signal amplitudes and fast reaction times for the immobilization signals, because of the electrostatically supported reaction. The hybridization signals in this case, however, were very slow and not detectable most of the time. We assume that mainly steric hindrances of the probe molecules at the chip surface reduced the hybridization efficiency in these cases.

In a second approach, we established more defined interface architecture at the FET sensor surfaces. Here we used the same silanization procedure as in the electrostatic immobilization with additional steps. We used succinic anhydride as a cross-linker molecule, which confirmed a covalent attachment of the amino end-functionalized probe ssDNA to the chip surface. We characterized this attachment procedure step-by-step, confirming the covalent attachment of the probe ssDNA to the chip surface (results were not shown here). By using this procedure and a time-dependent potentiometric readout, the reaction times and the signal amplitudes for immobilization and hybridization were comparable. For all the experiments, a differential readout using a reference FET led to a more stable sensor operation and reliable signals. We immobilized fully mismatch probe ssDNA sequences onto this reference channel. The reference channel can either be localized on a second, electrically identical chip or be part of the array on the same chip. Such an on-chip reference can be realized with an aligned microspotting procedure. However, the selectivity of the obtained results for in situ readouts was so far not high enough to enable the detection of SNPs. In our experiments, two mismatches of the probe molecules were detectable by the use of an on-chip reference.

An alternative approach is the use of impedimetric readout of the transistor channels. We used an ex situ assay, which resulted in a more reliable detection of the hybridization signals. For the impedimetric readout, the transfer function characteristics of each transistor channel were scanned individually. In our experiments, this procedure resulted in most reliable signals with selectivity high enough to enable the detection of SNPs. Further advantage of the transfer function method is a much quicker sensor readout, because the signals can be sampled with 2- to 3-kHz in our current design. In summary it can be said that the fully electronic, label-free detection of the DNA hybridization with FET devices is currently a growing research field. Nonetheless, so far the theoretical description of field effect–based DNA detection is still under discussion. More detailed studies into this direction will be mandatory to enable the application of these methods in modern DNA bioassays.

ACKNOWLEDGMENTS

We thank M. Krause, W. Staab, R. Stockmann, A. Steffens, and S. Schaal for the design, fabrication and encapsulation of the silicon chips and N. Wolters, R. Otto, H. Bousack, and D. Strobl for the design, optimization, and fabrication of the amplifier system. We thank D. Lomparski for the implementation of the readout software and O. Belinskyy, M.R. Sakkari, and M.M. Kottuppallil for assembly of the custom made, aligned microspotting unit. The experimental results presented here were achieved in the theses projects of F. Uslu and Y. Han. We thank F. Nakamura for the establishment of the covalent attachment procedure for the probe ssDNA. In addition, we thank M.J. Schöning and A. Poghossian for valuable discussions. This project was funded by the Helmholtz Association of German Research Centres.

REFERENCES

Antonisse, M.M.G., Snellink-Ruel, B.H.M., Lugtenberg, R.J.W., Engbersen, J.F.J., van den Berg, A. and Reinhoudt, D.N. (2000). Membrane characterization of anion-selective CHEMFETs by impedance spectroscopy. Anal. Chem. 72, 343–348.

Barbaro, M., Bonfiglio, A. and Raffo, L. (2006). A charge-modulated FET for detection of biomolecular processes: conception, modeling, and simulation. IEEE Trans. Electron Devices 53, 158–166.

Bergveld, P. (1970). Development of an ion-sensitive solid-state device for neurophysiological measurements. IEEE Trans. Biomed. Eng. BM17, 70.

Bergveld, P. (1991). A critical-evaluation of direct electrical protein-detection methods. Biosens. Bioelectron. 6, 55–72.

Bergveld, P. (1996). The future of biosensors. Sens. Actuator A-Phys. 56, 65–73.

Bergveld, P., Vandenberg, A., Vanderwal, P.D., Skowronskaptasinska, M., Sudholter, E.J.R. and Reinhoudt, D.N. (1989). How electrical and chemical-requirements for refets may coincide. Sensors and Actuators 18, 309–327.

Bergveld, P., Wiersma, J. and Meertens, H. (1976). Extracellular potential recordings by means of a field-effect transistor without gate metal, called Osfet. IEEE Trans. Biomed. Eng. 23, 136–144.

Bousse, L., Derooij, N.F. and Bergveld, P. (1983). Operation of chemically sensitive field-effect sensors as a function of the insulator-electrolyte interface. IEEE Trans. Electron Dev. 30, 1263–1270.

Cai, W., Peck, J.R., van der Weide, D.W. and Hamers, R.J. (2004). Direct electrical detection of hybridization at DNA-modified silicon surfaces. Biosens. Bioelectron. 19, 1013–1019.

Chovelon, J.M., Jaffrezicrenault, N., Fombon, J.J. and Pedone, D. (1991). Monitoring of isfet encapsulation aging by impedance measurements. Sens. Actuator B-Chem. 3, 43–50.

Dong, Y., Pappu, S.V. and Xu, Z. (1998). Detection of local density distribution of isolated silanol groups on planar silica surfaces using nonlinear optical molecular probes. Anal. Chem. 70, 4730–4735.

Fritz, J., Cooper, E.B., Gaudet, S., Sorger, P.K. and Manalis, S.R. (2002). Electronic detection of DNA by its intrinsic molecular charge. Proc. Natl. Acad. Sci. U. S. A. 99, 14142–14146.

Han, Y., Mayer, D., Offenhäusser, A. and Ingebrandt, S. (2006a). Surface activation of thin silicon oxides by wet cleaning and silanization. Thin Solid Films 510, 175–180.

Han, Y., Offenhäusser, A. and Ingebrandt, S. (2006b). Detection of DNA hybridization by a field-effect transistor with covalently attached catcher molecules. Surf. Interface Anal. 38, 176–181.

Ingebrantt, S., Han, Y., Sakkari, M.R., Stockmann, R., Belinskyy, O., Offenhäusser, A., 2005. Semiconductor Materials for sensing, vol. 828. Materials Research Society, Warrendale, pp. 307–312.

Ingebrandt, S., Han, Y., Nakamura, F., Schöning, M.J., Poghossian, A. and Offenhäusser, A. (2006a). Label-free detection of single nucleotide polymorphisms utilizing the differential transfer function of field-effect transistors. Biosens. Bioelectron. 22, 2834–2840.

Ingebrandt, S. and Offenhäusser, A. (2006b). Label-free detection of DNA using field-effect transistors. physica status solidi (a) 203, 3399–3411.

Ingebrandt, S., Yeung, C.K., Krause, M. and Offenhäusser, A. (2001). Cardiomyocyte-transistor-hybrids for sensor application. Biosens. Bioelectron. 16, 565–570.

Katz, E. and Willner, I. (2003). Probing biomolecular interactions at conductive and semiconductive surfaces by impedance spectroscopy: Routes to impedimetric immunosensors, DNA-Sensors, and enzyme biosensors. Electroanalysis 15, 913–947.

Kharitonov, A.B., Wasserman, J., Katz, E. and Willner, I. (2001). The use of impedance spectroscopy for the characterization of protein-modified ISFET devices: application of the method for the analysis of biorecognition processes. J. Phys. Chem. B 105, 4205–4213.

Kharitonov, A.B., Zayats, M., Lichtenstein, A., Katz, E. and Willner, I. (2000). Enzyme monolayer-functionalized field-effect transistors for biosensor applications. Sens. Actuator B-Chem. 70, 222–231.

Kim, D.S., Jeong, Y.T., Park, H.J., Shin, J.K., Choi, P., Lee, J.H. and Lim, G. (2004). An FET-type charge sensor for highly sensitive detection of DNA sequence. Biosens. Bioelectron. 20, 69–74.

Kricka, L.J. (2002). Stains, labels and detection strategies for nucleic acids assays. Ann. Clin. Biochem. 39, 114–129.

Kruise, J., Rispens, J.G., Bergveld, P., Kremer, F.J.B., Starmans, D., Haak, J.R., Feijen, J. and Reinhoudt, D.N. (1992). Detection of charged proteins by means of impedance measurements. Sens. Actuator B-Chem. 6, 101–105.

Lahav, M., Kharitonov, A.B. and Willner, I. (2001). Imprinting of chiral molecular recognition sites in thin TiO2 films associated with field-effect transistors: novel functionalized devices for chiroselective and chirospecific analyses. Chem.-Eur. J. 7, 3992–3997.

Landheer, D., Aers, G., McKinnon, W.R., Deen, M.J. and Ranuarez, J.C. (2005). Model for the field effect from layers of biological macromolecules on the gates of metal-oxide-semiconductor transistors. J. Appl. Phys. 98,

Lugtenberg, R.J.W., Egberink, R.J.M., van den Berg, A., Engbersen, J.F.J. and Reinhoudt, D.N. (1998). The effects of covalent binding of the electroactive components in durable CHEMFET membranes—impedance spectroscopy and ion sensitivity studies. J. Electroanal. Chem. 452, 69–86.

Macanovic, A., Marquette, C., Polychronakos, C. and Lawrence, M.F. (2004). Impedance-based detection of DNA sequences using a silicon transducer with PNA as the probe layer. Nucleic Acids Res. 32, e201-e207.

Marquette, C.A., Lawrence, I., Polychronakos, C. and Lawrence, M.F. (2002). Impedance based DNA chip for direct T-m measurement. Talanta 56, 763–768.

Offenhäusser, A., Sprössler, C., Matsuzawa, M. and Knoll, W. (1997). Field-effect transistor array for monitoring electrical activity from mammalian neurons in culture. Biosens. Bioelectron. 12, 819–826.

Poghossian, A., Cherstvy, A., Ingebrandt, S., Offenhäusser, A. and Schöning, M.J. (2005). Possibilities and limitations of label-free detection of DNA hybridization with field-effect-based devices. Sens. Actuator B-Chem. 111, 470–480.

Poghossian, A. and Schoning, M.I. (2004). Detecting both physical and (bio-)chemical parameters by means of ISFET devices. Electroanalysis 16, 1863–1872.

Pogorelova, S.P., Bourenko, T., Kharitonov, A.B. and Willner, I. (2002). Selective sensing of triazine herbicides in imprinted membranes using ion-sensitive field-effect transistors and microgravimetric quartz crystal microbalance measurements. Analyst 127, 1484–1491.

Pogorelova, S.P., Kharitonov, A.B., Willner, I., Sukenik, C.N., Pizem, H. and Bayer, T. (2004). Development of ion-sensitive field-effect transistor-based sensors for benzylphosphonic acids and thiophenols using molecularly imprinted TiO2 films. Anal. Chim. Acta 504, 113–122.

Pouthas, F., Gentil, C., Cote, D. and Bockelmann, U. (2004). DNA detection on transistor arrays following mutation-specific enzymatic amplification. Appl. Phys. Lett. 84, 1594–1596.

Pouthas, F., Gentil, C., Cote, D., Zeck, G., Straub, B. and Bockelmann, U. (2004). Spatially resolved electronic detection of biopolymers. Phys. Rev. E 70, 0319061-0319068.

Sakata, T. and Miyahara, Y. (2005). Potentiometric detection of single nucleotide polymorphism by using a genetic field-effect transistor. Chembiochem 6, 703–710.

Sallacan, N., Zayats, M., Bourenko, T., Kharitonov, A.B. and Willner, I. (2002). Imprinting of nucleotide and monosaccharide recognition sites in acrylamidephenylboronic acid-acrylamide copolymer membranes associated with electronic transducers. Anal. Chem. 74, 702–712.

Schasfoort, R.B.M., Streekstra, G.J., Bergveld, P., Kooyman, R.P.H. and Greve, J. (1989). Influence of an Immunological Precipitate on Dc and Ac Behavior of an Isfet. Sensor Actuator 18, 119–129.

Souteyrand, E., Cloarec, J.P., Martin, J.R., Wilson, C., Lawrence, I., Mikkelsen, S. and Lawrence, M.F. (1997). Direct detection of the hybridization of synthetic homo-oligomer DNA sequences by field effect. J. Phys. Chem. B 101, 2980–2985.

Sponer, J., Leszczynski, J. and Hobza, P. (2001). Electronic properties, hydrogen bonding, stacking, and cation binding of DNA and RNA bases. Biopolymers 61, 3–31.

Sprössler, C., Denyer, M., Britland, S., Knoll, W. and Offenhäusser, A. (1999). Electrical recordings from rat cardiac muscle cells using field-effect transistors. Phys. Rev. E 60, 2171–2176.

Sprössler, C., Richter, D., Denyer, M. and Offenhäusser, A. (1998). Long-term recording system based on field-effect transistor arrays for monitoring electrogenic cells in culture. Biosens. Bioelectron. 13, 613–618.

Tsuruta, H., Matsui, S., Hatanaka, T., Namba, T., Miyamoto, K. and Nakamura, M. (1994). Detection of the products of polymerase chain-reaction by an elisa system based on an ion sensitive field-effect transistor. J. Immunol. Methods 176, 45–52.

Uslu, F., Ingebrandt, S., Mayer, D., Böcker-Meffert, S., Odenthal, M. and Offenhäusser, A. (2004). Labelfree fully electronic nucleic acid detection system based on a field-effect transistor device. Biosens. Bioelectron. 19, 1723–1731.

Wang, J. (2002). Electrochemical nucleic acid biosensors. Anal. Chim. Acta 469, 63–71.

Yeow, T.C.W., Haskard, M.R., Mulcahy, D.E., Seo, H.I. and Kwon, D.H. (1997). A very large integrated pH-ISFET sensor array chip compatible with standard CMOS processes. Sens. Actuator B-Chem. 44, 434–440.

Zayats, M., Raitman, O.A., Chegel, V.I., Kharitonov, A.B. and Willner, I. (2002). Probing antigen-antibody binding processes by impedance measurements on ion-sensitive field-effect transistor devices and complementary surface plasmon resonance analyses: Development of cholera toxin sensors. Anal. Chem. 74, 4763–4773.

Part B

Protein-Based Nanobioelectronics

The term protein comes from the Greek word *protas*, meaning *of primary importance*. A protein is a complex organic compounds, typically of high molecular mass (up to 3000000 Da). The proteins can be assimilated to polymers built from 20 different L-α-amino acids, joined by peptide bonds. The details of their sequence are stored in the code of a gene: Genes are transcribed into RNA, which is then subject to posttranscriptional modification and control, resulting in mature mRNA. The mRNA is translated by ribosomes, originating like proteins. The amino-acid chain constituting a protein has two terminus, the carboxy, or C-terminus, and the amino, or N-terminus.

The three-dimensional structure is a peculiar property of each protein. The shape into which a protein naturally folds is known as its native state, which is determined by its sequence of amino acids. In biochemistry there exists a peculiar terminology related to four distinct aspects of the protein's structure: the *primary structure*, which is exactly the amino acid sequence; the *secondary structure*, related to the segments of the chain that assumes particular shapes such as alpha helix and beta sheet, formed by hydrogen bonding; the *tertiary structure*, the overall shape of each protein molecule and is formed mainly by hydrophobic interactions, with also a contribution of hydrogen bonds, ionic interactions and disulfide bonds; and finally, the *quaternary structure*, the shape resulting from the fusion of more than one protein molecule, called a protein complex. It is worth noting that proteins can deviate from those levels of structure in performing their biological functions. The functional rearrangements are usually referred as conformations, and transitions between them are called conformational changes. The natural folding process of a protein occurs in less than a minute after they are formed: This self-assembling process brings the protein from primary to higher structures. As a general rule, there exist more than one stable folded conformation corresponding to a precise biological activity and only one is considered to be the active one.

X-ray diffraction is used to gain precise information on the protein structure and folding. To this purpose proteins have to be turned into crystals, which is not a straightforward process. Because the crystal is made up of regular repeats of the protein, it scatters the radiation into thousands of rays. The direction of these rays tells us about the way in which the protein is packed inside the crystal. The brightness tells us about the structure of each protein molecule. Today, due to advances in synchrotron X-ray sources and detectors, protein crystal structures can be calculated in just hours with a resolution of few angstroms.

Molecules and ions able to chemically bind to a specific site onto a protein are called ligands, and the sites are called binding sites. The strength of ligand–protein binding characterizes the chemical specificity of the binding and is called affinity.

Proteins are involved in every function performed by a cell; therefore, by controlling the protein's activity it is possible to regulate these functions. The regulation can go through so-called allosteric modulation, involving binding of ligands, or covalent modulation, involving the protein's modification.

As mentioned, the activity of a protein is directly related to its structure. Small amino acid chains, lacking tertiary structure, also occur in nature: They are referred as polypeptides; for example, the hormones.

Proteins are generally classified as soluble, filamentous, or membrane-associated. Enzymes are soluble proteins that catalyze biochemical reactions. Antibodies also belong to this class. Membrane-associated proteins are usually in charge to do mass transport across the cell membrane, but they do not modify the chemical structure of the shifted mass or substrate. Another kind of membrane proteins are the receptors, which change their shape upon interaction with specific ligands. Filamentous proteins constitute the cytoskeleton of cells and are the basic components of several animal tissues, such as tubulin, actin, collagen, and keratin, which are building blocks of skin, hair, and cartilage. Another special class of proteins consists of motor proteins, such as myosin, kinesin, and dynein. These proteins are nanometer-sized molecular motors and are able to generate physical force that can move organelles, cells, and entire muscles. It is worth noting that there is considerable interest in the nanotechnological application of such proteins bound to a surface for the building of bio-inspired nanomachines.

Proteins are involved in every function performed by a cell, including regulation of important cellular functions such as signal transduction and metabolism. It is also possible to group the proteins into functional classes such as:

1. Enzymes, which catalyze metabolism reactions
2. Structural proteins, such as tubulin, or collagen
3. Regulatory proteins, such as transcription factors or cyclins, which regulate the cell cycle
4. Signaling molecules and their receptors, such as hormones and their receptors
5. Defensive proteins, such as antibodies and toxins

Proteins are sensitive to their environment. They result to be active in their native state, over a small pH range, and in a solution with small quantities of electrolytes. A protein that is not in its native state is said to be denatured or unfolded, because they have no well-defined secondary structure.

The sequence and hence the structure and other properties of a protein can be modified by methods belonging to genetic engineering, whereas the term protein engineering refers to a rather new research branch aimed at the design and creation of proteins with entirely new properties or functions.

As clearly comes out from this brief description of protein characteristics, the self-assembly process resulting in the folding of a protein is very attractive for the potential uses in nanotechnology, indeed learning about the mechanisms at the base of protein folding would allow the design and building of human-made machines on the nanoscale.

For their peculiar properties and functionalities proteins are considered with great interest in bioelectronics. Moreover their sizes and the potentiality offered by protein engineering made them suitable candidates for the development of protein-based nano-bioelectronics. This is actually the subject of the second part of the book. This is in turn divided in two sections. The first is focused on the exploitation of specific proteins for the implementation of nanoelectronic devices and nanoparticle arrays. The second section is instead related to the recent development in the field of biosensing and proteomics due to the progress in the today electronics and the related technological methods.

"Nanoelectronics Devices Based on Proteins" is the title of the first chapter of the first section. Here the authors go in the fascinating world of these new hybrid devices, starting with a short discussion of transport mechanisms in proteins, and introducing all the strategies and methods that have been developed to probe and interconnect proteins. They overview all of the SPM-based techniques to probe and measure a single protein, and present a recent program in the field of advanced and alternative nanolithographies to fabricate the nanoscale electrodes to be interfaced to proteins. Then they present material concerning the measurements performed by several groups on prototype protein nanodevices. One of the main concerns about the reliability of these hybrid devices is their lifetime and reproducibility, along with the major objection on the "bio" side concerning the retention of protein functionality (actually related to folding properties, as elucidated) in environments different from the living organism where they usually live, which is the object of the last part of this chapter.

The second chapter brings us into the fascinating world of two-dimensional bacterial surface layer proteins. These proteins are isolated from prokaryotic organisms and represent a valid alternative to patterning methods at the nanoscale due to their self-assembling properties. Starting from the description of S-layers and their surface lattice symmetry, the authors explain the potentiality of these biomaterials in the implementation of lithographic templates. These templates can be used as masks for further substrate processing or as linking sites for immobilization of different nanostructures; alternatively, they can act as molds for the wet chemical synthesis of nanoparticles. CdS, gold, and platinum nanoparticles have been produced in this way. Finally, the authors discuss the use of S-layers as scaffolds for preformed nanoparticle binding.

The first chapter of the second section brings our attention to the role of redox proteins and enzymes in nanoscale biolectronics. The biological action of these macromolecules, such as charge transfer and catalysis, is converted into electrical signal by means of electrochemistry methods. The progress of technology has made it possible to conduct these studies at the level of a single molecule, by using a combination of scanning tunneling microscopy and protein assembly on solid surfaces. This aspect is a common task in molecular electronics and represents the first step toward the implementation of any bio-inspired nanodevice. The authors review the theoretical concepts that underlie the development of these methods, i.e., electrochemical electron transfer and redox processes in electrochemical STM. Then experimental approaches and results are discussed. Particular attention is given to protein assembly, which is the key process for the successful and safe conduction of experiments. Real cases and significant results are presented: cytochrome c, azurin, nitrite reductase, and cytochrome c_4. For each case experimental results are exhaustively compared with theoretical predictions and simulations.

"Ion Channels in Tethered Bilayer Lipid Membranes on Au Electrodes" is the title of the second chapter of this section. Membrane proteins represent one of the most studied and interesting classes of proteins for the diverse biological functions that they perform. A tethered bilayer lipid membrane is a biomimetic support, which is as close as possible to the real cell membrane, for the incorporation and study of membrane proteins. Most of the recent developments of biosensing devices at the single molecule scale are based on this architecture. Anchoring the biomimetic membrane to a solid support offers the possibility of integrating these systems into microelectronic readout systems. As the authors point out, the development of anchor lipids for grafting onto the surface and the implementation of a uniform lipid bilayer are the first fundamental steps of these studies. Electrochemical impedance spectroscopy and surface plasmon resonance spectroscopy are the techniques routinely used to detect the surface uniformity of the layers. Proteins can then be fused into the layers. The authors present experimental results on three membrane proteins—valinomycin, gramicidin, and M2δ. These examples drive the interest of the reader toward the implementation of devices for drug screening or detection of toxins, based on such engineered biological systems, which if down-scaled, should allow the measurement of single channel fluctuations of embedded membrane proteins.

Finally, the last chapter reviews the properties of hybrid diads composed by colloidal nanocrystals and proteins, recently realized by different groups. Colloidal nanocrystals present intrinsic properties, like very wide spectral range of emission tunability, due to their size, and long-lasting efficient fluorescence, that make them suitable as versatile bioprobes for advanced molecular and cellular imaging techniques. The authors, after the presentation of these unique characteristics, review the synthetic approaches developed for the production of colloidal nanocrystals of different sizes and materials. One of the major critical issues in working with nanocrystals is the necessity to functionalize their surface to make them water soluble and link them other biomolecules, such as proteins, DNA, and viruses. The binding of few or a single protein or enzyme to a nanocrystal can be detected by using fluorescence resonance energy transfer (FRET) and bioluminescence resonance energy transfer (BRET) techniques. An exhaustive description of all the implementation and studies of these hybrid nanosystems, along with their potential use in the field of nanobioelectronics and biosensing, conclude and complete this part of the book.

Protein-Based Nanoelectronics

5

Nanoelectronic Devices Based on Proteins

Giuseppe Maruccio and Alessandro Bramanti

1 PROTEINS IN NANOELECTRONICS

Acquiring the capability of handling and assembling structures at the molecular scale is the challenging target of present research in nanotechnology and the key to the forthcoming breakthrough in the field. Yet, this formidable task is accomplished in nature with extraordinary efficiency. Remarkably complex biological molecules such as DNA and proteins are built with atomic precision and, even more interesting to nanotechnology, possess self-assembly properties. The latter can be defined as "the self-organization of one or more entities without any external source of information about the structure to be formed as the total energy of the system is minimized to result in a stabler state". As a consequence, biomolecules are interesting materials for a large number of applications. Behind self-organization there lie: (a) driving mechanisms (electric field, diffusion); and (b) recognition capabilities. The former bring reactants into contacts; the latter provide that bonds form selectively, depending on the matching of compatible bio-linkers (Bashir 2001). The inherent potential for the fabrication of new functional nanodevices is evident (Keren, Krueger, et al. 2002).

A. Offenhäusser and R. Rinaldi (eds.), *Nanobioelectronics - for Electronics, Biology, and Medicine*,
DOI: 10.1007/978-0-387-09459-5_6, © Springer Science+Business Media, LLC 2009

2 OVERVIEW AND THEORY OF CHARGE TRANSPORT MECHANISMS IN PROTEINS

The rationale behind considering proteins as materials for bioelectronics relies in their recognized electron transfer (ET) capabilities, which are pivotal to the vital processes of both animals and plants. In fact, neither aerobic respiration in microbes and animals, nor photosynthesis—two basic processes sustaining life—could even subsist without ET reactions. Among the most studied proteins in the literature as to ET, let us mention: metalloprotein azurin—one of whose commonest variants bears a copper ion in the active site and is involved in the respiratory cycle of bacterium *Pseudomonas aeruginosa*—whose popularity results from its good ET performance and high stability; myoglobin, involved in the storage and transportation of oxygen in muscles; ferritin, a ion-storage protein; and cytochromes, including iron atoms in the redox-active sites.

Research activity spanning more than half a century has shown that protein charge transfer takes place between redox cofactors (a donor and an acceptor) separated by distances in the order of several tens of Å (whereby long-range ET). In particular, ET was in past usually studied by spectroscopic and electrochemical methods with ET rate often dependent from the employed technique (e.g., in the case of azurin, $300\,s^{-1}$ and $150–200\,s^{-1}$ from AC impedance and ER techniques against $4–12\,s^{-1}$ from cyclic voltammetry studies). Then, more recently, STM and CP-AFM (see Section 3) made ET probing possible at the single-molecule level, providing information previously buried in the statistical averaging of conventional techniques (Tao 1996).

To gain some theoretical insight, consider a generic (intramolecular or intermolecular) ET process between a donor (D) and an acceptor (A), at fixed distance. In some cases this reaction may take place by a superexchange mechanism (consisting of direct quantum tunneling between the donor and acceptor) or subsequent tunneling steps between intermediate ions and/or molecular sites; in other cases a sequential (incoherent) hopping between adjacent sites must be invoked to explain the long distance or, equivalently, the short time (Kuhn, Rupasov et al. 1996).

The major factors influencing ET (Farver and Pecht 1992) are: (1) the driving force; (2) the distance between the two redox centers (on which ET decreases exponentially for tunneling and is inversely proportional for hopping, respectively); (3) the nature of the microenvironment (mediating the virtual state or providing an intermediate state); and (4) the reorganization energy λ required for the structural adjustments in the reactants and surrounding molecules, which are necessary to make electron transfer possible.

In a general semiclassical picture, the ET rate (Gray and Winkler 2005) (charge transferred per unit time) for vibration-induced electron tunneling (Broo and Larsson 1991) is given by:

$$k = \frac{(2\pi)^2}{h} H_{da}^2 \, FC \tag{1}$$

where FC is the nuclear (Franck–Condon) factor and $H_{da}{}^2$ is the electronic factor.

Observe first that the hypothesis underlying (Eq 1) is the Franck–Condon principle, i.e., the assumption that the electron's velocity is high enough to allow considering nuclei as fixed during the ET process. However, the condition for ET to occur is that fluctuations bring the involved molecules into a transition state configuration, in which the reactant and product states are degenerate.

It should be pointed out that the process is electronically nonadiabatic: Because of the weak coupling between sites, the transition state will be reached several times before the electron is transferred. Also, it is worth mentioning that non-Franck–Condon effects have been reported, depending on the D-A distance, tunneling energy and pathway structure, temperature, and relaxation of the solvent molecules (Gray and Winkler 2005; Skourtis, Balabin et al. 2005). Details are well beyond our scope, but analytical correction factors have been obtained for the effect of solvent (Gray and Winkler 2005 and refs. therein).

Analyzing (Eq 1) more deeply, the FC factor involves a balance between the driving forces and the reorganization energy in order to reach the transition state and in the Marcus model is approximately proportional to $\exp(-\Delta G^*/k_B T)$, where the activation energy:

$$\Delta G^* = \frac{\lambda}{4}\left(1 - \frac{\Delta G^0}{\lambda}\right)^2$$

is determined by the reorganization energy λ and the free energy of the reaction, $-\Delta G^0$ (Larsson, Broo et al. 1995). According to this expression, the ET rate is maximum when the nuclear factor is optimized, i.e., if $\Delta G° = \lambda$. The reorganization energy λ, in turn, is strongly influenced by the protein folding characteristics. Qualitatively, in the simple case of equal D and A (self-exchange), the more similar are the structures of the reduced (D) and oxidized (A) protein, the lower is λ expectedly. In particular, in the case of Azurin, λ is minimized and the electron transfer favored since the copper site geometry is essentially unchanged in the Cu(II) and Cu(I) state.

Concerning the electronic term of (EQ 1), instead it is the tunneling matrix element from donor to acceptor, depending on the interaction between reactants and products at the nuclear configuration of the transition state and primarily on the distance R between D and A ($H_{da}^2 = C\exp(-\beta R)$), for example, $\beta \cong 1.1$ Å from Azurin protein studies (Ramamurthy and Schanze 1998). The pre-exponential term C depends very much on the character of the coupling between D and A, and the intervening microenvironment separating D and A, which mediates the virtual state or provides intermediate states. (If D and A are in contact, H_{da} is roughly proportional to the overlap between their orbitals.)

More generally, the bridge between D and A may be a sequence of heterogeneous links such as covalent bonds, hydrogen bonds, and through-space jumps, characterized by different individual couplings; then the total coupling is the product of the different couplings for the individual links. Now the nature of an ET pathway is clearly determined by the structural details of the proteins involved and possibly of the surrounding medium and it is of crucial importance to identify the electron tunneling pathway. In this respect, it is commonly accepted that chemical bonds are predominantly used, hydrogen bonds are used subsidiarily, and through-space routes are used rarely (Beratan 1987). In protein, specifically, the influence of α-helices and β-sheets in ET is of particular importance, since these are two of the most common secondary structure motifs. The former are tightly packed, 0.54-nm–wide helical arrangement of amino acids, held together by periodic hydrogen bonds between N-H and C=O groups. The latter are parallel amino acid chains, parallel and aside to each other with an alternation in the sequence such that hydrogen bonds can form; although the overall shape is roughly flat (whereby the term *sheet*), when looked at more closely they reveal to be very loosely bound helices. There is experimental evidence that β-sheets are generally more ET efficient, although the difference to α-helices may become very little at small D-A distances. The larger ET rates can be attributed to the different strength of hydrogen bonds, proportional to the electron coupling efficiency of the latter (Gray and Winkler 1996). Understanding the structural issues that are most relevant to ET may help choose and, in the future, even engineer proteins for molecular electronics application.

Hopping has already been mentioned as another important actor in ET transport beside tunneling. Evidence for its role is provided, among others, by the observation that ET in some redox enzyme assemblies actually occurs over far longer distances than a reasonable space limit for tunneling (around 2 nm), and still on a proper time scale for the related biological functions (milliseconds to microseconds). Furthermore, a crossover from tunneling to hopping was reported by Bernd Giese and co-workers (Giese, Wessely et al. 1999; Giese, Amaudrut et al. 2001) who studied the charge transfer between G bases, separated by adenine-thymine $(A\bullet T)_n$ bridges of various lengths in double DNA strands. (The switch between these conduction mechanisms occurs because the tunneling rates decrease considerably as the distance increases.) Theoretical work (Kim, Park et al. 1996) has evidenced that hopping may be favored through sequences of electrophilic sites, such as dangling H bonds.

So far, our brief survey has considered protein-mediated ET in a somewhat fundamental perspective, focusing on basic phenomena and their dependence on structural characteristics. However, when it comes to employing proteins in molecular electronic devices, tone must consider the crucial issue of their organization in solid state.

3 PROBING AND INTERCONNECTING MOLECULES/ PROTEINS

The main difficulty to achieve the final goal of molecular-scale electronics (i.e., the use of single molecules/proteins as active elements in nanodevices) is related to the extremely small size of the molecular building blocks, which is (on the other hand) the main advantage in miniaturization. In particular, reliable methods to interconnect and probe transport in molecules/proteins at the nanoscale are needed.

In this respect, the first two-terminal conductance measurements at the molecular scale were carried out in molecular tunnel junctions with the invention of the scanning probe microscope (SPM). In fact, besides enabling molecular scale resolution imaging and manipulation, scanning tunneling microscopy (STM) and scanning tunneling spectroscopy (STS) allow measuring the tunneling current across molecules on conductive substrates (Datta, Tian et al. 1997) and mapping the electronic density of states (and wave functions) in nanostructures (Lu, Grobis et al. 2003; Maltezopoulos, Bolz et al. 2003). The typical experiment performs SPM measurements on proteins immobilized in a monolayer. However, it is worth noting that, in the STM, the position of the tip and conductivity of the sample are coupled and it is thus not easy to extrapolate the conductivity of molecules from STM data. Conductive probe atomic force microscopy (CP-AFM) allows circumventing this difficulty by exploiting the force feedback to control the position of tip with respect to the substrate. (CP-AFM also enables transport studies on highly resistive samples.) In a recent work, conductive AFM measurements on ferritin (Xu, Watt et al. 2005) have shown currents as high as some pA across a single holoferritin molecule (ferritin filled with ferrihydrite). Recently, a new method to measure the resistance of single molecules by the repeated formation of many molecular break-junctions was reported by Xu and Tao (2003), who collected current vs. distance curves by driving an STM tip (held at a fixed bias) into and out the sample and observed well-defined peaks in conductance histograms at integer multiples of fundamental conductance values, identified with the conductance of the single molecule (Fig. 5.1).

Concerning redox proteins (and redox molecules in general), it is worth mentioning also the electrochemical scanning tunneling microscope (EC-STM) (Fig. 5.2) (Liu, Fan et al. 1986; Sonnenfeld and Hansma 1986), which permits STM experiments in an electrochemical cell, independently controlling the electrochemical potentials of the substrate and tip toward a common reference electrode. The probed sample is the working electrode and a bias potential is applied between the tip and the sample for tunneling. The tunneling current is determined by both the bias potential (at a fixed electrochemical potential of one of the electrodes relative to the reference electrode) and the overpotential of both electrodes

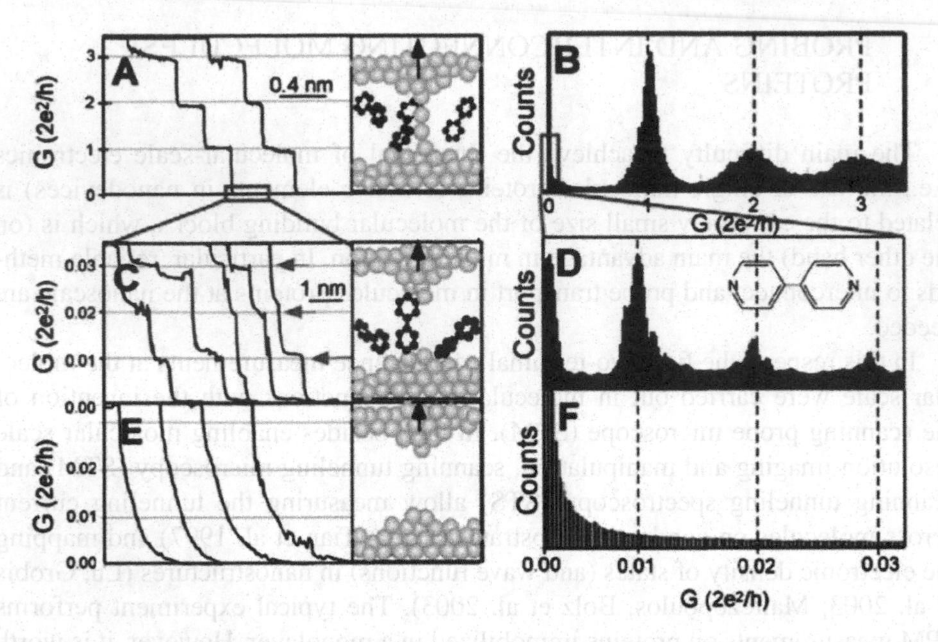

FIG. 5.1. **A.** A contact between a gold STM tip and a gold substrate shows quantized variation of conductivity, as the tip is raised from the substrate. **B.** The corresponding histogram from 1000 conductance curves evidences peaks near integer multiples of $G_0=2e^2/h$. **C.** Once the contact is broken conductivity begins to vary in $0.01G_0$ steps, as from the histogram in *(D)*, related to the presence of molecules such as 4,4′ bypiridine in the solution and how many are trapped within the contact. As a counterproof, no conductance steps *(E)* or peaks *(F)* are observed in the absence of molecules (Reproduced with permission from Xu B. and Tao N.J. (2003). Science 301, 1221–1223. Copyright 2003 American Association for the Advancement of Science.)

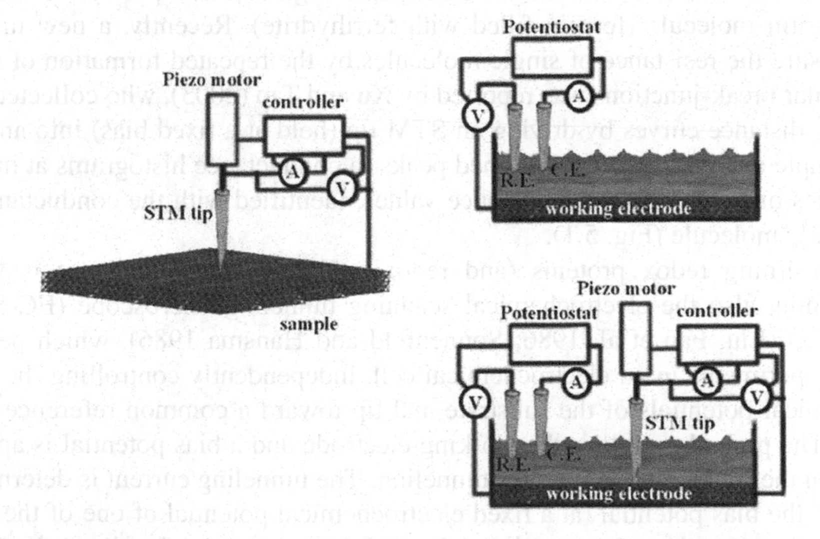

FIG. 5.2. Experimental setup of: **A.** an STM. **B.** An electrochemical cell. **C.** An EC-STM.

(for a given bias voltage). Although the experimental apparatus introduces considerable variation with respect to the electrochemical process at a protein–metal interface, much insight is provided on the fine tunneling mechanisms involved as well as the effect of metal contacts, a crucial issue for biomolecular electronic devices. As demonstrated by the Ulstrup group, EC-STM allows gaining considerable informations about electron transfer (ET) in biological macromolecules at the nanoscale and probing whether proteins immobilized in SAMs retain their functional properties (Zhang, Grubb et al. 2003). Analogous studies on azurin in air and liquid (Alessandrini, Gerunda et al. 2003) have also shown the expected resonance in conductivity, proving that ET is mediated by metal in the active site. Furthermore, all of these studies confirm the capability of immobilized proteins of retaining their fundamental ET properties, which is encouraging as to their biomolecular electronic applications; further confirms have come from reliability measurements on device-like configurations, described in the following.

Despite the remarkable possibility to study ET by scanning probe methods, however, a major goal is to fabricate nanometer spaced electrodes bridged by single molecules/proteins in order to demonstrate electronic devices working at the nanoscale. For this purpose, due to the intrinsic limitations of optical lithography, research is focused worldwide on the development of postoptical lithographies, with major efforts in engineering reliable interconnection schemes compatible with the new materials recently proposed.

The main drawback of standard photolithography lies in its resolution limit, ultimately determined by the radiation wavelength. The Rayleigh resolution criterion defines the critical dimension (the smallest half-pitch) of a feature on an integrated circuit (IC): $\approx \kappa_1 \lambda / N_A$, where λ is the wavelength, N_A is the numerical aperture of the imaging system and κ_1 is a value between 0.25 and 1 that depends on the configuration of the illumination system and the optical response of the photoresist.

Among the numerous, different technologies for patterning below the 100-nm scale, electron beam lithography (EBL) is of course the usually employed, mainly because of its versatility, which permits the fabrication of a variety of different nanodevices, combining EBL with other processes such as lift-off, etching, and electrodeposition/electroplating. With EBL, the resist is exposed by a thin electron beam, slowly moved and selectively turned on and off. The EBL resolution is limited by large-angle electron scattering events leading to backscattering and additional exposure in the resist (the proximity effect). As a consequence, in order to fabricate pairs of nanoelectrodes with separation in the few-nanometer range, postprocessing techniques are typically used such as electrodeposition, electromigration, and mechanically controllable break junctions. In particular, nanogaps ranging from 20 to 3.5 nm were reproducibly fabricated by Kervennic and co-workers by monitoring the conductance between two free-standing EBL-fabricated electrodes during

Pt electrodeposition (from an aqueous solution of 0.1 mol of K_2PtCl_4 and 0.5 mol of H_2SO_4) stopping the electrodeposition process at predefined conductance values (Kervennic, Van der Zant et al. 2002).

Electromigration (usually at cryogenic temperature) (Park, Pasupathy et al. 2002) and mechanically controllable break junctions (realized by the gentle fracture of an electrode due to an induced mechanical deformation in the supporting substrate) (Reed, Zhou et al. 1997; Champagne, Pasupathy et al. 2005) (Fig. 5.3) have been also exploited to achieve nanometer-spaced electrodes. Since all these approaches do not completely ensure the reproducibility of the fabricated gap, the real separation has to be typically measured by SEM. Motivated by this, Zandbergen and co-workers (Zandbergen, van Duuren et al. 2005) developed an e-beam sculpting technique (Fig. 5.4) to fabricate nanojunctions in a TEM by reshaping a metallic bridge in a controllable manner until the creation of a nanogap. Finally, it is worth noting as the robustness of the electrodes and possible failure mechanisms during operation also have to be investigated (Maruccio, Visconti et al. 2003; Coura, Legoas et al. 2004).

Besides the scattering-related limitation in resolution, electron bream lithography suffers from high costs and comparatively long processing times. Among the other lithographic techniques, we should mention:

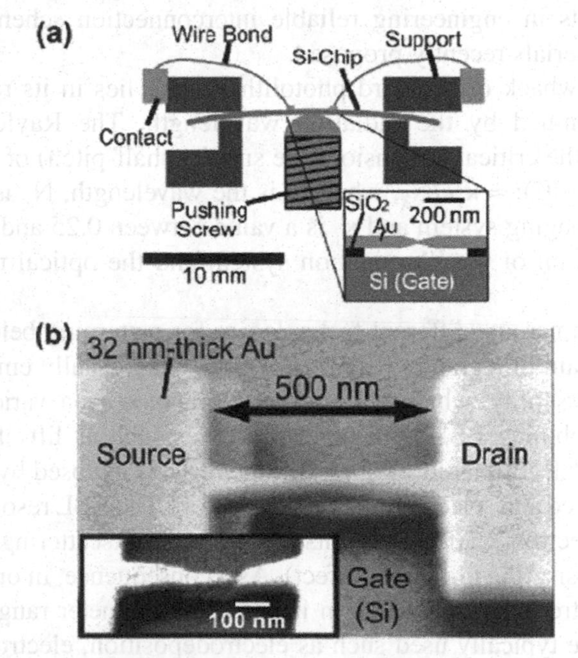

FIG. 5.3. **A.** Schematic of the mechanically adjustable and electrically gated nanojunction by Champagne and co-workers, SEM-imaged in *(B)* at a of 78° tilt angle. **B.** inset: an electromigration-broken nanojunction (Reproduced with permission from Champagne A.R. et al. (2005). Nano Letts. 5, 305–308. Copyright 2005 American Chemical Society.)

FIG. 5.4. Successive moments in the fabrication of a nanogap by the TEM sculpting technique (Reproduced with permission from Zandbergen H.W. et al. (2005). Nano Letts. 5, 549–553. Copyright 2005 American Chemical Society.)

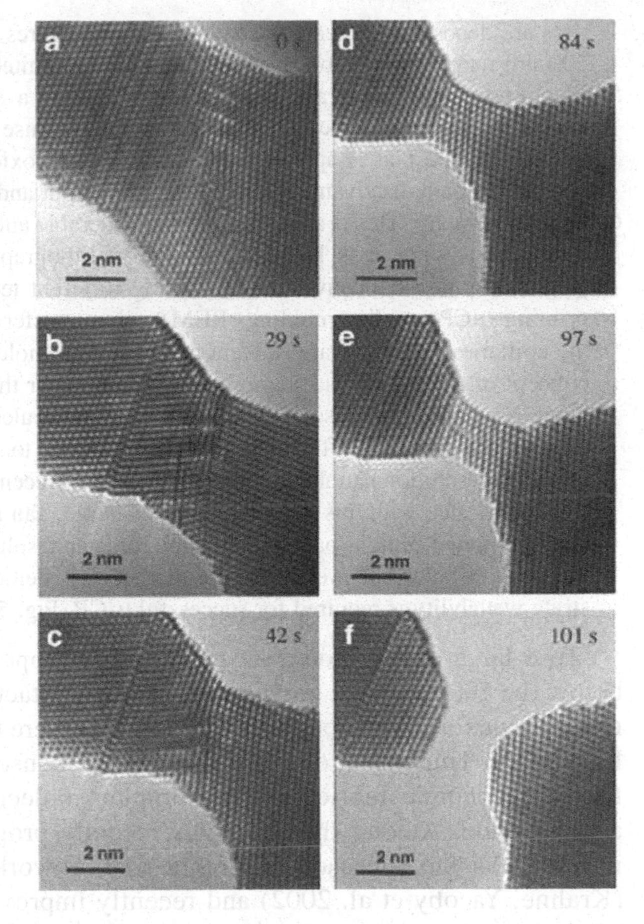

1. Ion beam lithography: A technique closely related to EBL and exploiting light high-energy ions, like protons, instead of electrons.
2. X-ray lithography: (using 0.1–10 nm radiation) (Silverman 1997), which uses collimated x-rays to expose a resist in a parallel replication process. The increased lateral resolution comes from both the extremely short wavelength and the high penetration ability. A mask is employed to define the pattern into a resist layer, such mask is usually the most crucial element. X-ray lithography is expensive to perform because of the need for a synchrotron radiation facility.
3. Interference lithography: Simple and maskless. Exposure is obtained from the light of two or more interfering coherent laser beams. The interference patterns define the final geometry, and this is the major drawback of this technique, since only highly symmetrical structures can be obtained. Conversely, a large area can be exposed simultaneously, which results in higher speed.
4. Shadow masks techniques: Exploit a very thin shadow-mask, whose nanometric holes play the role of the resist. The mask is placed very close to the substrate and then mate-

rials are deposited to create the desired nanostructures on the substrate. Features in the 400-nm range can be fabricated, with resolution limited by the mask thickness.

5. Scanning probe lithography (SPL): Exploits (a) a scanning probe for scratching, nanoindentation, and local heating, or (b) the intense electric field in the vicinity of the tip (when a bias is applied) for field enhanced oxidation and electron exposure of resists. Its main disadvantages are slow throughput and small patternable area.
6. Soft lithography: This is also attracting considerable attention, especially in bio-inspired applications since it is biocompatible. Soft lithography (Xia and Whitesides 1998) (Fig. 5.5A) is a collective term including different techniques, such as microcontact printing (μCP), replica molding (REM), microtransfer molding (μTM), micromolding in capillaries (MIMIC), and solvent-assisted micromolding (SAMIM) sharing the basic concept of an elastomeric stamp or mold to transfer the pattern to the substrate. Their name results from the use of flexible organic molecules and materials rather than rigid inorganic materials. Soft-lithography has potential to fulfill the demands of low-cost and high-resolution nanolithographic techniques. Recently, a number of protein patterns were fabricated with this technique. For example, Tan and co-workers (Tan, Tien et al. 2002) reported protein patterns with micrometer resolution by μCP and examined the influence of substrate wettability on the process, demonstrating that a minimum substrate wettability is required for successful μ-CP (Fig. 5.5B).

To a large extent, however, most of the proposed methods for patterning below the 100-nm scale are appropriate for contacting only single devices and not suitable for mass production because of severe technological and economic limitations. Thus, great efforts are currently focused on developing techniques for the economic fabrication of complex molecular electronics circuits on a large scale. Among the strategies recently proposed, we want to mention in particular the approach of Krahne and co-workers (Maruccio et al. 2007) (Krahne, Yacoby et al. 2002) and recently improved by G. Maruccio and co-workers to define networks of nanojunctions by only optical lithography and wet etching of an AlGaAs/GaAs quantum-well structure without the need of expensive e-beam systems. In this method, the thickness of the quantum well

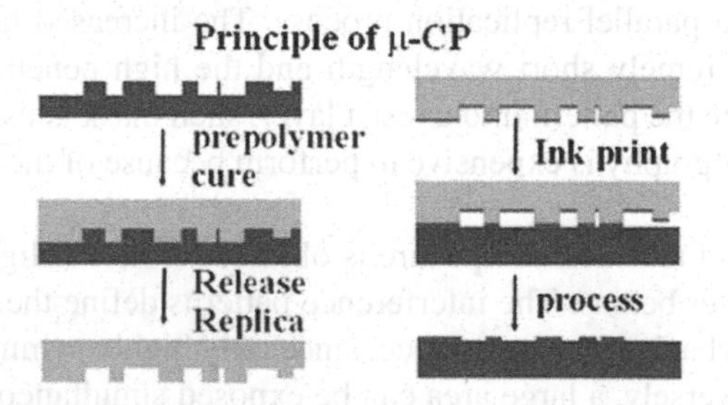

FIG. 5.5. **A.** μ-CP printing. **B.** Pattern of fluorescently labeled proteins by μ-CP (Reproduced with permission from Tan J.L. et al. (2002). Langmuir 18, 519–523. Copyright 2002 American Chemical Society.)

and the deposited metal layer control the gap size with subnanometer precision. The leakage current through the semiconductor layer can be trimmed down selectively oxidizing the two AlGaAs barriers, above and below the QW (G. Maruccio and co-workers, (Maruccio et al. 2007) (Fig. 5.6). The approach is economical and suitable for the simultaneous production of wafer-scale arrays of electrodes, mainly because the electrode pattern is defined only by photolithography; it thus represents an important step toward low-cost massive fabrication of nanodevices.

Before concluding this discussion on methods to study transport in single molecules/proteins, we have to point out its extreme sensitivity on the nature and quality of the contacts. This is a primary problem in molecular electronics. Ideally, the contacts should not interfere in the measurement of the molecular conductivity. Therefore, they should be ohmic and of low resistance, since any potential barrier at the interface deeply influences charge transfer and any contribution to the characteristic from the molecule–contact interface should be reduced to minimum. Xu and Tao also recently demonstrated how the conductance of single molecules is very sensitive to the microscopic details of the contacts, by reporting two different single-molecule conductance values corresponding to two possible molecular–electrode contact geometries (Xu and Tao 2003). It is thus of crucial importance to demonstrate methods to create ideal Ohmic and reproducible molecule–metal contacts. Motivated by this, T. Dadosh and co-workers (Dadosh et al. 2005) (Fig. 5.7) proposed to fabricate single-molecules devices by means of a dimer-based contacting scheme, which makes use of two colloidal gold particles bridged by a dithiolated molecule to control the molecule–electrode interface.

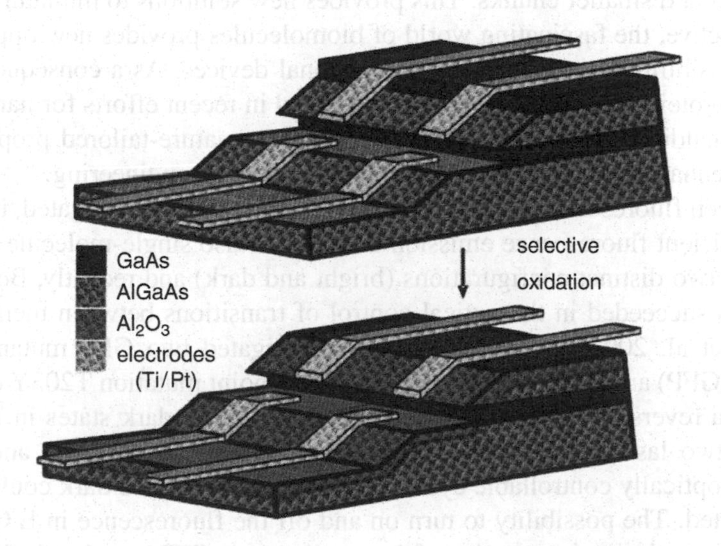

FIG. 5.6. Mesa nanojunctions without *(A)* and with *(B)* selective oxidation of the AlGaAs barriers.

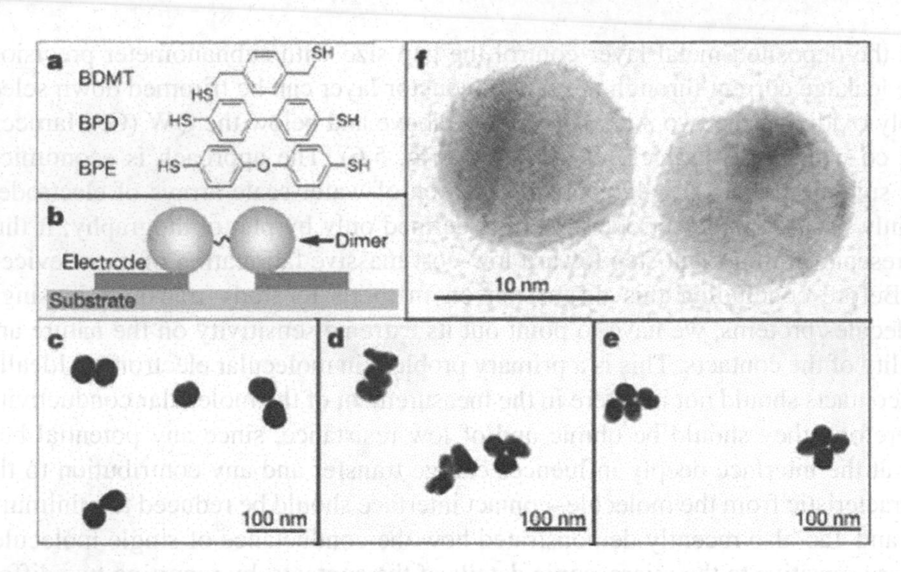

FIG. 5.7. **A.** Molecules investigated by Dadosh and Yacoby (2005). **B.** Contacted dimer. **C.–E.** Dimers, trimers and tetramers imaged by TEM. **F.** High-resolution SEM image of a dimer (Reprinted by permission from Dadosh et al. (2005). Nature 436, 677. Copyright 2005 Macmillan Publishers Ltd.)

4 EXPERIMENTAL RESULTS ON PROTEIN DEVICES

A bottom-up approach is very promising in order to build nanodevices from molecular building blocks instead of carving lithographically big pieces of matter into smaller and smaller chunks. This provides new solutions to miniaturization. In this perspective, the fascinating world of biomolecules provides new opportunities and the possibility to implement new functional devices. As a consequence, both DNA and proteins have been largely investigated in recent efforts for nanobioelectronics (Rinaldi and Maruccio 2006). Their unique nature-tailored properties can further be enhanced for specific applications by genetic engineering.

The green fluorescent protein (GFP) is one of the most investigated, because of its very efficient fluorescence emission that allows also single-molecule detection. It exists in two distinct configurations (bright and dark) and recently, Beltram and co-workers succeeded in the optical control of transitions between them (Cinelli, Pellegrini et al. 2001). Specifically, they investigated two GFP mutants: EGFP (enhanced GFP) and E^2GFP (obtained by a single point mutation T203Y of EGFP), achieving a reversible transition between the bright and dark states in E^2GFP by means of two laser beams with different wavelengths $\lambda = 476\,\mathrm{nm}$ and 350 nm. Unlimited optically controllable cycles between the bright and dark configurations were reported. The possibility to turn on and off the fluorescence in E^2GFP could open the way to the implementation of dense volumetric GFP-based optical memories exploiting photoconversion from the dark to bright state by irradiation at 350 nm (WRITE step), fluorescence emission following weak excitation at 476 nm (READ step), and photobleaching (switching off of the emission, as ERASE step).

Another interesting protein is Bacteriorhodopsin (BR), a transmembrane protein found in the cellular membrane of *Halobacterium salinarium*, which functions as a light-driven proton pump and thus is interesting as photonic material. Motivated by its peculiar properties, its use as the active component in memory devices was investigated. (For a review on Bacteriorhodopsin-based applications see Birge, Gillespie et al. 1999.) In particular, two different applications were proposed: (1) holographic associative memories based on a Fourier transform optical loop and exploiting the real-time holographic properties of a BR films; and (2) branched-photocycle three-dimensional optical memories, in which parallel write, read, and erase processes are performed exploiting a sequential multiphoton process and an unusual branching reaction that creates a long-lived photoproduct. The field of proposed applications for BR is very broad and ranges from electronics to optoelectronics and computing (e.g., random access thin-film memories, neural-type logic gates, photon counters and photovoltaic converters, artificial retinas, picosecond photodetectors, multilevel logic gates optical computing, and different kinds of memories.) (See Birge, Gillespie et al. 1999 for further details.) Biomolecular photodiodes based on different proteins have been recently reported by various groups worldwide (Choi, Nam et al. 2001; Manoj and Narayan 2003; Choi and Fujihira 2004; Frolov, Rosenwaks, Carmeli, and Carmeli, 2005). A typical approach makes use of protein monolayers/heterolayers such as those consisting of cytochrome c and GFP used as electron acceptors and sensitizers by Choi and Fujihira, who turned on and off photocurrent using intermittent illumination at 488 nm by an Ar ion laser. Exploiting a similar approach and a bicomponent SAM made of two helical peptides with different (selectively activatable) chromophores and opposite directions of dipole moments (when immobilized on gold), the photocurrent direction in a molecular photodiode was switched from anodic to cathodic by Yasutomi and co-workers (Yasutomi, Morita et al. 2004), changing the wavelength of irradiating light from 351 nm to 459 nm (Fig. 5.8). In fact, since at a certain applied-potential range, the direction of the dipole moment determines the photocurrent flow, this last one can be switched by alternating photoirradiation.

Still concerning the exploitation of proteins in optoelectronic devices, it is worth mentioning the work of Das and co-workers (Das, Kiley et al. 2004), who investigated the integration of photosynthetic protein complexes (both reaction centers [RCs] from the purple bacterium *Rhodobacter sphaeroides* and Photosystem I (PSI) isolated from spinach chloroplasts) in photodetectors and photovoltaic cells. They found the peptides to enhance the average open circuit voltage by a factor of 2 to 3 and, in a conservative estimate, provided an internal quantum efficiency of 12% for the hybrid devices under short-circuit conditions (Fig. 5.9).

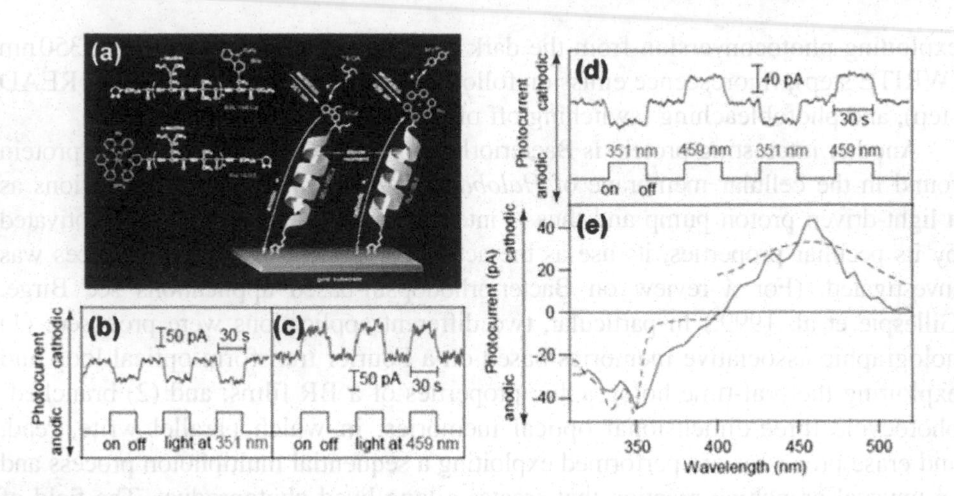

FIG. 5.8. **A.** Molecular structures of the hexadecapeptides SSL16ECz and Rul16SS. When co-assembled in highly ordered bicomponent SAMs (linked via Au-S bridge on a gold substrate), their oppositely directed dipole moments determine the switching of photocurrent from anodic to cathodic upon selective photoexcitation of the sensitizer (an ECz or Ru group, respectively). **B,C.** Periodic photocurrent generation by the SSL16ECz (Rul16SS) SAM, irradiated at 351 nm (459 nm) in a 50-mM TEOA (MV^{2+}) aqueous solution. **D.** Photocurrent switching vs. time under alternating photoirradiation at 351 and 459 nm. **E.** The action spectrum *(purple solid line)*, in the anodic and cathodic photocurrent regions, respectively, agrees with the absorption spectra of SSL16ECz (blue dashed line) and Rul16SS *(red dashed line)* in ethanol, respectively (Reprinted with permission from Yasutomi S. et al. (2004). Science 304, 1944–1947. Copyright 2004 American Association for the Advancement of Science.)

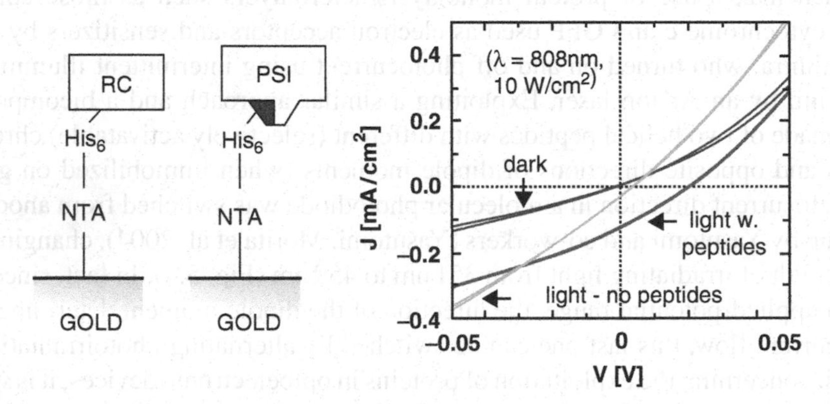

FIG. 5.9. **A,B.** Oriented immobilization of photosynthetic protein complexes (RC and PSI) on gold surfaces functionalized with DTSSP and (then) Ni^{2+}-NTA. **C.** Stabilization of RC complexes with A6K/V6D peptides improves the internal quantum efficiency of the devices to 12% under short circuit conditions (Reprinted with permission from Das et al. (2004), Nano Letts. 4, 1079–1083. Copyright 2004 American Chemical Society) See original paper for further details.

Finally, exploiting the redox properties and the functional groups of the blue copper protein Azurin, a protein-based field effect transistor was recently demonstrated by Maruccio and co-workers (Maruccio, Biasco et al. 2005). The β-barrel structure of Azurin contains two potential redox centers: a blue copper site coordinated directly to amino acid residues, and a disulfide bridge (Cys3-Cys26) at the opposite end of the molecule, at a distance of ≈2.6 nm from the copper site (Farver, Lu et al. 1999). The disulfide bridge can be employed to covalently immobilize the protein in monolayers. On the other hand, the Azurin ET capability depends on the equilibrium between the two possible oxidation states of the copper site -Cu(I) and Cu(II)- by means of the reversible redox reaction, which converts continuously the Cu(II) copper oxidized state into the Cu(I) reduced state and vice versa. In their prototype structure based on a planar nanojunction fabricated by EBL and a silver gate electrode, a protein monolayer was obtained by means of a two-step procedure involving: (a) the self-assembly of 3-mercaptopropyltrimethoxysilane, and (b) the reaction of the free thiol groups of 3-MPTS with the (unique) surface disulfide bridge of Azurin, thus resulting in an oriented monolayer. The current-voltage characteristic (I_{ds}-V_{ds}) of the Pro-FET under forward drain-source bias (V_{ds}) exhibited a low-current plateau at low field and then rose up to hundreds of pA (Fig. 5.10), whereas a pronounced resonance centered at V_g=1.25 V was observed in the transfer characteristic. From an electronic viewpoint, their device switched from a n-MOS FET behavior before resonance to a p-MOS FET after resonance. Although among all the fabricated nanodevices, only a limited group exhibited a clear gate effect over a number of gate sweeps, whereas the others failed during the first few sweeps—the aging of nanodevices is a general issue of molecular electronics (Park, Pasupathy et al. 2002; Kagan, Afzali et al. 2003; Lee, Lientschnig et al. 2003; Maruccio, Visconti et al. 2003)—this result could pave the way to fabricate both p- and n-type devices on the same chip and thus exploit the advantages of complementary logic, such as the decrease of the logic gate occupation area, the reduction in the fabrication complexity and, notably, a decay of power consumption, because as opposed

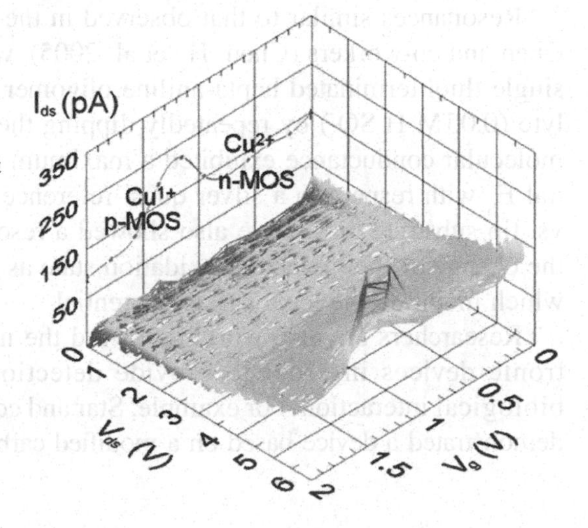

FIG. 5.10. Protein FET, drain-source current (I_{ds}) characteristic vs. the drain-source bias V_{ds} and gate bias V_g, in the dark, at room temperature. No significant leakage current was observed between the planar electrodes and the back gate. The pronounced resonance peak is centered at V_g = 1.25 V. In the projection, the same resonance is visible in a transfer characteristic at V_{ds} = 5.5 V (Reprinted with permission from Maruccio G. et al. (2005). Adv. Mater. 17, 816–822.) Copyright 2005 Wiley-VCH.

to unipolar inverters that consume power in the low state, CMOS devices consume power only when switching. The authors ascribed the unique transport mechanism of their biomolecular devices to hopping from one reduced (Cu(I)) protein to an adjacent oxidized (Cu(II)) protein, with the azurin redox state regulated by V_g: The higher is V_g, the greater is the fraction of reduced azurins. In this scenario, the ET is maximized when (at a particular value of V_g) the fraction of oxidized molecules equal that of reduced molecules. Beyond numerical simulations, the key role of the copper atom in electron transfer was also supported by the significantly lower currents and the absence of modulation in devices based on two azurin variants (in which the Cu atom is absent (Apo-form) or replaced by a Zn atom). Xu and co-workers (Xu, Watt et al. 2005) carried out a similar comparison on the ferritin protein by CP-AFM, again finding a lower current in the apo form.

Maruccio and co-workers also studied transport in molecular tunnel-junctions based on the blue-copper protein Azurin at the single-molecule level under ambient conditions by improving the mesa-gap technique for the interconnection of individual molecules and the fabrication of large-scale nanojunction-arrays (Maruccio et al. 2007). In such devices, they observed negative differential resistance and their results about conduction through the Azurin protein were consistent with transport via the electron pathway proposed by Farver and connecting the two different redox-active sites in Azurin (namely, the Cu(II) center and the disulfide bridge RSSR linking cysteines-3 and -26). Notably, such results are in good agreement (for both current values and peak positions) with the signatures of resonant tunneling (and NDR) recently reported in the current-voltage characteristics of Cu-Azurin metalloprotein CP–AFM junctions at low applied forces at Oxford University (Davis, Morgan et al. 2005). A systematic study of conduction in Azurin by conductive AFM and as a function of the compressional force employed to create the contact was also carried out by Davis and co-workers (Zhao, Davis et al. 2004) (Fig. 5.11).

Resonances similar to that observed in the azurin transistor were reported by Chen and co-workers (Chen, He et al. 2005), who measured the conductance of a single thiol-terminated hepta-aniline oligomer under potential control in electrolyte (0.05 M H_2SO_4) by repeatedly dipping their STM tip into the substrate. The molecular conductance exhibited a maximum as a function of the surface potential E_S with respect to a silver quasi-reference electrode (Fig. 5.12). The current vs. tip-substrate bias curve also showed a resonance that the authors ascribed to the change of the molecular oxidation state as a consequence of the applied bias, which modified the local surface potential.

Researchers are also working toward the integration of proteins in nanoelectronic devices in order to provide detection capabilities based on specific biological interaction. For example, Star and co-workers (Star, Gabriel et al. 2003) demonstrated a device based on a modified carbon nanotube FET (Fig. 5.13A) and

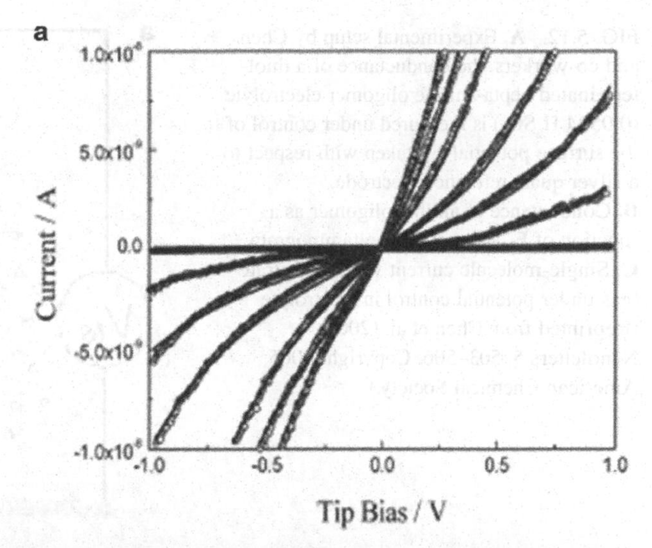

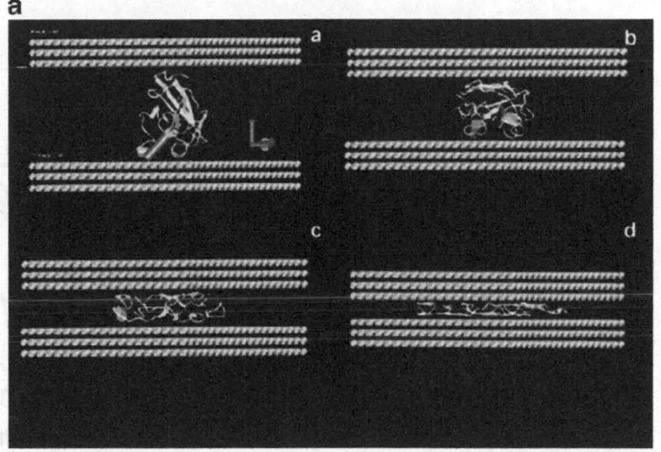

FIG. 5.11. **A.** Typical I–V curves for forces (experimental data, *open circle*) varying between 10.4 and 56.4 nN, interpolated by Simmons model *(solid line)*. **B.** Structural evolution of azurin on a surface of CH_4 molecules, at interfacial separation distances of *(A)* 4.0 nm, *(B)* 2.7 nm, *(C)* 1.7 nm, and *(D)* 1.0 nm (Reprinted with permission from Zhao J.W. et al. (2004), J. Amer. Chem. Soc. 126, 5601–5609. Copyright 2004 American Chemical Society).

able to detect protein binding (namely, biotin-streptavidin binding). Specifically, the nanotube was coated with a thin layer of polymer material, which was previously biotinylated and successively exposed to streptavidin. They observed a clear change in the device characteristic after exposure to streptavidin in response to biotin-streptavidin binding, as shown in Fig. 5.13. Control experiments supported the specificity of such biosensor. This approach could open the way to the fabrication of complex microarrays for the electronic detection of specific bio-interactions. This would be a considerable step forward in this field, where the readout schemes are usually optical (involving the use of fluorescent dyes). In fact, an electronic readout has the potential to increase significantly the number of different probe sites per unit area and thus the speed and efficiency of analysis.

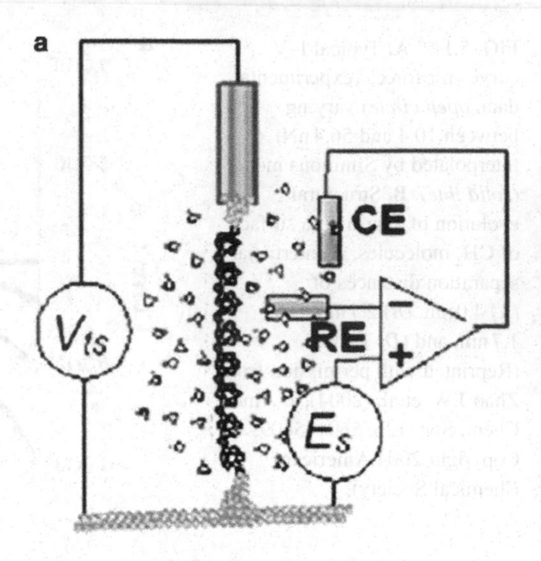

FIG. 5.12. **A.** Experimental setup by Chen and co-workers, the conductance of a thiol-terminated hepta-aniline oligomer electrolyte (0.05 M H_2SO_4) is measured under control of the surface potential E_S, taken with respect to a silver quasi-reference electrode.
B. Conductance of aniline oligomer as a function of E_S and a cyclic voltammogram.
C. Single-molecule current vs. tip-substrate bias under potential control in electrolyte (Reprinted from Chen et al. (2005). Nanoletters 5, 503–506. Copyright 2005 American Chemical Society.)

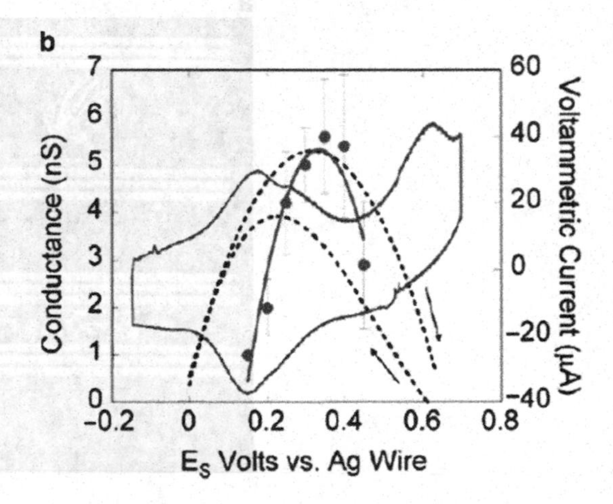

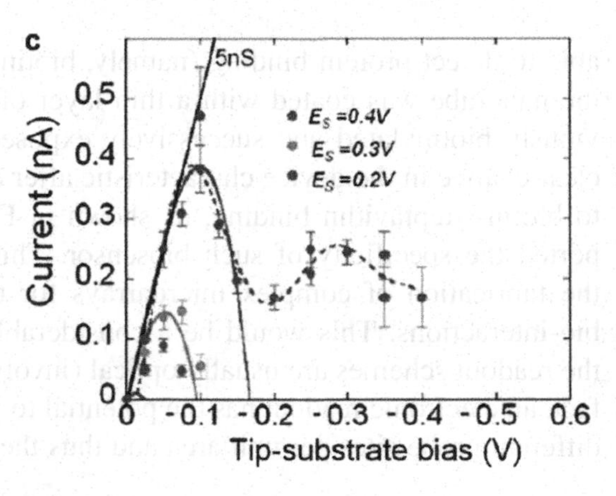

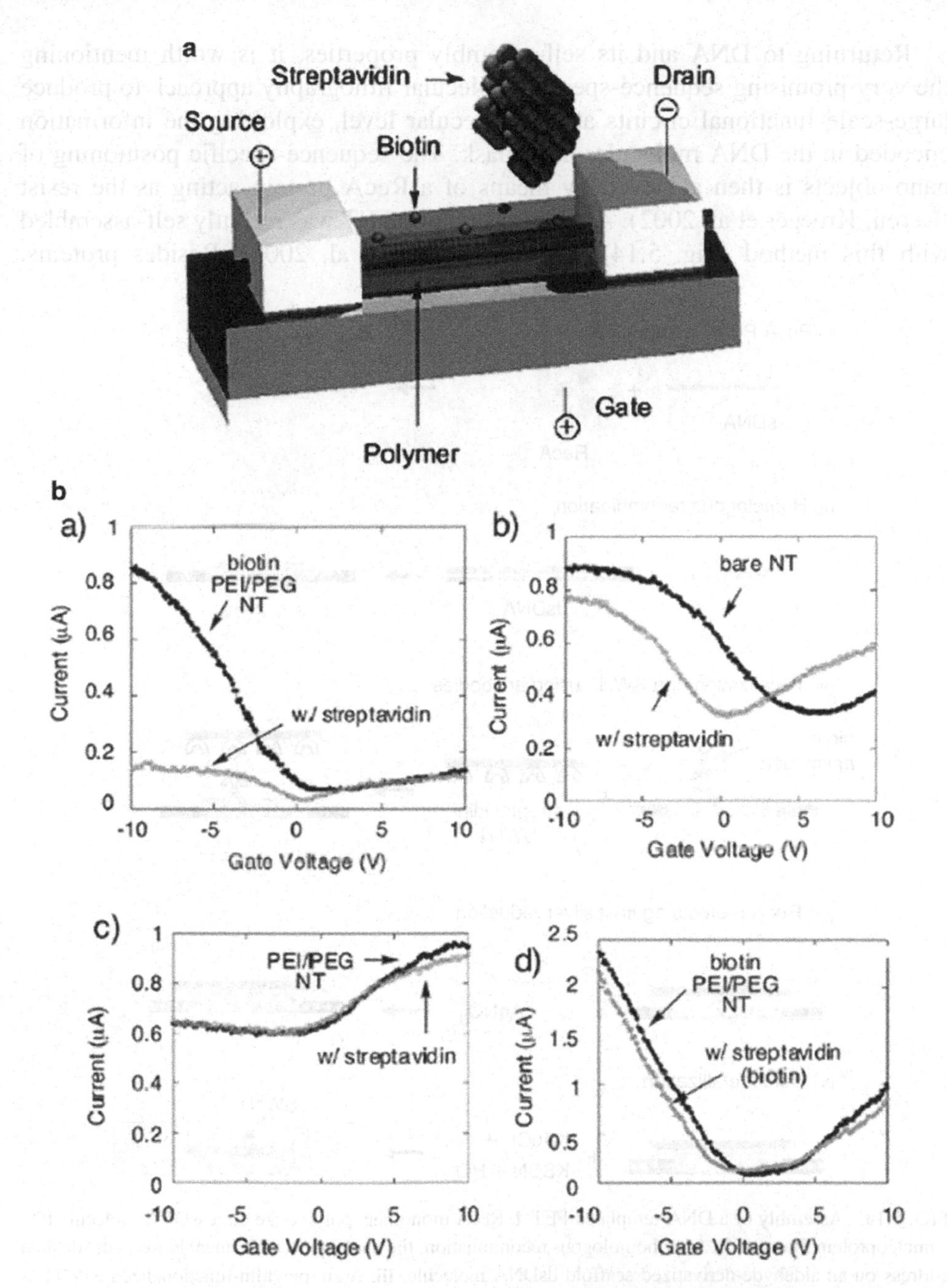

FIG. 5.13. **Top.** Schematic of the modified carbon nanotube (CNT) FET, functionalized with a molecular receptor (biotin) and successively exposed to streptavidin. **Bottom.** Change of the device characteristic due to biotin-streptavidin binding. **A.** Response—in the absence and presence of streptavidin—of a biotinylated, polymer-coated CNT-FET. **B.** A bare CNT-FET. **C.** A polymer-coated CNT-FET device. **D.** Response of a biotinylated, polymer-coated CNT-FET in the absence and presence of streptavidin previously incubated with biotin (Reprinted with permission from Star A. et al. (2003), Nano Letts. 3, 459–463. Copyright 2003 American Chemical Society).

Returning to DNA and its self-assembly properties, it is worth mentioning the very promising sequence-specific molecular lithography approach to produce large-scale functional circuits at the molecular level, exploiting the information encoded in the DNA molecules as a mask. The sequence-specific positioning of nano-objects is then achieved by means of a RecA protein acting as the resist (Keren, Krueger et al. 2002). A carbon nanotube FET was recently self-assembled with this method (Fig. 5.14) (Keren, Berman et al. 2003). Besides proteins,

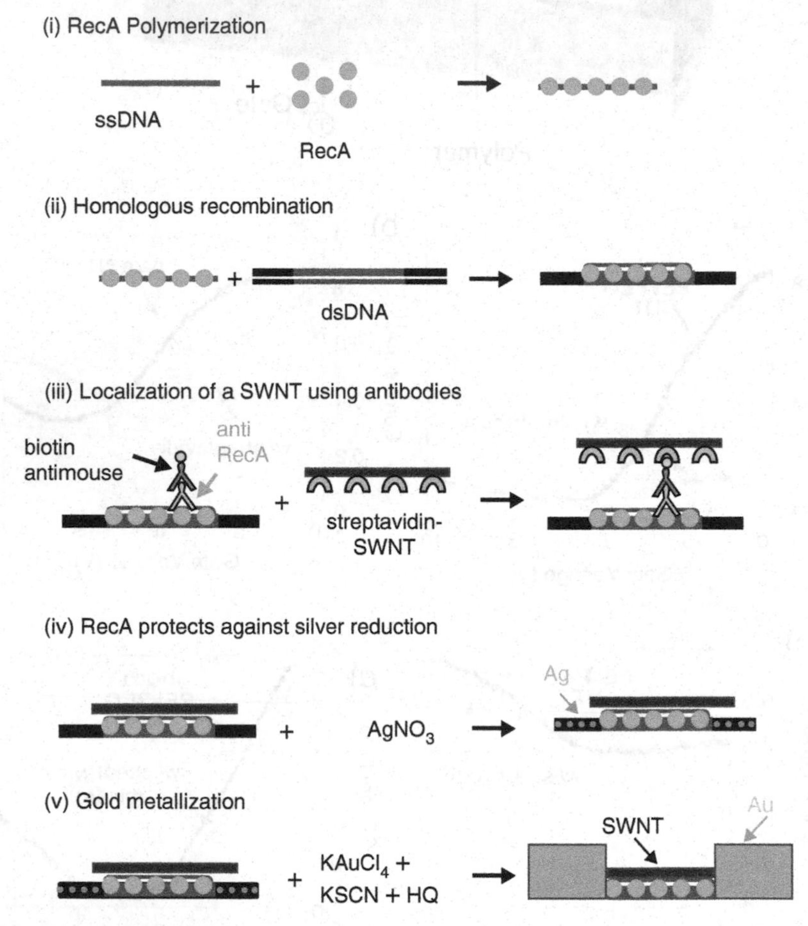

FIG. 5.14. Assembly of a DNA-templated FET. **i.** RecA monomers polymerize on a ssDNA molecule into a nucleoprotein filament. **ii.** After homologous recombination, the nucleoprotein filament binds at the desired address on an aldehyde-derivatized scaffold dsDNA molecule. **iii.** A streptavidin-functionalized SWNT is localized by the DNA-bound RecA, using a primary antibody to RecA and a biotin-conjugated secondary antibody. **iv.** After incubation in an $AgNO_3$ solution, silver clusters form on the segments unprotected by RecA. **v.** Electroless gold deposition, using the silver clusters as nucleation centers, leads to the construction of two DNA-templated gold wires, which in turn contact the SWNT bound at the gap (Reprinted with permission from Keren K. et al. (2003). Science 302, 1380–1382. Copyright 2003 AAAS.)

genetically engineered viruses have also been employed to order nanostructures (Lee, Mao et al. 2002).

5 RELIABILITY OF PROTEIN-BASED ELECTRONIC DEVICES (AGING OF PROTEINS IN AMBIENT CONDITION AND UNDER HIGH ELECTRIC FIELDS, ETC.)

One of the main concerns about the feasibility of novel devices is the resilience of the molecular materials to the usual working conditions of devices. An important issue is the surviving of proteins in solid state, at "reasonably practical" conditions of temperature, pressure, humidity etc. Finally, a third major concern in nanoscaled devices is the remarkable intensity of the electric fields, which could be expected to produce permanent structural alteration. In fact, to obtain appreciable current in conducting molecules, working voltages of several volts are usually necessary, resulting in electric fields as large as $10^7 \div 10^8$ V/m in the case of contacts spaced a few tens of nanometers. Although thermodynamical considerations on the expected stability of macromolecules, per se, apparently reinforce a doubtful attitude towards proteins as reliable, long-lasting materials, experimental evidence supports the feasibility of using protein in nanodevices.

In particular, the protein functionality in monolayers has been investigated by voltammetry, electrochemical impedance spectroscopy (EIS), electrochemical STM and X-ray photoelectron spectroscopy (XPS) (Chi, Zhang et al. 2000). In this respect, in their comprehensive investigations (Chi, Zhang et al. 2000), Ulstrup and co-workers demonstrated how suitable functional groups (e.g., the Azurin disulfide group [Cys3-Cys26]), can be exploited to anchor proteins on a substrate of interest (e.g., Au[111]) avoiding denaturation (i.e., loss of functionality). Specifically, in their work, the presence of a functional protein adlayer was first evidenced by their voltammetric results, displaying a clear azurin signal associated with oxidation and reduction of the copper center (and absent for the three reference molecules investigated, i.e., cysteine, cystine, and Zn-azurin). Adsorption via sulfur groups was supported by XPS data and they also investigated the interfacial electron transfer between the copper center and the electrode surface, measuring a rate constant of ca. 30 s^{-1} by means of electrochemical impedance spectroscopy. Finally, individual proteins in a dense monolayer were observed by EC-STM.

The same protein was also studied by Alessandrini and co-workers (2003), who investigated a mixture of Cu- and Zn-Azurin chemisorbed on Au(111) by EC-STM in buffer solution. A resonance (as a function of the substrate potential) was clearly observed for the tunneling in Cu-Azurins, demonstrating the preserved functionality of the immobilized proteins. Moreover, their approach also allowed them to demonstrate that the Cu ion in the active site mediates the electron tunneling through azurin, since no such resonance was observed in the case of Zn-Azurins (Fig. 5.15).

Investigation on the structural integrity of proteins is often performed also in terms of fluorescence spectroscopy, exploiting the emitting properties of natural chromoproteins or artificially dye-labeled proteins (Fidy, Balog et al. 1998; Kohler, Friedrich et al. 1998; Park, Andrews et al. 1999; Park, Thomas et al. 2000; Cohen, McAnaney et al. 2002). The fluorescence spectrum depends on the energy

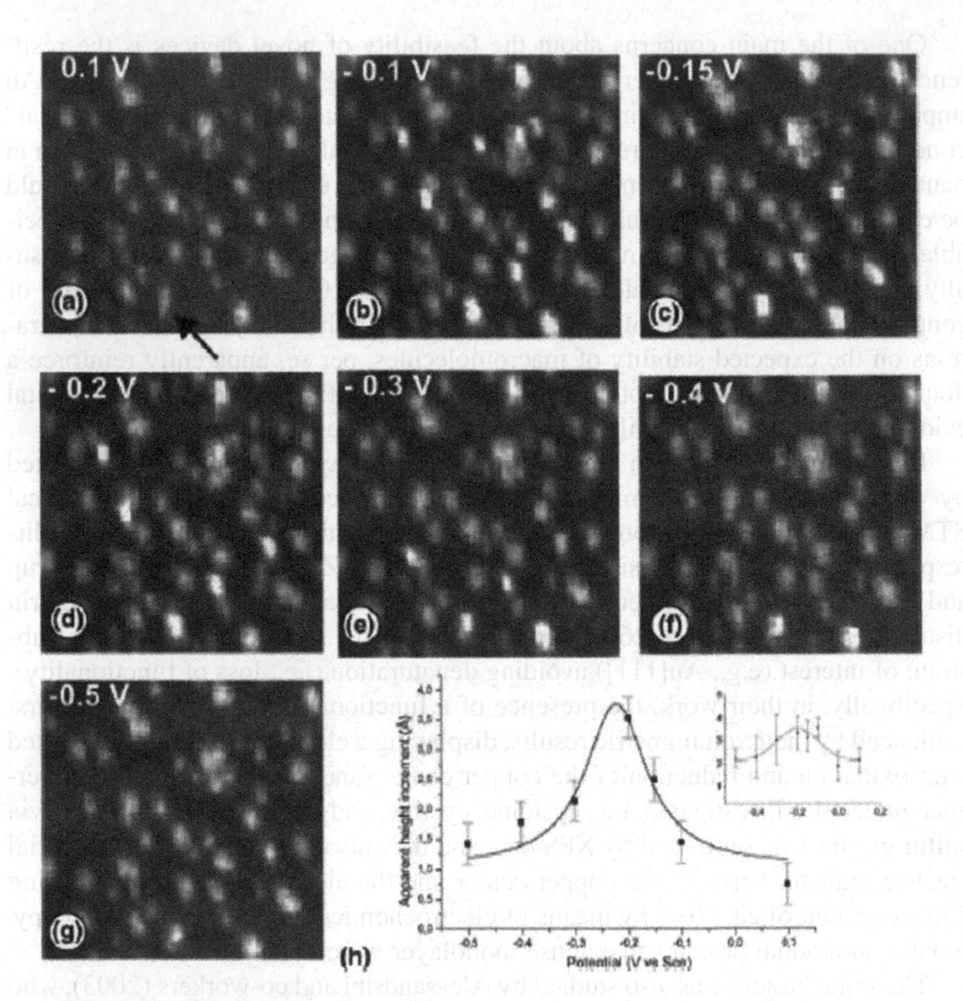

FIG. 5.15. **A.–G.** Sequence of ECSTM images of mixed Cu– and Zn–azurin molecules on a Au(111) substrate, at variable substrate potential (0.1–0.5 V vs. SCE). The scan range is 130 nm. Although some molecules do not change their apparent height other apparently do. (See the spot evidenced in *A*.) The apparent height increment of the same spot, taken with respect to the nonvarying ones, is reported in *(H)* along with the apparent height increment averaged over six molecules *(inset)* (Reproduced with permission from Alessandrini, Gerunda et al. (2003). Chem. Phys. Letts. 376, 625–630.)

levels involved in the transition, which are in turn generally sensitive to folding. Hence, alteration of the structural integrity reflects on the fluorescence spectrum of the protein, provided that the two (ground and excited) energy states in a transition are shifted differently.

In spectroscopic measurements azurin proves, once again, a very good molecular probe. Its intrinsic fluorescence is entirely due to the triptophan amino acid residue Trp48 and shows to be a sensitive indicator of its conformation. The native spectrum is characterized by a peak at an unusually short wavelength (~308 nm), which red-shifts toward 355 nm for denatured protein (Fig. 5.16A). Interestingly, all varieties of azurin show the same emission and excitation spectra, independently of the presence and type of metal ion, which demonstrates that there are no significant structural differences apart from the active site itself. Conversely, there is considerable difference in the quantum yield, owing to metal-related fluorescence quenching. This makes measurements with copper azurin more difficult than with zinc azurin and, especially, with apoazurin. Since it has been proved that apoazurin is the least stable variety of the three (Leckner, Bonander et al. 1997; Svergun, Richard et al. 1998; Merzel and Smith 2002; Danielewicz-Ferchmin, Banachowicz et al. 2003), it can be used as probe of the

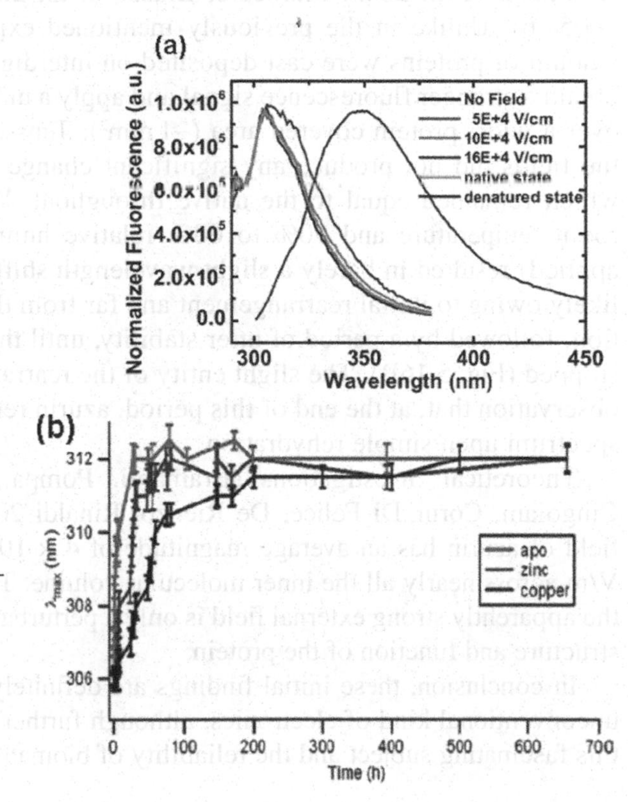

FIG. 5.16. (a) Fluorescence spectra of native apoazurin in solution and on SiO_2 without and with an applied electric field. As a reference, the red-shifted emission spectrum of denatured apoazurin in buffer is reported. (b) Ageing of the different protein derivatives at ambient conditions probed by intrinsic fluorescence experiments."

Azurin functionality, since its retention of the native conformation in a given condition guarantees on the structural integrity of the other types in the same environment.

In the past, most fluorescence spectroscopy experiments on proteins were held in liquid, but fluorescence spectra from immobilized proteins are needed to test their integrity in self-assembled monolayers. Because of the small quantity of protein, a sensitive measurement is required in order to detect the weak fluorescence signal. In the case of Azurins immobilized in monolayers, the spectra definitely showed retention of the native conformation, with apoazurin exhibiting only a slight red-shift (~2–3 nm) and zinc azurin featuring practically unchanged emission (Pompa, Biasco et al. 2004).

As discussed, a second challenge is evaluating the response of proteins to electric fields and other stressing factors. In this respect, molecular dynamics simulations (Xu, Phillips et al. 1996) had shown that fields as strong as 10^8 V/m, whether applied in brief pulses or in a sinusoidal periodical fashion, had no permanent structural effects on a single molecule of bovine pancreatic trypsin inhibitor (BPTI) in vacuum; more recently, experimental evidence confirmed the results. The effect of intensive electric fields and week-long aging were investigated by fluorescence spectroscopy on solid state proteins at ambient conditions (Bramanti et al. 2005; Maruccio, Biasco et al. 2005; Pompa, Bramanti et al. 2005a,b). Unlike in the previously mentioned experiments, a relatively large amount of proteins were cast deposited on interdigitated electrodes, in order to obtain a stronger fluorescence signal and apply a uniform ~10^7 V/m electric field over a wide, protein covered area (~1 mm^2). Tens-of-minutes–long exposure to the fields did not produce any significant change in the fluorescence spectra, which remained equal to the native throughout. Weeks of exposure to air, at room temperature and 50% to 60% relative humidity (without electric field applied) resulted in barely a slight wavelength shift in the first few hours, most likely owing to initial rearrangement and far from the known effect of denaturation, followed by a period of utter stability, until the experiment was arbitrarily stopped (Fig. 5.16B). The slight entity of the rearrangement is confirmed by the observation that, at the end of this period, azurin returned to the native state and spectrum upon simple rehydration.

Theoretical investigations (Bramanti, Pompa, Maruccio, Calabi, Arima, Cingolani, Corni,.Di Felice, De Rienzo, Rinaldi 2005) proved that the internal field of azurin has an average magnitude of 4×10^9 V/m and is larger than 10^7 V/m across nearly all the inner molecular volume. This backs the hypothesis that the apparently strong external field is only a perturbation, which cannot impair the structure and function of the protein.

In conclusion, these initial findings are definitely encouraging for this novel, unconventional kind of electronics, although further investigations are needed on this fascinating subject and the reliability of biomaterials in general.

6 OUTLOOK: USEFULNESS OF PROTEINS IN FUTURE ROBUST MOLECULAR DEVICES, CAPABILITY OF REACTING TO BIOLOGICAL ENVIRONMENT (BIOSENSORS), AND POTENTIAL COMMERCIAL APPLICATIONS

The discussed results back the conviction that proteins are interesting materials and that, thanks to their structural and functional versatility, they are potentially suited to a wide range of applications. Further development toward commercial applications can be envisaged as (at least) threefold.

On the edge of molecular electronics protein–based circuits can be conceived of as an interesting solution, once appropriate techniques for patterning and interconnecting devices are made available. The nanoscopic size of the computing units—down to a single molecule—and their possibly low cost can be the strengths of new molecular machines. Conversely, new architectures and computing paradigms may be required to meet the constraints of this novel biohardware.

Somehow related to the previous point are protein-based biosensors. In these—likely hybrid—devices, the usefulness of proteins mainly results from their capability of specifically reacting to the biological environment. In those medical applications in which continuous monitoring of physical parameters is mandatory and the biosensor is most often the weakest ring of the chain, technology can greatly benefit from this innovative use of these naturally biocompatible and bioreactive materials.

Patterning at the nanoscale is the third, but not less important potential development. Using the self-assembly properties of proteins and possibly other molecules (e.g., DNA) can lead to very complex macroscopic structures, achieving finer-than-lithographic resolution without use of traditional lithography. This is, in fact, what is needed to build memories and complex computing circuits; that is, this may be the milestone marking the beginning of the true post-silicon era.

ACKNOWLEDGMENTS

We are thankful for the invaluable support and exciting collaboration by various colleagues. We would like to thank Ross Rinaldi, Roberto Cingolani, Pier Paolo Pompa, Franco Calabi, Valentina Arima, Adriana Biasco, Eliana D'Amone, Paolo Visconti, Antonio Della Torre at NNL in Lecce (Italy), Elisa Molinari, Rosa Di Felice, Stefano Corni at S^3-INFM research center in Modena (Italy), Gerard Canters and Martin Verbeet at Leiden University (NL). Financial support by NNL, IIT, EC through SAMBA and SpiDME project, Italian MIUR (FIRB Molecular Devices) is gratefully acknowledged.

REFERENCES

Alessandrini, A., Gerunda, M. et al. (2003). Electron tunneling through azurin is mediated by the active site Cu ion. Chem. Phys. Letts. 376(5–6), 625–630.

Bashir, R. (2001). DNA-mediated artificial nanobiostructures: state of the art and future directions. Superlattices Microstructures 29(1), 1–16.

Beratan, D.N., Onuchic, J.N., Hopfield, J.J. (1987). Electron tunneling through covalent and non-covalent pathways in proteins. J. Chem. Phys. 86, 4488–4498.

Birge, R.R., Gillespie, N.B. et al. (1999). Biomolecular electronics: protein-based associative processors and volumetric memories. J. Phys. Chem. B 103(49), 10746–10766.

Bramanti, A., Pompa, P.P., Maruccio, G., Calabi, F., Arima, V., Cingolani, R., Corni, S., Di Felice, R., De Rienzo, F., and Rinaldi, R. (2005). Azurin for biomolecular electronics: a reliability study. Jpn. J. Appl. Phys. 44(9A), 6864–6866.

Broo, A. and Larsson S. (1991). Electron-transfer in azurin and the role of aromatic side groups of the protein. J. Phys. Chem. 95(13), 4925–4928.

Champagne, A. R., Pasupathy, A. N. et al. (2005). Mechanically adjustable and electrically gated single-molecule transistors. Nano Letts. 5(2), 305–308.

Chen, F., He, J. et al. (2005). A molecular switch based on potential-induced changes of oxidation state. Nano Letts. 5(3), 503–506.

Chi, Q.J., Zhang, J.D. et al. (2000). Molecular monolayers and interfacial electron transfer of *Pseudomonas aeruginosa* azurin on Au(111). J. Amer. Chem. Soc. 122(17), 4047–4055.

Choi, J.W. and Fujihira, M. (2004). Molecular-scale biophotodiode consisting of a green fluorescent protein/cytochrome c self-assembled heterolayer. Appl. Phys. Letts. 84(12), 2187–2189.

Choi, J.W., Nam, Y.S. et al. (2001). Rectified photocurrent of the protein-based bio-photodiode. Appl. Phys. Letts. 79(10), 1570 –1572.

Cinelli, R.A.G., Pellegrini, V. et al. (2001). Green fluorescent proteins as optically controllable elements in bioelectronics. Appl. Phys. Letts. 79(20), 3353–3355.

Cohen, B.E., McAnaney, T.B. et al. (2002). Probing protein electrostatics with a synthetic fluorescent amino acid. Science 296(5573), 1700–1703.

Coura, P.Z., Legoas, S.B. et al. (2004). On the structural and stability features of linear atomic suspended chains formed from gold nanowires stretching. Nano Letts. 4(7), 1187–1191.

Dadosh, T., Gordin, Y., Krahne, R., Khivirich, I., Mahalu, D., Frydman, V., Sperling, J., Yacoby, A. and Bar-Joseph, I., (2005). Measurement of the conductance of single conjugated molecules. Nature 436, 677.

Danielewicz-Ferchmin, I., Banachowicz, E. et al. (2003). Protein hydration and the huge electrostriction. Biophys. Chem. 106(2), 147–153.

Das, R., Kiley, P.J. et al. (2004). Integration of photosynthetic protein molecular complexes in solid-state electronic devices. Nano Letts. 4(6), 1079–1083.

Datta, S., Tian, W.D. et al. (1997). Current-voltage characteristics of self-assembled monolayers by scanning tunneling microscopy. Phys. Rev. Letts. 79(13), 2530–2533.

Davis, J.J., Morgan, D.A. et al. (2005). Molecular bioelectronics. J. Mater. Chem. 15(22), 2160–2174.

Farver, O., Lu, Y. et al. (1999). Enhanced rate of intramolecular electron transfer in an engineered purple CUA azurin. Proc. Natl. Acad. Sci. U S A 96(3), 899–902.

Farver, O. and Pecht I. (1992). Long-range intramolecular electron-transfer in azurins. J. Amer. Chem. Soc. 114(14), 5764–5767.

Fidy, J., Balog, E. et al. (1998). Proteins in electric fields and pressure fields, experimental results. Biochim. et Biophys Acta Prot. Struct. Mol. Enzymol. 1386(2), 289–303.

Frolov, L., Rosenwaks, Y., Carmeli, C., and Carmeli, I. (2005). Fabrication of a photoelectronic device by direct chemical binding of the photosynthetic reaction center protein to metal surfaces. Adv. Mater. 17, 2434.

Giese, B., Amaudrut, J. et al. (2001). Direct observation of hole transfer through DNA by hopping between adenine bases and by tunnelling. Nature 412(6844), 318–320.

Giese, B., Wessely, S. et al. (1999). On the mechanism of long-range electron transfer through DNA. Angewandte Chemie Int. Ed. 38(7), 996–998.

Gray, H.B. and Winkler, J.R. (1996). Electron transfer in proteins. Annu. Rev. Biochem. 65, 537–561.

Gray, H.B. and Winkler, J.R. (2005). Long-range electron transfer. Proc. Natl. Acad. Sci. U S A 102(10), 3534–3539.

Kagan, C.R., Afzali, A. et al. (2003). Evaluations and considerations for self-assembled monolayer field-effect transistors. Nano Letts. 3(2), 119–124.

Keren, K., Berman, R.S. et al. (2003). DNA-templated carbon nanotube field-effect transistor. Science 302(5649), 1380–1382.

Keren, K., Krueger, M. et al. (2002). Sequence-specific molecular lithography on single DNA molecules. Science 297(5578), 72–75.

Kervennic, Y.V., Van der Zant, H.S.J. et al. (2002). Nanometer-spaced electrodes with calibrated separation. Appl. Phys. Letts. 80(2), 321–323.

Kim, K.S., Park, I. et al. (1996). The nature of a wet electron. Phys. Rev. Letts. 76(6), 956.

Kohler, M., Friedrich, J. et al. (1998). Proteins in electric fields and pressure fields, basic aspects. Biochim. et Biophys Acta Prot. Struct. Mol. Enzymol. 1386(2), 255–288.

Krahne, R., Yacoby, A. et al. (2002). Fabrication of nanoscale gaps in integrated circuits. Appl. Phys. Letts. 81(4), 730–732.

Kuhn, O., Rupasov, V. et al. (1996). Effective bridge spectral density for long-range biological energy and charge transfer. J. Chem. Phys. 104(15), 5821–5833.

Larsson, S., Broo, A. et al. (1995). Connection between structure, electronic-spectrum, and electron-transfer properties of blue copper proteins. J. Phys. Chem. 99(13), 4860–4865.

Leckner, J., Bonander, N. et al. (1997). The effect of the metal ion on the folding energetics of azurin, a comparison of the native, zinc and apoprotein. Biochim. et Biophys Acta Prot. Struct. Mol. Enzymol. 1342(1), 19–27.

Lee, J.O., Lientschnig, G. et al. (2003). Absence of strong gate effects in electrical measurements on phenylene-based conjugated molecules. Nano Letts. 3(2), 113–117.

Lee, S.W., Mao, C.B. et al. (2002). Ordering of quantum dots using genetically engineered viruses. Science 296(5569), 892–895.

Liu, H., Fan, F.F. et al. (1986). Scanning electrochemical and tunneling ultramicroelectrode microscope for high-resolution examination of electrode surfaces in solution. J. Amer. Chem. Soc. 108, 3838.

Lu, X.H., Grobis, M. et al. (2003). Spatially mapping the spectral density of a single C-60 molecule. Phys. Rev. Letts. 90(9), art. no. 096802.

Maltezopoulos, T., Bolz, A. et al. (2003). Wave-function mapping of InAs quantum dots by scanning tunneling spectroscopy. Phys. Rev. Letts. 91(19), art. no. 196804.

Manoj, A.G. and Narayan, K.S. (2003). Voltage-controlled spectral tuning of photoelectric signals in a conducting polymer-bacteriorhodopsin device. Appl. Phys. Letts. 83(17), 3614–3616.

Maruccio, G., Biasco, A. et al. (2005). Towards protein field-effect transistors, Report and model of prototype. Adv. Mater. 17(7), 816.

Maruccio, G., Visconti, P. et al. (2004). Nano-scaled biomolecular field-effect transistors: prototypes and evaluations. Electroanalysis 16(22), 1853–1862.

Maruccio, G., Visconti, P. et al. (2003). Planar nanotips as probes for transport experiments in molecules. Microelectr. Eng. 67–68, 838–844.

Maruccio G, Marzo P, Krahne R, Passaseo A, Cingolani R, Rinaldi R (2007), Protein conduction and negative differential resistance in large-scale nanojunction arrays, SMALL 3 (7), 1184–1188

Merzel, F. and Smith, J.C. (2002). Is the first hydration shell of lysozyme of higher density than bulk water? Proc. Natl. Acad. Sci. U S A 99(8), 5378–5383.

Park, E.S., Andrews, S.S. et al. (1999). Vibrational stark spectroscopy in proteins, a probe and calibration for electrostatic fields. J. Phys. Chem. B 103(45), 9813–9817.

Park, E.S., Thomas, M.R. et al. (2000). Vibrational Stark spectroscopy of NO bound to heme: effects of protein electrostatic fields on the NO stretch frequency. J. Amer. Chem. Soc. 122(49), 12297–12303.

Park, J., Pasupathy, A.N. et al. (2002). Coulomb blockade and the Kondo effect in single-atom transistors. Nature 417(6890), 722–725.

Pompa, P.P., Biasco, A. et al. (2004). Structural stability study of protein monolayers in air. Phys. Rev. E 69(3), art. no. 032901.

Pompa, P.P., Bramanti, A. et al. (2005a). Retention of nativelike conformation by proteins embedded in high external electric fields. J. Chem. Phys. 122(18), art. no. 181102.

Pompa, P.P., Bramanti, A. et al. (2005b). Aging of solid-state protein films: behavior of azurin at ambient conditions. Chem. Phys. Letts. 404(1–3), 59–62.

Ramamurthy, V. and Schanze, K.S. (1998). Organic and Inorganic Photochemistry, Marcel Dekker, New York.

Reed, M.A., Zhou, C. et al. (1997). Conductance of a molecular junction. Science 278(5336), 252–254.

Rinaldi R. and Maruccio G. (2006) Nano-Bio Electronics, published by WILEY-VCH · Weinheim · Berlin as a chapter in the book entitled Series on Nanotechnology for Life Sciences - Vol 4 (Nanodevices for Life Sciences, ISBN-10: 3-527-31384-2, ISBN-13: 978-3-527-31384-6, (2006) approx 400pp with 175 figs) edited by Challa Kumar

Silverman, J.P. (1997). X-ray lithography, status, challenges and outlook for 0.13 um. J. Vacuum Sci. Technol. B 15, 2117.

Skourtis, S.S., Balabin, I.A. et al. (2005). Protein dynamics and electron transfer: electronic decoherence and non-Condon effects. Proc. Natl. Acad. Sci. U S A 102(10), 3552–3557.

Sonnenfeld, R. and Hansma, P.K. (1986). Atomic-Resolution Microscopy in Water Science 232, 211.

Star, A., Gabriel, J.C.P. et al. (2003). Electronic detection of specific protein binding using nanotube FET devices. Nano Letts. 3(4), 459–463.

Svergun, D.I., Richard, S. et al. (1998). Protein hydration in solution: experimental observation by x-ray and neutron scattering. Proc. Natl. Acad. Sci. U S A 95(5), 2267–2272.

Tan, J.L., Tien, J. et al. (2002). Microcontact printing of proteins on mixed self-assembled monolayers. Langmuir 18(2), 519–523.

Tao, N.J. (1996). Probing potential-tuned resonant tunneling through redox molecules with scanning tunneling microscopy. Phys. Rev. Letts. 76(21), 4066–4069.

Xia, Y.N. and Whitesides, G.M. (1998). Soft lithography. Angewandte Chemie-Int. Ed. 37(5), 551–575.

Xu, B.Q. and Tao, N.J.J. (2003). Measurement of single-molecule resistance by repeated formation of molecular junctions. Science 301(5637), 1221–1223.

Xu, D., Phillips, J.C. et al. (1996). Protein response to external electric fields, Relaxation, hysteresis, and echo. J. Phys. Chem. 100(29), 12108–12121.

Xu, D.G., Watt, G.D. et al. (2005). Electrical conductivity of ferritin proteins by conductive AFM. Nano Letts. 5(4), 571–577.

Yasutomi, S., Morita, T. et al. (2004). A molecular photodiode system that can switch photocurrent direction. Science 304(5679), 1944–1947.

Zandbergen, H.W., van Duuren, R. et al. (2005). Sculpting nanoelectrodes with a transmission electron beam for electrical and geometrical characterization of nanoparticles. Nano Letts. 5(3), 549–553.

Zhang, J.D., Grubb, M. et al. (2003). Electron transfer behaviour of biological macromolecules towards the single-molecule level. J. Phys. Cond. Mater. 15(18), S1873–S1890.

Zhao, J.W., Davis, J.J. et al. (2004). Exploring the electronic and mechanical properties of protein using conducting atomic force microscopy. J. Amer. Chem. Soc. 126(17), 5601–5609.

6

S-Layer Proteins for Assembling Ordered Nanoparticle Arrays

Dietmar Pum and Uwe B. Sleytr

1 INTRODUCTION

The current challenge in the development of the next generation of nanoscale systems will require entirely new materials, fabrication technologies, and devices. The ultimately high requirements in the synthesis of molecular functional units, in particular of nanoelectronic devices, can only be met when novel concepts based on state-of-the-art top-down lithographic procedures are combined with bottom-up approaches using principles learned from nature. Thus, the aim of current research is the development of tools and techniques for a novel combined top-down/bottom-up fabrication technology for nanostructures with critical dimensions below 100 nm that can be used in a flexible way for a variety of devices and architectures.

The bioinspired synthesis of inorganic materials, such as metallic or semiconducting nanoparticles, has already attracted much attention over the last two decades. In particular, the broad base of knowledge about the binding of biological molecules has paved the way for investigating the potential of S-layer proteins (Sleytr and Beveridge 1999) and their self-assembly products as catalysts, templates, and scaffolds for the generation of ordered nanoparticle arrays (Sleytr, Egelseer, Pum, and Schuster 2004; Sára, Pum, Schuster, and Sleytr 2005; Sleytr, Sára, Pum, and Schuster 2005).

A. Offenhäusser and R. Rinaldi (eds.), *Nanobioelectronics - for Electronics, Biology, and Medicine*, 167
DOI: 10.1007/978-0-387-09459-5_7, © Springer Science+Business Media, LLC 2009

2 DESCRIPTION OF S-LAYERS

Crystalline bacterial surface layer (S-layer) proteins represent the outermost cell envelope component of a broad spectrum of bacteria and archaea (Fig. 6.1A) (Sleytr and Beveridge 1999; Sleytr, Egelseer, Pum, and Schuster 2004; Sára, Pum, Schuster, and Sleytr 2005; Sleytr, Sára, Pum, and Schuster 2005). S-layers are monomolecular arrays composed of a single protein or glycoprotein species (M_w 40–200 kDa) and exhibit oblique (p1, p2), square (p4), or hexagonal (p3, p6) lattice symmetry with unit cell dimensions in the range of 3 to 30 nm. Depending on the lattice symmetry one morphological unit (= unit cell) consists of one, two, three, four, or six identical S-layer proteins. S-layers are highly porous protein lattices (30–70% porosity) with pores of uniform size and morphology in the 2- to 8-nm range (Fig. 6.1B). S-layers are generally 5 to 10 nm thick.

S-layers are also highly anisotropic structures with respect to their physicochemical surface properties. Generally, in Bacillacaea the outer S-layer face is less corrugated than the inner one and bears no net charges, whereas the inner face is either net positively or negatively charged. Functional groups (e.g., carboxyl, amino, or hydroxyl groups) or genetically incorporated functional domains (e.g., streptavidin sites for biotin binding, fused with the self-assembling part of the protein) are repeated with the periodicity of the S-layer lattice at a distance resembling the lattice constants, leading to regular arrays of bound functional molecules or nanoparticles. Studies on the structure–function relationship of different S-layers from Bacillaceae revealed the existence of specific (lectin type) binding domains on the N-terminal part of S-layer proteins for heteropolysaccharides (secondary cell wall polymers; SCWPs) covalently linked to the peptidoglycan matrix of the cell wall (Sára 2001; Sleytr, Sára, Mader, Schuster, and Unger 2001).

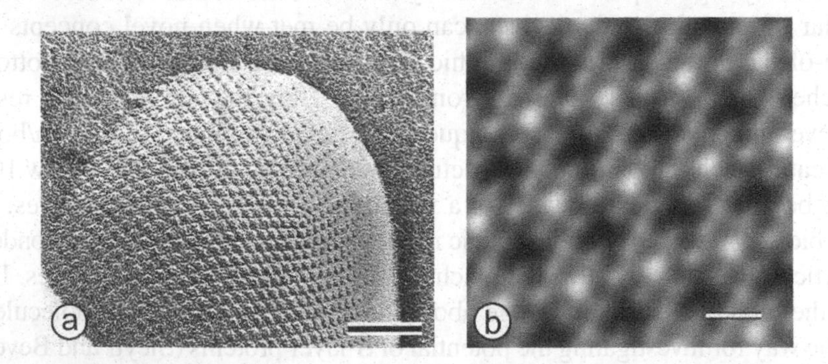

FIG. 6.1. **A.** Transmission electron microscopical image of a freeze-etched and metal (Pt/C) shadowed preparation of a bacterial cell. The S-layer shows hexagonal lattice symmetry. Bar, 100 nm. **B.** Digital image reconstructions of a scanning force microscopical image of an S-layer protein monolayer reassembled on a silicon surface. The S-layer lattice shows square (p4) lattice symmetry. Bar, 10 nm.

One of the most fascinating properties of isolated S-layer proteins is their capability to form free floating self-assembly products in solution (e.g., flat sheets, cylinders) (Sleytr and Messner 1989), to recrystallize into extended monomolecular layers on solid supports (Pum and Sleytr 1995; Györvary, Stein, Pum, and Sleytr 2003), at the air–water interface, and on lipid films (Schuster and Sleytr 2000) and to cover liposomes (Küpcü, Sára, and Sleytr 1995; Mader, Küpcü, Sleytr, and Sára 2000) and nanocapsules (Toca-Herrera, Krastev, Bosio, Küpcü, Pum, Fery, Sára, and Sleytr 2005). Depending on the S-layer protein species used and the environmental conditions, double layers in back-to-back orientation may be formed. The reassembly occurs after removal of the disrupting agent used in the dissolution and isolation procedure. In general, a complete disintegration of S-layer lattices in the constituent (glyco)protein subunits on bacterial cells can be achieved using high concentrations of chaotropic agents (e.g., guanidine hydrochloride, urea), by lowering or raising the pH, or by applying metal-chelating agents (e.g., EDTA, EGTA) or cation substitution. The formation of self-assembled arrays is only determined by the amino acid sequence of the polypeptide chains and consequently the tertiary structure of the S-layer protein species. In various S-layer proteins from Bacillacaea it was shown that significant portions of the C- or N-terminal part can be deleted without losing the capability of the subunits for lattice formation (Jarosch, Egelseer, Huber, Moll, Mattanovich, Sleytr, and Sára 2001).

3 METHODS, MATERIALS, AND RESULTS

3.1 NANOPARTICLE FORMATION BY SELF-ASSEMBLY ON S-LAYER PATTERNED SUBSTRATES

The first report on the use of S-layers as lithographic templates was published by Douglas and coworkers almost two decades ago (Douglas, Clark, and Rothschild 1986). In a three-step process, S-layer fragments from *Sulfolobus acidocaldarius* were deposited onto a smooth 20-nm thick carbon coated substrate, metal coated by evaporation with tantalum/tungsten (Ta/W), and then ion milled. This S-layer shows hexagonal lattice symmetry with a center-to-center spacing of the morphological units of 22 nm, a thickness of 10 nm, and pores 5 nm in diameter. The thickness of the metal film was ~1.2 nm. Under ion milling this composite protein–metal structure exhibited differential metal removal leading to 15-nm sized holes periodically arranged in a hexagonal lattice resembling the lattice geometry of the underlying S-layer lattice. Because of the shadowing angle of 40° from the normal to the substrate the coated unmilled S-layers exhibited thicker metal deposits along the edges of the hexagonally symmetric holes. The thinnest metal thickness was found in the S-layer holes shadowed by their front edges relative to the incident beam. Subsequent ion milling (normal incidence, 30 s) removed the metal in the holes first and led to a film of nearly uniform thickness perforated by a periodic array of oval holes and metal strips running along the most heavily coated molecular rows. Although significant fluctuations in the shape and size of the holes in the metal film were observed, this first approach laid the foundation for using S-layers as patterning elements for generating ordered nanostructured materials. Later on, in a basically similar approach, this group used fast-atom beam milling to fabricate 10 nm diameter sized holes in a 3.5-nm-thick titanium oxide (TiO_2) layer (Douglas, Devaud, and Clark 1992). Scanning force microscopy was used to study the surface profile of these protein–metal structures and the patterning of the supporting substrate.

A decade later, Douglas and co-workers made use of the S-layer based lithographic approach described in the preceding to fabricate a precisely ordered and located lattice of 5-nm diameter titanium nanoclusters (Winningham, Gillis, Choutov, Martin, Moore, and Douglas 1998). The crystalline S-layer protein fragments (1–2 µm in extent) were deposited onto a hydrophilic silicon oxide surface and coated with a thin layer of titanium by evaporation (thickness ~1.2 nm). Subsequently, the titanium film was oxidized in air (thickness of the TiO_2 layer 3.5 nm) exhibiting 6-nm diameter metallized pores. Low energy electron enhanced etching (LE4) in a DC hydrogen plasma was used for etching such a thin delicate mask since only LE4 avoided the damage commonly inflicted by standard ion bombardment. The isotropic etching widened the holes to a diameter of 18 nm and transferred their pattern into the silicon. After etching, the mask was completely removed and the patterned surface oxidized in oxygen plasma. After deposition of a further (1.2-nm) titanium layer an ordered array of 5-nm sized metal nanoclusters was formed at the etched hole positions. This superlattice of titanium nanoclusters resembled the lattice geometry of the underlying S-layer patches.

The lithographic approach of using S-layers as nanometric masks was also exploited to fabricate magnetic nanostructures. The hexagonally packed intermediate (HPI) S-layer of *Deinococcus radiodurans* was used for patterning ferromagnetic films (Panhorst, Brückl, Kiefer, Reiss, Santarius, and Guckenberger 2001). HPI displays hexagonal (p6) lattice symmetry with a center-to-center spacing of 18 nm, a thickness of 6.5 nm, and 6-nm wide pores. Accordingly, a hexagonal pattern of uniform 10-nm sized dots and a lattice spacing of 18 nm was fabricated from 2.5-nm thick sputter coated Co, FeCo, Fe, CoNi, and NiFe films. These patterns occurred after dry etching with Ar ions. The most critical parameter for a successful formation of nanodot arrays were etching time, energy, and density of the Ar ions.

Recently, regular arrays of nanometer sized magnetic dots were also generated using the S-layer of *Sulfolobus acidocaldarius* as templates (Malkinski, Camley,

Celinski, Winningham, Whipple, and Douglas 2003). In this work, a thin layer of chromium (Cr) was deposited onto 1- to 2-μm sized S-layer fragments from *Sulfolobus acidocaldarius* at an angle of 60°. The topography of the protein lattice produced a shadow at each pore leading to an ordered array of holes in the Cr layer. Subsequently, the pattern of this ordered array was transferred into the silicon by plasma treatment yielding an identical pattern of ~4-nm sized holes in the silicon subtrate. An assembly of Fe/Pd dots with an average dot size of 10 nm in diameter, 6.5 nm in height and a lattice spacing of 22 nm (according to the lattice parameters of the underlying S-layer) was fabricated by molecular beam epitaxy. The dots consisted of a sandwich of four 1-nm thick Fe and 0.4-nm thick Pd layers in which the terminating Pd layer was 1.5 nm thick. Finally, after the metal deposition the protein layer was removed. The magnetic properties of these dot arrays were different from those of equivalent continuous films (Malkinski, Camley, Celinski, Winningham, Whipple, and Douglas 2003).

In addition to the described thermal evaporation techniques cross-beam pulsed laser deposition was used to transfer the nanomorphology of S-layers into metallic (Pt/C) overlayers (Gorbunov, Mertig, Kirsch, Eichler, Pompe, and Engelhardt 1997; Neubauer, Pentzien, Reetz, Kautek, Pum, and Sleytr 1997). The observed surface corrugations of the metallized S-layer sheets showed higher corrugations compared to the uncoated sample. Thus it was concluded that the protein layer had been mechanically stabilized by the metal film.

3.2 WET CHEMICAL SYNTHESIS OF NANOPARTICLES

Based on the investigation of mineral formation by bacteria in natural environments (Douglas and Beveridge 1998), S-layer lattices can be used in wet chemical processes for the precipitation of metal ions from solution, as well (Sleytr, Bayley, Sára, Breitwieser, Küpcü, Mader, Weigert, Unger, Messner, Jahn-Schmid, Schuster, Pum, Douglas, Clark, Moore, Winningham, Levy, Frithsen, Pankovc, Beagle, Gillis, Choutov, and Martin 1997). In this approach self-assembled S-layer structures were exposed to metal-salt solutions followed by slow reaction with a reducing agent such as hydrogen sulfide (H_2S). Nanoparticle superlattices were formed according to the lattice spacing and symmetry of the underlying S-layer. Furthermore, since the precipitation of the metals was confined to the pores of the S-layer, the nanoparticles also resembled the morphology of the pores.

The first example exploiting this technique was the precipitation of cadmium sulfide (CdS) on S-layer lattices composed of the S-layer protein from *Geobacillus stearothermophilus* NRS 2004/3a variant 1 (NRS), and the S-layer protein of *Bacillus sphaericus* CCM2177 (SbpA) (Shenton, Pum, Sleytr, and Mann 1997). NRS reassembles into mono- and double-layered self-assembly products in solution and into monolayers on solid supports (Sára, Pum, Küpcü, Messner, and Sleytr 1994). NRS shows oblique (p1) lattice symmetry with unit cell dimensions of a = 9.8 nm, b = 7.5 nm, and a base angle of 80°. The thickness of this S-layer is 4.5 nm and pores show a diameter of 4 to 5 nm. SbpA forms monolayer self-assembly products in suspension and on solid supports (Györvary, Stein, Pum, and Sleytr 2003). The S-layer exhibits square lattice symmetry with a lattice constant of 13.1 nm, and a thickness of the S-layer of 8 nm. Pores in this S-layer are 4 to 5 nm wide. After incubation of the S-layer self-assembly products with a $CdCl_2$ solution for several hours the hydrated samples were exposed to H_2S for at least 1 to 2 days. The generated CdS nanoparticles were 4 to 5 nm in size and their superlattices resembled the oblique lattice symmetry of SbsB, or the square lattice symmetry of SbpA, respectively (Shenton, Pum, Sleytr, and Mann 1997). In addition, mineralization of the dispersed nanostructures formed in suspension of NRS produced characteristic stripe patterns of organized CdS nanoparticles. The most common stripe pattern consisted of fringes 16 nm in width and 32 nm spaced apart. They were aligned parallel to the longer unit cell vector of the underlying S-layer lattice. These stripe patterns were Moiré fringes, which originated from the superposition of two oblique CdS/S-layer lattices aligned in back-to-back orientation along the longer base vector. In this particular arrangement, the two associated S-layers were in perfect register every fourth lattice row.

Further on, a superlattice of 4- to 5-nm sized gold particles was formed by using SbpA (with previously induced thiol groups) as template for the precipitation of a tetrachloroauric (III) acid ($HAuCl_4$) solution (Fig. 6.2) (Dieluweit, Pum, and Sleytr 1998). Gold nanoparticles were formed either by reduction of the metal salt with H_2S or under the electron beam in a transmission electron microscope. The latter approach is technologically important because it allows the definition of areas in which nanoparticles are eventually formed (Dieluweit, Pum, and Sleytr 1998; Wahl, Mertig, Raff, Selenska-Pobell, and Pompe 2001). As determined by electron diffraction the gold nanoparticles were crystalline but their ensemble was not crystallographically aligned. Later on, the wet chemical approach was used in the formation of Pd- (salt: $PdCl_2$), Ni- ($NiSO_4$), Pt- ($KPtCl_6$), Pb- ($Pb(NO_3)_2$), and Fe- ($KFe(CN)_6$) nanoparticle arrays (unpublished results). Recently, small spot X-ray photoelectron emission spectroscopy (XPS) was used to characterize the elemental composition of the nanoclusters. XPS demonstrated that they consisted primarily of elemental gold (Dieluweit, Pum, Sleytr, and Kautek 2005).

In a similar approach arrays of platinum nanoparticles were fabricated on the S-layer of *Sporosarcina ureae* (Mertig, Kirsch, Pompe, and Engelhardt 1999). *Sp. ureae* exhibits square (p4) lattice symmetry with a lattice spacing of 13.2 nm. Platinum cluster deposition was achieved by precipitation of platinum from K_2PtCl_4 solution and subsequent reduction with NaN_3. Transmission electron microscopical investigations revealed the formation of well-separated metal clusters with an average diameter of 1.9 ± 0.6 nm. Seven clusters per unit cell were formed. They were found in

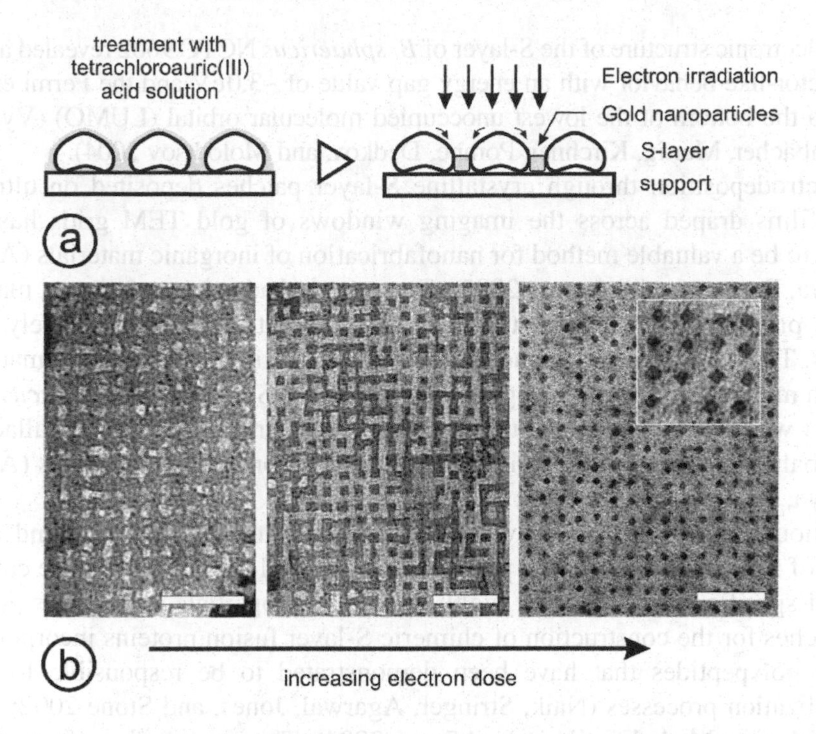

FIG. 6.2. **A.** Schematic drawing of the S-layer templated wet chemical synthesis of gold nanoparticles. The S-layer is incubated with a tetrachloroauric (III) acid solution. **B.** Upon irradiation with an electron beam a superlattice of gold nanoparticles with square lattice symmetry is formed. The nanoparticles resemble the morphology of the S-layer pores (inset). Bars, 50 nm. (Reprinted from Dieluweit, Pum, and Sleytr 1998, with permission from Elsevier.).

the pores and indentations in the S-layer lattice. Moreover, UV-VIS spectroscopy was able to stress the role of the S-layer in the process of cluster deposition as templates with a very high density of specific affinity sites in which nucleation takes place. In subsequent work, the formation of palladium nanoparticles on the S-layer of *Bacillus sphaericus* NCTC 9602 and of *Sp. ureae* was shown as well (Pompe, Mertig, Kirsch, Wahl, Ciachi, Richter, Seidel, and Vinzelberg 1999; Mertig, Wahl, Lehmann, Simon, and Pompe 2001; Wahl, Engelhardt, Pompe, and Mertig 2005; Hüttl, Ullrich, Wolf, Kirchner, Mertig, and Pompe 2006). The S-layer or *B. sphaericus* NCTC 9602 shows square (p4) lattice symmetry with a lattice constant of 12.5 nm. Palladium cluster deposition was achieved after activation of the S-layer with K_2PdCl_4. Upon electron irradiation in the TEM 5- to 7-nm sized metallic nanoparticles forming regular arrays resembling the lattice geometry of the S-layer were formed. However, nanoparticle formation was only observed in flattened S-layer cylinders in which Pt and Pd complexes were physically accumulated because of the particular geometry of the resulting S-layer double layer. In addition to the precipitation of nanoparticles for applications in molecular electronics, it must be noted here that recent investigations of the electronic structure of the S-layer of *B. sphaericus* NCTC 9602 revealed a semiconductor-like behavior with an energy gap value of ~3.0eV and the Fermi energy close to the bottom of the lowest unoccupied molecular orbital (LUMO) (Vyalikh, Danzenbächer, Mertig, Kirchner, Pompe, Dedkov, and Molodtsov 2004).

Electrodeposition through crystalline S-layer patches deposited on ultrathin AuPd films draped across the imaging windows of gold TEM grids has also proved to be a valuable method for nanofabrication of inorganic materials (Allred, Sarikaya, Baneyx, and Schwartz 2005). Contrary to shadowing techniques, material growth proceeds from the substrate outward without the need of entirely open regions. Thus, it offers the unique prospect of fabricating nanopatterned materials through multilayered crystalline protein patches. The S-layer of *Deinococcus radiodurans* was used for making cuprous oxide (Cu_2O), nickel, platinum, palladium, and cobalt nanostructured materials on electrically conducting substrates (Allred, Sarikaya, Baneyx, and Schwartz 2005).

Although native S-layers have clearly demonstrated the presence and availability of functional sites for the precipitation of metal ions, a much more controlled and specific way of making highly ordered nanoparticle arrays uses genetic approaches for the construction of chimeric S-layer fusion proteins incorporating unique polypeptides that have been demonstrated to be responsible for biomineralization processes (Naik, Stringer, Agarwal, Jones, and Stone 2002; Naik, Jones, Murray, McAuliffe, Vaia, and Stone 2004). The precipitation of metal ions or binding of metal nanoparticles (see the following) is then confined to specific and precisely localized positions in the S-layer lattice. Currently several silver and cobalt precipitating peptides are under investigation (Naik, Stringer, Agarwal, Jones, and Stone 2002; Naik, Jones, Murray, McAuliffe, Vaia, and Stone 2004). First results are promising and have demonstrated the feasibility to genetically engineer S-layer fusion proteins incorporating metal binding peptides capable of forming monolayers on technologically important substrates such as silicon, glass, gold, or polymeric surfaces. In particular, the possibility of using S-layer streptavidin fusion proteins as patterning elements for the immobilization of biotinylated functional molecules recently has been demonstrated (Moll, Huber, Schlegel, Pum, Sleytr, and Sára 2002; Huber, Liu, Egelseer, Moll, Knoll, Sleytr, and Sára 2006).

3.3 BINDING OF PREFORMED NANOPARTICLES

Although wet chemical methods lead to crystalline arrays of nanoparticles with spacing in register with the underlying S-layer lattice, they do not allow us to precisely control particle size and hence the contact distances of neighboring particle surfaces, both of which are important for studying and exploiting quantum phenomena.

Thus, the binding of preformed, often core-shell, nanoparticles into regular arrays on S-layers has significant advantages for the development of nanoscale electronic devices. Based on the work on binding biomolecules, such as enzymes or antibodies, onto S-layers (Sleytr, Egelseer, Pum, and Schuster 2004; Sára, Pum, Schuster, and Sleytr 2005; Sleytr, Sára, Pum, and Schuster 2005), it has already been demonstrated that metallic and semiconducting nanoparticles can be bound in regular arrangements on S-layers. This is because contrary to conventional carriers, in which location, local density, and orientation of functional groups are only approximately known, with S-layers the properties of a single constituent unit are replicated with the periodicity of the lattice and thus define the characteristics of the whole two-dimensional array. The pattern of bound molecules frequently reflects the lattice symmetry, size of the morphological units, and physicochemical properties of the array. Specific binding of molecules on S-layer lattices may be induced by different noncovalent forces. For example, the distribution of net negatively charged domains on S-layers could be visualized by electron microscopical methods after labelling with positively charged topographical markers, such as polycationic ferritin (PCF; diameter, 12 nm) (Fig. 6.3). The regular arrangement of free carboxylic acid groups on the hexagonal S-layer lattice from *Thermoproteus tenax* was clearly demonstrated in this way (Messner, Pum, Sára, Stetter, and Sleytr 1986).

Recently, gold and amino functionalized CdSe had been bound onto S-layer protein monolayers and self-assembly products of SbpA (Fig. 6.4) (Györvary, Schroedter, Talapin, Weller, Pum, and Sleytr 2004). SbpA monolayers recrystallized on hydrophobic silicon surfaces expose the outer S-layer face toward the

FIG. 6.3. Transmission electron microscopic image of a freeze-dried and metal (Pt/C) shadowed preparation of a bacterial cell wall sacculus onto which polycationic ferritin molecules (PCF; diameter 12 nm) had been bound in an ordered hexagonal arrangement. Bar, 100 nm.

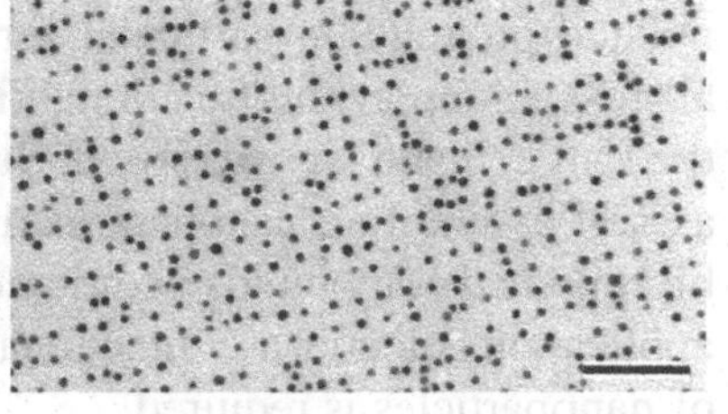

FIG. 6.4. Transmission electron microscopic image of preformed gold nanoparticles (diameter 5 nm) bound on an S-layer with square (p4) lattice symmetry. The superlattice of the gold nanoparticles resembles the lattice geometry of the underlying S-layer. Bar, 50 nm.

environment. Amino-functionalized 4-nm sized CdSe particle were bound at 1-ethyl-3,3'(dimethylaminopropyl) carbodiimide (EDC) activated carboxyl groups at the outer S-layer face in register with the underlying square S-layer lattice. On hydrophilic silicon surfaces SbpA forms double layers where the inner S-layer surfaces are facing each other and thus, again, expose their outer S-layer face towards the environment. The inner face is only accessible where the double layers are incomplete. Citrate stabilized negatively charged gold nanoparticles of 5 nm in diameter were bound by electrostatic interactions at the inner S-layer face forming extended superlattices (Györvary, Schroedter, Talapin, Weller, Pum, and Sleytr 2004).

The hexagonally packed intermediate (HPI) S-layer of *Deinococcus radiodurans* was used for the self-assembly of preformed gold nanoparticles into superlattices commensurate with the underlying S-layer lattice (Hall, Shenton, Engelhardt, and Mann 2001; Bergkvist, Mark, Yang, Angert, and Batt 2004). Each hexamer is in the form of a hollow cone-shaped protrusion with a positively charged central channel. High-resolution TEM studies showed that negatively charged monodisperse gold nanoparticles with mean sizes of ~8 nm and ~5 nm, respectively, were electrostatically bound at these sites forming micrometer sized crystalline domains. Additional experiments with 20-nm sized positively charged as well as with 5-nm sized negatively charged gold nanoparticles demonstrated that superlattices were not formed and thus not templated by the S-layer. The nanoparticles were either negatively charged because of surface citrate ions or positively charged because of surface coating with poly-L-lysine.

A major breakthrough in the regular binding of metallic and semiconducting nanoparticles was achieved by the successful design and expression of S-layer-streptavidin fusion proteins which allowed a specific binding of biotinylated ferritin molecules into regular arrays (Fig. 6.5) (Moll, Huber, Schlegel, Pum, Sleytr, and Sára 2002; Huber, Liu, Egelseer, Moll, Knoll, Sleytr, and Sára 2006).

The fusion proteins had the inherent ability to self-assemble into monomolecular protein lattices. The fusion proteins and streptavidin were produced independently in *Escherichia coli*, isolated, purified, and mixed to refold into heterotetramers of 1:3 stoichiometry. Self-assembled chimeric S-layers could be formed in suspension, on liposomes, silicon wafers, and accessory cell wall polymer–containing cell wall fragments. The two-dimensional protein crystals displayed streptavidin in defined repetitive spacing, and they were capable of binding D-biotin and biotinylated proteins, in particular ferritin. Further on, it could be demonstrated that all fused streptavidin functionalities had the same position and orientation within the unit cell and were exposed. Such chimeric S-layer protein lattices can be used as self-assembling nanopatterned molecular affinity matrices capable to arrange biotinylated compounds in ordered arrays on surfaces (Fig. 6.6). In addition, it has application potential as a functional coat of liposomes when a spherical arrangement of nanoparticles is required.

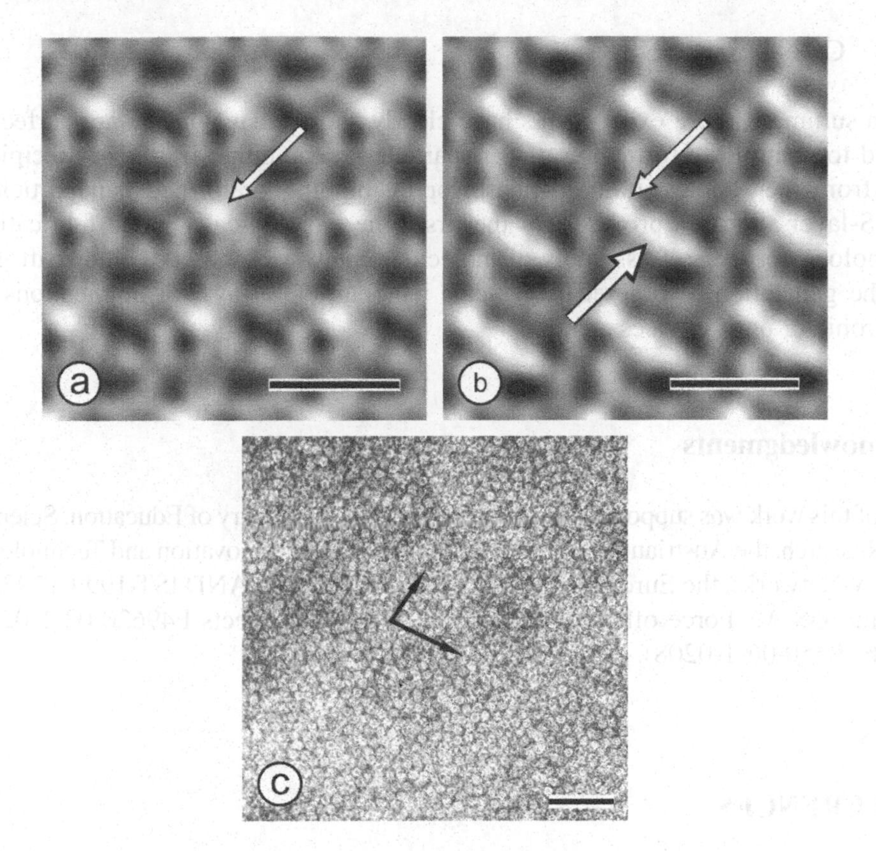

FIG. 6.5. **A.** Digital image reconstructions of transmission electron microscopical images of negatively stained preparations of native SbsB S-layer protein *(A)* and Streptavidin – SbsB S-layer fusion protein *(B)*. The region of highest protein mass in the SbsB lattice is marked in *(A)* and *(B)* by corresponding arrows while the second arrow in *(B)* points toward the additional streptavidin mass. Bars, 10 nm. **C.** Bound biotinylated ferrtin molecules are reflecting the geometry of the underlying SbsB S-layer lattice. The vector pair indicates the orientation of the oblique (p1) lattice. Bar, 100 nm. (Reprinted from Moll, Huber, Schlegel, Pum, Sleytr, and Sára 2002, with permission from the National Academy of Sciences USA.).

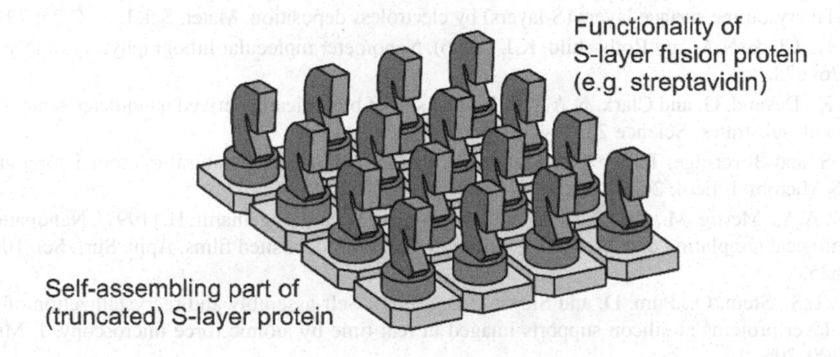

FIG. 6.6. Schematic drawing of S-layer fusion proteins with p1 lattice symmetry. All fused functionalities (e.g., streptavidin) shown as knights here exhibit the same position and orientation within the S-layer lattice.

4 CONCLUSIONS

In summary, these experiments have clearly shown that S-layers are perfectly suited to control the formation of nanoparticle arrays, either by direct precipitation from the vapour or liquid phase, or by binding preformed nanoparticles. The S-layer approach provides for the first time a biologically based fabrication technology for the self-assembly of molecular catalysts, templates, and scaffolds for the generation of ordered large-scale nanoparticle arrays for applications in electronic or optic devices.

Acknowledgments

Part of this work was supported by the Austrian Federal Ministry of Education, Science and Research, the Austrian Federal Ministry of Transport, Innovation and Technology (MNA-Network), the European Commission (Project, BIOAND IST-1999-11974), and the US Air Force office of Scientific Research (Projects F49620-03-1-0222 and FA9550-06-1-0208).

REFERENCES

Allred, D.B., Sarikaya, M., Baneyx, F. and Schwartz, D.T. (2005). Electrochemical nanofabrication using crystalline protein masks. Nano Lett. 5, 609–613.

Bergkvist, M., Mark, S.S., Yang, X., Angert, E.R. and Batt, C.A. (2004). Bionanofabrication of ordered nanoparticle arrays: effect of particle properties and adsorption conditions. J. Phys. Chem. B 108, 8241–8248.

Dieluweit, S., Pum, D. and Sleytr, U.B. (1998). Formation of a gold superlattice on an S-layer with square lattice symmetry. Supramol. Sci. 5, 15–19.

Dieluweit, S., Pum, D., Sleytr, U.B. and Kautek, W. (2005). Monodisperse gold nanoparticles formed on bacterial crystalline surface layers (S-layers) by electroless deposition. Mater. Sci. Eng. C 25, 727–732.

Douglas, K., Clark, N.A. and Rothschild, K.J. (1986). Nanometer molecular lithography. Appl. Phys. Lett. 48, 676–678.

Douglas, K., Devaud, G. and Clark, N.A. (1992). Transfer of biologically derived nanometer-scale patterns to smooth substrates. Science 257, 642–644.

Douglas, S. and Beveridge, T.J. (1998). Mineral formation by bacteria in natural microbial communities. FEMS Microbiol. Ecol. 26, 79–88.

Gorbunov, A.A., Mertig, M., Kirsch, R., Eichler, H., Pompe, W. and Engelhardt, H. (1997). Nanopatterning by biological templating and laser direct writing in thin laser deposited films. Appl. Surf. Sci. 109/110, 621–625.

Györvary, E.S., Stein, O., Pum, D. and Sleytr, U.B. (2003). Self-assembly and recrystallization of bacterial S-layer proteins at silicon supports imaged in real time by atomic force microscopy. J. Microsc. 212, 300–306.

Györvary, E., Schroedter, A., Talapin, D.V., Weller, H., Pum, D. and Sleytr, U.B. (2004). Formation of nanoparticle arrays on S-layer protein lattices. J. Nanosci. Nanotech. 4,115–120.

Hall, S.R., Shenton, W., Engelhardt, H. and Mann, S. (2001). Site-specific organization of gold nanoparticles by biomolecular templating. Chem. Phys. Phys. Chem. 3, 184–186.

Huber, C., Liu, J., Egelseer, E.M., Moll, D., Knoll, W., Sleytr, U.B. and Sára, M. (2006). Heterotetramers formed by an S-layer-streptavidin fusion protein and core-streptavidin as a nanoarrayed template for biochip development. Small 2, 142–150.

Hüttl, R., Ullrich, F., Wolf, G., Kirchner, A., Mertig. M. and Pompe, W. (2006). Calorimetric methods for catalytic investigations of novel catalysts based on metallized S-layer preparations. Thermochim. acta 440, 13–18.

Jarosch, M., Egelseer, E.M., Huber, C., Moll, D., Mattanovich, D., Sleytr, U.B. and Sára, M. (2001). Analysis of the structure-function relationship of the S-layer protein SbsC of Bacillus *stearothermophilus* ATCC 12980 by producing truncated forms. Microbiology 147, 1353–1363.

Küpcü, S., Sára, M. and Sleytr, U.B. (1995). Liposomes coated with crystalline bacterial cell surface protein (S-layers) as immobilization structures for macromolecules. Biochim. Biophys. Acta 1235, 263–269.

Mader, C., Küpcü, S., Sleytr, U.B and Sára, M. (2000). S-layer-coated liposomes as a versatile system for entrapping and binding target molecules. Biochim. Biophys. Acta 1463, 142–150.

Malkinski, L., Camley, R.E., Celinski, Z., Winningham, T.A., Whipple, S.G., and Douglas, K. (2003). Hexagonal lattice of 10-nm magnetic dots. J. Appl. Phys. 93, 7325–7327.

Mertig, M., Kirsch, R., Pompe, W. and Engelhardt, H. (1999). Fabrication of highly oriented nanocluster arrays by biomolecular templating. Eur. Phys. J. D 9, 45–48.

Mertig, M., Wahl, R., Lehmann, M., Simon, P., and Pompe, W. (2001). Formation and manipulation of regular metallic nanoparticle arrays on bacterial surface layers: an advanced TEM study. Eur. Phys. J. D 16, 317–320.

Messner, P., Pum, D., Sára, M., Stetter, K.O., and Sleytr, U.B. (1986). Ultrastructure of the cell envelope of the archaebacteria Thermoproteus tenax and Thermoproteus neutrophilus. J. Bacteriol. 166, 1046–1054.

Moll, D., Huber, C., Schlegel, B., Pum, D., Sleytr, U.B. and Sára, M. (2002). S-layer-streptavidin fusion proteins as template for nanopatterned molecular arrays. Proc. Natl. Acad. Sci. USA 99, 14646–14651.

Naik, R.R., Stringer, S.J., Agarwal, G., Jones, S.E. and Stone, M.O. (2002). Biomimetic synthesis and patterning of silver nanoparticles. Nature Mater. 1, 169–172.

Naik, R.R., Jones, S.E., Murray, C.J., McAuliffe, J.C., Vaia, R.A., and Stone, M.O. (2004). Peptide templates for nanoparticle synthesis derived from polymerase chain reaction-driven phage display. Adv. Funct. Mater. 14, 25–30.

Neubauer, Pentzien, S., Reetz, S., Kautek, W., Pum, D. and Sleytr, U.B. (1997). Pulsed-laser metal contacting of biosensors on the basis of crystalline enzyme-protein layer composites. Sens. Actuat. B 40, 231–236.

Panhorst, M., Brückl, H., Kiefer, B., Reiss, G., Santarius, U. and Guckenberger, R. (2001). Formation of metallic surface structures by ion etching using a S-layer template. J. Vac. Sci. Technol. B 19, 722–724.

Pompe, W., Mertig, M., Kirsch, R., Wahl, R., Ciachi, L.C., Richter, J., Seidel, R. and Vinzelberg, H. (1999). Formation of metallic nanostructures on biomolecular templates. Zeitschrift für Metallkunde 90, 1085–1091.

Pum, D. and Sleytr, U.B. (1995). Monomolecular reassembly of a crystalline bacterial cell surface layer (S-layer) on untreated and modified silicon surfaces. Supramol. Sci. 2, 193–197.

Sára, M., Pum, D., Küpcü, S., Messner, P. and Sleytr, U.B. (1994). Isolation of two physiologically induced variant strains of *Bacillus stearothermophilus* NRS 2004/3a and characterization of their S-layer lattices. J. Bacteriol. 176, 848–860.

Sára, M. (2001). Conserved anchoring mechanisms between crystalline cell surface S-layer proteins and secondary cell wall polymers in Gram-positive bacteria. Trends Microbiol. 9, 47–49.

Sára, M., Pum, D., Schuster, B. and Sleytr, U.B. (2005). S-layers as patterning elements for application in nanobiotechnology. J. Nanosci. Nanotech. 5, 1939–1953.

Schuster, B. and Sleytr, U.B. (2000). S-layer-supported lipid membranes. Rev. Mol. Biotechnol. 74, 233–254.

Shenton, W., Pum, D., Sleytr, U.B. and Mann, S. (1997). Biocrystal templating of CdS superlattices using self-assembled bacterial S-layers. Nature 389, 585–587.

Sleytr, U.B. and Messner P. (1989). Self-assemblies of crystalline bacterial cell surface layers. In H. Plattner (Ed.): Electron Microsc. of subcellular dynamics. CRC Press, Boca Raton, FL, pp 13–31.

Sleytr, U.B., Bayley, H., Sára, M., Breitwieser, A., Küpcü, S., Mader, C., Weigert, S., Unger, F.M., Messner, P., Jahn-Schmid, B., Schuster, B., Pum, D., Douglas, K., Clark, N.A., Moore, J.T., Winningham, T.A., Levy, S., Frithsen, I., Pankovc, J., Beagle, P., Gillis, H.P., Choutov, D.A. and Martin, K.P. (1997). Applications of S-layers. FEMS Microbiol. Rev. 20, 151–175.

Sleytr, U.B. and Beveridge, T.J. (1999). Bacterial S-layers. Trends Microbiol. 7, 253–260.

Sleytr, U.B., Sára, M., Mader, C., Schuster, B. and Unger, F.M. (2001). Use of a secondary cell wall polymer of prokaryotic microorganisms, Int. Patent WO 01/81425.

Sleytr, U.B., Egelseer, E.M., Pum, D. and Schuster, B. (2004). S-Layers. In C. Niemeyer, and C. Mirkin (Eds.), Nanobiotechnology. John Wiley-VCH Verlag, Weinheim, pp 77–92.

Sleytr, U.B., Sára, M., Pum, D. and Schuster, B. (2005). Crystalline bacterial cell surface layers (S-layers): a versatile self-assembly system. In A. Ciferri (Ed.): Supramolecular Polymers. Taylor & Francis, Boca Raton, FL, pp 583–616.

Toca-Herrera, J.L., Krastev, R., Bosio, V., Küpcü, S., Pum, D., Fery, A., Sára, M. and Sleytr, U.B. (2005). Recrystallization of bacterial S-layers on flat polyelectrolyte surfaces and hollow polyelectrolyte capsules. Small 1, 339–348.

Vyalikh, D.V., Danzenbächer, S., Mertig, M., Kirchner, A., Pompe, W., Dedkov, Y.S. and Molodtsov, S.L. (2004). Electronic structure of regular bacterial surface layers. Phys. Rev. Lett. 93, 238103.

Wahl, R., Mertig, M., Raff, J., Selenska-Pobell, S. and Pompe, W. (2001). Electron-beam induced formation of highly ordered palladium and platinum nanoparticle arrays on the S-layer of *Bacillus sphaericus* NCTC 9602. Adv. Mat. Sci. Technol. 13, 736–740.

Wahl, R., Engelhardt, H., Pompe, W. and Mertig, M. (2005). Multivariate statistical analysis of two-dimensional metal cluster arrays grown in vitro on a bacterial surface layer. Chem. Mater. 17, 1887–1894.

Winningham, T.A., Gillis, H.P., Choutov, D.A., Martin, K.P., Moore, J.T., and Douglas, K. (1998). Formation of ordered nanocluster arrays by self-assembly on nanopatterned Si(100) surfaces. Surf. Sci. 406, 221–228.

Electronics for Proteomics

7

Electrochemical Biosensing of Redox Proteins and Enzymes

Qijin Chi, Palle S. Jensen, and Jens Ulstrup

1 INTRODUCTION

As a crucial part of nanoscale science and technology, molecular electronics is rooted in the concept of molecular charge transfer (Aviram and Ratner 1974; Jortner and Ratner 1997; Hush 2003). Two essential steps involved in bottom-up fabrication for molecular electronic devices are organizing molecules into nanoscale structures, and interfacing such nanostructures with macroscopically addressable components (e.g. metal and semiconductor electrodes). The function of molecular electronic devices thus rests fundamentally on design and control of charge transfer through target molecules and across the interfaces between molecules and macroscopic electrodes (Nitzan and Ratner 2003; Flood, Stoddart, Steuerman, and Heath 2004). As a result, detailed understanding of the fundamental charge transfer process becomes a central issue for designing efficient molecular devices.

Integration of biological molecules with inorganic nanomaterials (e.g. metal nanoparticles, nanoscale electrodes, and carbon nanotubes) has opened new avenues to assemble bioelectronic nanostructures for a variety of applications, ranging from drug delivery to biosensors (Katz and Willner 2004a, b). Redox proteins and enzymes with vital roles in biological respiration and photosynthesis

A. Offenhäusser and R. Rinaldi (eds.), *Nanobioelectronics - for Electronics, Biology, and Medicine*,
DOI: 10.1007/978-0-387-09459-5_8, © Springer Science+Business Media, LLC 2009

are among the most attractive biomacromolecules for bioelectronics, because the biological action (e.g. electron transfer and catalysis) is converted into electric signals that can be detected directly by sensitive but economically attractive means such as electrochemistry (Armstrong and Wilson 2000; Zhang, Chi, Kuznetsov, Hansen, Wackerbarth, Christensen, Andersen, and Ulstrup 2002; Davis, Morgan, Wrathmell, Axford, Zhao, and Wang 2005). Understanding of protein charge transfer has, further, mostly been based on average results of macroscopic measurements. With the development of supersensitive microscopies such as scanning tunneling microscopy (STM) and surface spectroscopies such as surface enhanced Raman (SERS) and fluorescence spectroscopy, however, the time has come that protein-based electronic and energy transport can be approached at the nanoscale and single-molecule level.

In order for electrochemical approaches to protein charge transfer at the nanoscale and single-molecule level, protein molecules must be organized in appropriate configurations by which experimental measurements and theoretical framing can become feasible. Figure 7.1 shows schematic representations of *three* possible configurations in which the protein molecule is confined between

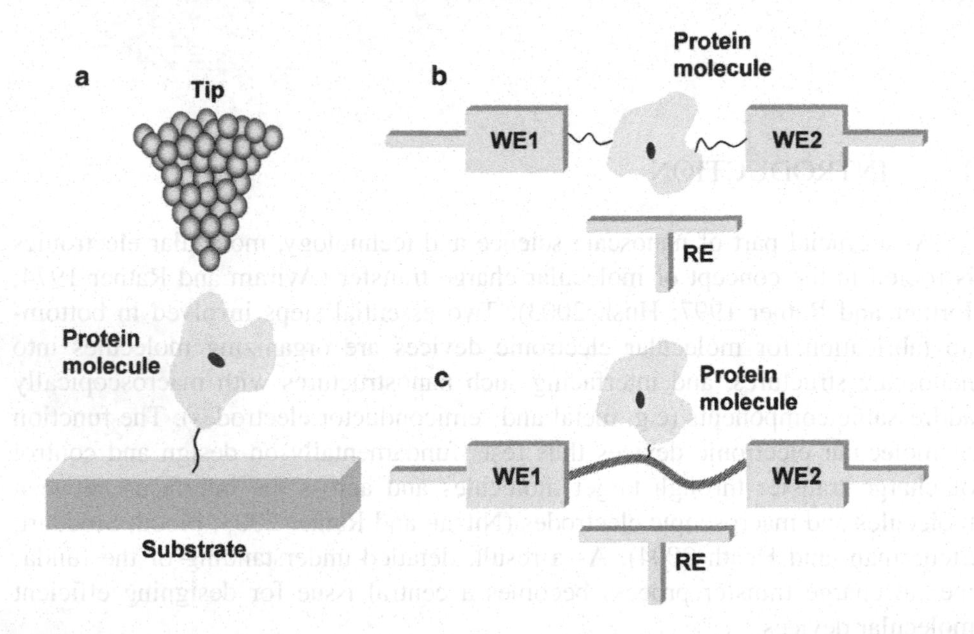

FIG. 7.1. Schematic illustration of three possible configurations that could be applied in electrochemical approaches to nanoscale and single-molecule charge transfer of proteins. ***a.*** Protein molecule assembled between a STM tip and a substrate. ***b.*** Direct organization in the nanogap between two working electrodes. ***c.*** Similar to *(b)* but with a nanowire or nanorod as support. WE and RE denote working electrode and reference electrode, respectively. WE1, WE2, and RE correspond, respectively, to source, drain, and gate electrodes in a molecular transistor.

two enclosing electrodes. In *the first configuration* (Fig. 7.1a), the protein molecule is assembled on a substrate with well-defined surface (e.g. a single-crystal electrode). A scanning probe (e.g. STM tip) can approach the target protein molecule for probing of local electronic structure and current–voltage correlations. This configuration can further be combined with electrochemistry, ascertaining that measurements are performed under electrochemical potential control in a biologically favorable medium. *The second configuration* (Fig. 7.1b) illustrates that the protein molecule is directly confined in a nanogap consisting of two nanoscale working electrodes. By using a bipotentiostat, charge transfer through the protein molecule and the protein–metal interfaces can in principle be controlled and measured. This configuration has not yet been established for proteins; however, although similar configurations have been extensively used in organic molecular transistor studies (Nitzan and Ratner 2003). The application of this configuration is currently hampered by challenges of assembling protein molecules in the nanogap and of detecting sensitivity. *The third configuration* (Fig. 7.1c) is similar to the second one, but a nanotube or nanowire is inserted in the nanogap as a support for the protein molecule assembly. This has made the configuration more practical, illustrated for example by glucose oxidase on a carbon nanotube connecting two nanoelectrodes (Besteman, Lee, Wiertz, and Dekker 2003). We have focused on interfacial and intramolecular charge transfer of redox-sensing metalloproteins using a broad variety of approaches over the past years. Most attention has been given to electrochemical and STM configurations, which constitute the major part of this chapter (Zhang Chi, Kuznetsov, Hansen, Wackerbarth, Christensen, Andersen, and Ulstrup 2002; Chi, Zhang, Jensen, Christensen, and Ulstrup 2006). In this report, we emphasize current status and future prospects of the field extending to concerted efforts of theoretical formalism and experimental approaches.

2 THEORETICAL CONSIDERATIONS

2.1 ELECTROCHEMICAL ELECTRON TRANSFER

Electrochemistry has offered a range of convenient approaches to many issues in (redox) chemistry. In contrast to electron transfer (ET) in homogeneous solution, electrochemical ET is complicated by the nature of the anisotropic and inhomogeneous conditions (or boundary conditions) at liquid/solid interfaces (Marcus and Sutin 1985; Kuznetsov and Ulstrup 1999). Theoretical formulation should in principle take into account all the boundary conditions, although in reality only approximations are feasible.

One of the advantages in electrochemical approaches is that the ET reaction can be directly monitored by the current flow. The current is a total outcome of contributions from all electronic levels of the electrode (Fig. 7.2a), but mostly dominated by those around the Femi level (Zhang, Chi, Albrecht, Kuznetsov, Grubb, Hansen, Wackerbarth, Welinder, and Ulstrup 2005; Chi, Zhang, Jensen, Christensen, and Ulstrup 2006). The redox level of the molecule, with the initial level at equilibrium above the Fermi level (ε_F), is strongly coupled to the environment. By nuclear configurational fluctuations the initial level is lowered to match populated levels in the electrode where ET proceeds, and subsequently the electron is trapped on the molecule, followed by relaxation to values well below the Fermi level of the electrode.

When the redox molecules are immobilized on the electrode surface, the following equations hold broad validity for the correlation between the (cathodic) current density and all factors affecting the process:

$$j_{cath}(\eta) = e\Gamma_{ox}^{(1-\alpha)}\Gamma_{red}^{\alpha}\int d\varepsilon\rho(\varepsilon)f(\varepsilon)k_{ET}(\varepsilon,\eta) \qquad (1)$$

$$k_{ET}(\varepsilon,\eta) = \kappa_{el}(\varepsilon,\eta)\frac{\omega_{eff}}{2\pi}\exp\left(-\frac{\left[E_r + e\eta - (\varepsilon - \varepsilon_F)\right]^2}{4E_r k_B T}\right) \qquad (2)$$

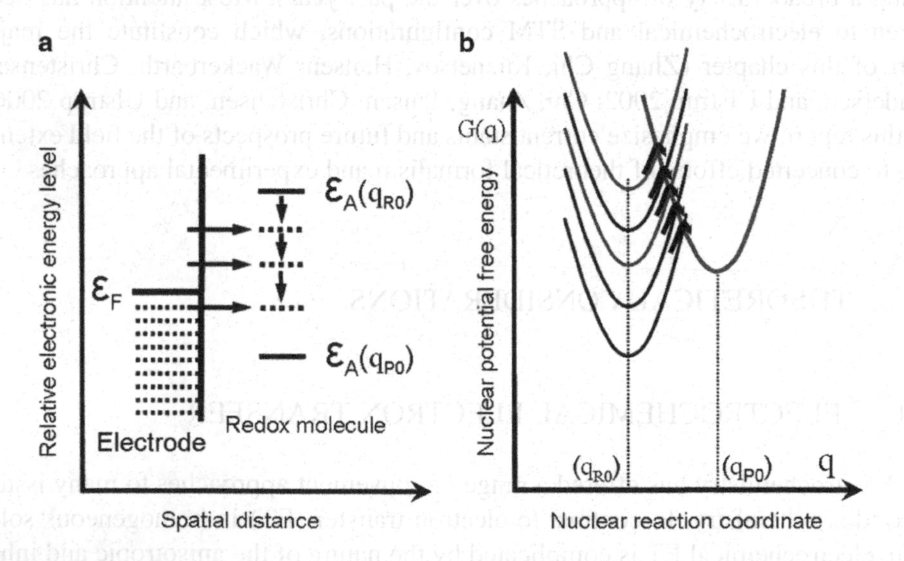

FIG. 7.2. One-dimensional schematic representation of electronic energy levels at molecule/electrode interfaces *(a)* and corresponding free energy surfaces *(b)* for electron transfer in electrochemical configuration (Reproduced with permission from ref Chi, Q., Zhang, J., Jensen, P.S., Christensen, H.E.M. and Ulstrup, J. (2006). Long-range interfacial electron of metalloproteins based on molecular wiring assemblies. Faraday Discussions 131, 181–195.)

$$\kappa_{el}(\varepsilon,\eta)=\left[T_{\varepsilon A}(\varepsilon,\eta)\right]^2\sqrt{\frac{4\pi^3}{E_r k_B T\hbar^2\omega_{eff}}} \tag{3}$$

where ε is the particular energy level at the metal electrode, η is the overpotential, Γ_{ox} and Γ_{red} are the population of adsorbed molecules in their oxidized and reduced form, respectively; $\rho(\varepsilon)$ is the electronic level density, $f(\varepsilon)$ is the Fermi function, ω_{eff} is the effective vibrational frequency of all the nuclear modes contributing to the nuclear reorganization free energy (E_r), α is the transfer coefficient, and $T_{\varepsilon A}(\varepsilon, \eta)$ is the electron exchange factor describing the electronic coupling between the energy level of the electrode and the molecular redox level. Other symbols have their usual meaning.

There are *three conspicuous differences* in comparison with ET in homogeneous conditions. *First*, the electrode incorporates a continuous manifold of electronic states (Fig. 7.2), whereas only a single pair of states is involved in homogenous ET. As a consequence, some phenomena (e.g. an inverted free energy region) are absent from electrochemical ET. Another consequence of this difference is that the electronic transmission coefficient, $\kappa_{el}(\varepsilon,\eta)$, is always infinitesimally small. However, a macroscopically large number of levels ($\approx \rho(\varepsilon_F)k_BT$) can contribute to the integral current form (Eq. 1) by the continuous nature of the electronic spectrum of the metal electrode. *Second*, the nature of the electronic wave functions is different. These are spatially localized for ET in homogeneous solution, but two-dimensionally delocalized in electrochemical ET. Therefore, the electrochemical tunneling factor can be conveniently approached by invoking electronic density rather than individual level wave functions. *Finally*, the driving force is tuned by the overpotential applied to the electrode, being zero at the equilibrium redox potential.

2.2 REDOX PROCESSES IN ELECTROCHEMICAL STM

A combination of electrochemistry and STM has provided a sensitive means to study the nature of interfacial ET, by which both molecular electronic structures and current–voltage relations can be probed. To this end, redox molecules are organized on the electrode surface with a suitable orientation. The theoretical framework for understanding of the STM redox process has been pursued over the past decade (Kuznetsov, Sommer-Larsen, and Ulstrup 1992; Schmickler and Widrig 1992; Friis, Kharkats, Kuznetsov, and Ulstrup 1998; Sumi 1998; Friis, Andersen, Kharkats, Kuznetsov, Nichols, Zhang, and Ulstrup 1999; Kuznetsov and Ulstrup 2000; Zhang, Chi, Kuznetsov, Hansen, Wackerbarth, Christensen, Andersen, and Ulstrup 2002; Zhang, Kuznetsov, and Ulstrup 2003). The two-step model, which can be viewed as an evolution of electrochemical ET notions to the

STM configuration, has emerged as one model suitable for interpretation of experimental observations for various systems. A schematic diagram is shown in Fig. 7.3 to illustrate the energy levels of the STM substrate (i.e. working electrode), the STM tip, and a redox molecule in the substrate-tip gap. The energy levels are all modified by changing either the substrate potential or the bias voltage. When the bias voltage is fixed, the substrate and tip energy levels are shifted in parallel. As a result, the redox level is tuned by the overpotential relative to the substrate and tip Fermi levels.

The bias voltage $V_{bias} = E_T - E_S$ (where E_S is the substrate and E_T the tip potential) can either be small (defined as $|e\gamma V_{bias}| < (E_r + e\xi\eta)/(1-\gamma)$) or large ($|e\gamma V_{bias}| > (E_r + e\xi\eta)/(1-\gamma)$). ξ is the fraction of the substrate-solution potential drop and γ the fraction of the bias voltage drop at the molecular redox center. In the case of small bias voltage (negative), an initially oxidized (vacant) level is above the tip Fermi level and an initially reduced (occupied) level is below both the substrate and tip Fermi levels (Fig. 7.3a). As for electrochemical ET, nuclear configurational fluctuations bring the redox level into the energy region close to the Fermi level of the electrodes. This induces a two-step ET process. Interfacial ET proceeds first from the tip to the vacant redox level and then from the temporarily occupied redox level to the substrate (Fig. 7.3a). A corresponding ET sequence in the opposite direction would occur at positive bias voltage. The tunneling current (I_t) is thus gated by the redox level, which can be directly displayed by changes either in STM image contrast or in the current-overpotential relations. In the case of large bias voltage (negative), the reduced level of the molecule at equilibrium

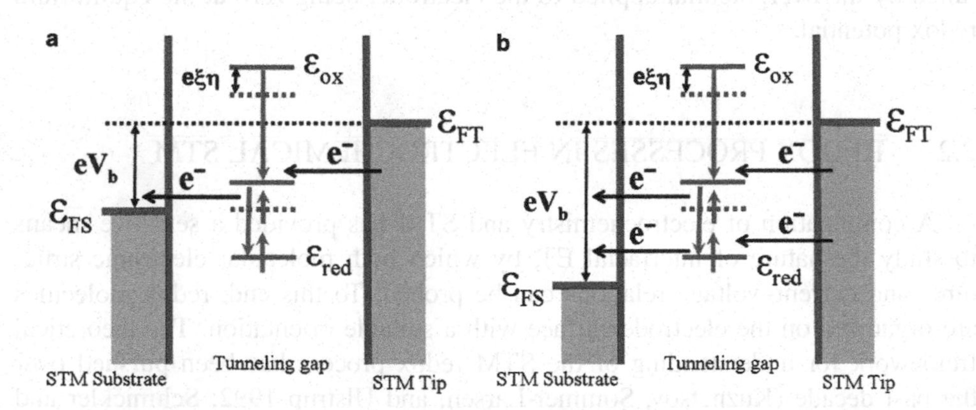

FIG. 7.3. Schematic energy diagrams of two-step ET mechanisms in electrochemical STM redox processes: *(a)* and *(b)* correspond, respectively, to the cases with small and large negative bias voltages applied between the STM substrate and tip (Reproduced with permission from ref Chi, Q., Zhang, J., Jensen, P.S., Christensen, H.E.M. and Ulstrup, J. (2006). Long-range interfacial electron of metalloproteins based on molecular wiring assemblies. Faraday Discussions 131, 181–195, with modifications.)

remains above the Fermi level of the substrate (Fig. 7.3b). The situation becomes more complicated because the large bias voltage notion is determined not only by the ratio of eV_{bias}/E_r but also by the overpotential. The tunneling current is expected first to rise with increasing overpotential, but the reduced level is still trapped in the region between the Fermi levels and transmits electrons in an activationless mode. Upon further increasing overpotential, both the oxidized and reduced levels are in the energy window with coherent and activationless electron tunneling, independent of both the overpotential and bias voltage. If the overpotential continues to rise, the reduced level will eventually traverse the substrate Fermi level and the current drop. The overall current-overpotential relation thus consists of an ascending branch at small overpotential and a descending branch at large overpotential, bridged by a tunnelling current plateau (Zhang, Grubb, Hansen, Kuznetrov, Boisen, Wackerbarth, and Ulstrup 2003).

The *nonadiabatic limit* of weak electronic coupling between the redox center and the enclosing electrodes generally applies for interfacial ET of proteins. Here our discussion is focused on the case of small bias voltages. The dependence of tunneling (I_t) on the *effective* overpotential ($\xi\eta$) and the bias voltage (V_{bias}) can be approximately expressed by two-step ET rate constants in a relatively simple combination.

$$I_t \approx e\frac{\vec{k}^{o/r}\vec{k}^{r/o}}{\vec{k}^{o/r}+\vec{k}^{r/o}} \tag{4}$$

where $\vec{k}^{o/r}$ and $\vec{k}^{r/o}$ are the rate constants for ET between the tip and the protein and between the protein and the substrate, respectively; an arrow indicates the electron flow direction following the pathway shown in Fig. 7.3A. $\vec{k}^{o/r}$ and $\vec{k}^{r/o}$ have the following forms:

$$\vec{k}^{o/r} = \kappa_t\rho_t\frac{\omega_{eff}}{2\pi}\frac{2k_BT}{\alpha_t}\exp\left(-\frac{(E_r - e\xi\eta - e\gamma V_{bias})^2}{4E_rk_BT}\right) \tag{5}$$

$$\vec{k}^{r/o} = \kappa_s\rho_s\frac{\omega_{eff}}{2\pi}\frac{2k_BT}{\alpha_s}\exp\left(-\frac{(E_r - eV_{bias} + e\xi\eta + e\gamma V_{bias})^2}{4E_rk_BT}\right) \tag{6}$$

where κ_t and κ_s are the electronic transmission coefficients, cf. Eqs. 1 to 3, for ET between the tip and the protein, and between the protein and the substrate, respectively; ρ_t and ρ_s are the electronic level densities of the tip and the substrate; α_t and α_s are the ET transfer coefficients, respectively, between the tip and the protein and between the protein and the substrate; ξ is the fraction of the substrate-solution potential drop and γ the fraction of the bias voltage drop at the molecular redox center. Other symbols are defined above or have their usual meaning. These equations can be used in theoretical computations in comparisons with experimental observations (Chi, Zhang, Jensen, Christensen, and Ulstrup 2006).

Equations 4 to 6 apply in the *nonadiabatic* limit of weak interactions between the (protein) molecule and the enclosing electrodes, $\kappa_{eff} \approx \kappa_i \rho k_{BT} << 1$ (i = s, t). This limit is often representative of protein interfacial ET. Two important differences emerge in the opposite limit, i.e., *adiabatic* limit of strong interactions. One is that the effective electronic transmission coefficient, κ_{eff}, approaches unity, leaving only the effective frequency $\omega_{eff}/2\pi$ in the pre-exponential factors of Equations 5 and 6. The other difference is that a large number of electrons, $n_{el} >> 1$ instead of only a single electron is transmitted during the relaxation of the redox level through the energy window between the Fermi levels of the substrate and tip. This can raise the current in a single STM electronic transmission even by orders of magnitude (see Section 4.2).

3 EXPERIMENTAL APPROACHES

Major considerations of experimental approaches to protein ET in electrochemical and STM configurations include: (a) establishment of an ultraclean environment; (b) choice of suitable target proteins; (c) preparation and handling of single-crystal electrodes; (d) assembly of protein molecules at liquid/solid interfaces; (e) fabrication of high-quality STM tips; and (f) development of sensitive accessory apparatus for electrochemical and STM measurements.

Target proteins could in principle cover all redox protein and enzymes, but heme proteins, blue copper proteins, and iron-sulphur proteins have been in focus. Here we present data for heme and blue copper proteins. Figure 7.4 shows three-dimensional crystallographic structures of *pseudomonas aeruginosa* azurin (a), the trimeric copper redox enzyme *Alcaligenes xylosoxidans* nitrite reductase (b), horse heart cytochrome *c* (c), and the two-center heme protein *Pseudomonas stutzeri* cytochrome C_4 (d). These proteins and enzyme represent increasing complexity in structure from *single* redox center to *multiple* redox centers, providing opportunity and challenges for systematic investigations.

3.1 MATERIALS AND REAGENTS

The Au(111) electrodes used in both electrochemical and STM experiments were prepared from polycrystalline gold wires (≥99.99%) by a hydrogen flame (Hamelin 1996; Chi, Zhang, Friis, Andersen, and Ulstrup 1999; Chi, Zhang, Nielsen, Friis, Chorkendorff, Canters, Andersen, and Ulstrup 2000; Zhang, Chi, Nielsen, Friis, Andersen, and Ulstrup 2000). The quality of the electrodes was

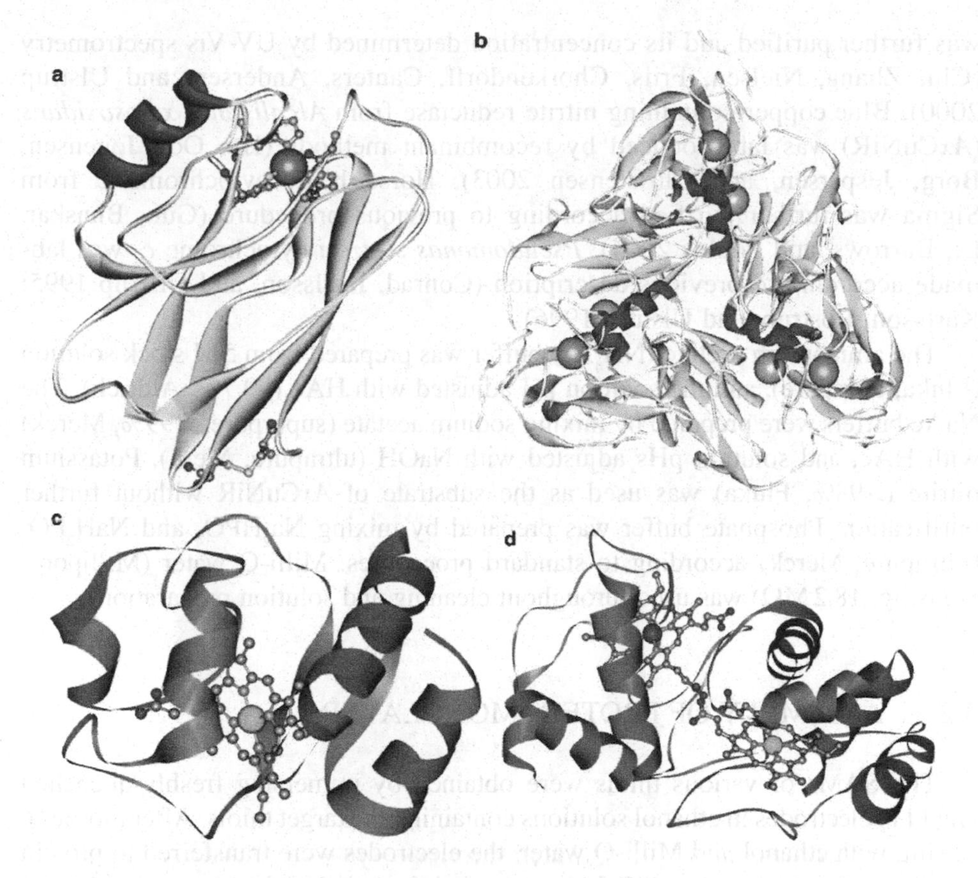

FIG. 7.4. Three-dimensional crystallographic structures of some target proteins. **a.** *Pseudomonas aruginosa* azurin. **b.** *Alcaligenes xylosoxidans* blue copper nitrite reductase. **c.** Horse hart cytochrome *c*. **d.** *Pseudomonas stutzeri* cytochrome c_4. The red and blue ribbons in cytochrome c_4 illustrate the C- and N-terminal domain, respectively.

regularly checked by cyclic voltammograms recorded in 0.5 M H_2SO_4 and by STM characterization. The electrodes were treated by annealing and quenching with a hydrogen flame before use (Zhang, Bilic, Reimers, Hush, and Ulstrup 2005; Zhang, Demetriou, Welinder, Albrecht, Nichols, and Ulstrup 2005). Organic molecules used in preparation of self-assembled monolayers (SAMs) as supporting layers were largely from commercial sources. For example, straight-chain alkanethiols $CH_3(CH_2)_{n-1}SH$ (n = 4–18) with the highest available purity, carboxyl-containing thiols, cystamine, and 4-methylbenzenethiol (MBT) were all obtained from Sigma and used as received.

Protein samples are either commercially available or lab-made using genetic engineering techniques. *Pseudomonas aeruginosa* azurin from Sigma was further purified and its concentration determined by UV-Vis spectrometry (Chi, Zhang, Nielsen, Friis, Chorkendorff, Canters, Andersen, and Ulstrup 2000). Blue copper-containing nitrite reductase from *Alcaligenes xylosoxidans* (*Ax*CuNiR) was lab-produced by recombinant methods (Ho, Ooi, Jøgensen, Borg, Jespersen, and Christensen 2003). Horse heart cytochrome *c* from Sigma was further purified according to previous procedure (Guo, Bhaskar, Li, Barrows, and Poulos 2004). *Pseudomonas stutzeri* cytochrome c_4 was lab-made according to previous description (Conrad, Karlsson, and Ulstrup 1995; Karlsson, Rostrup, and Ulstrup 1996).

The ammonium acetate (NH_4Ac) buffer was prepared from 5 M stock solution (Fluka, ultrapure), and the solution pH adjusted with HAc (99.7%, Aldrich). The NaAc buffers were prepared by mixing sodium acetate (superpure, > 99%, Merck) with HAc, and solution pHs adjusted with NaOH (ultrapure, Merk). Potassium nitrite (>98%, Fluka) was used as the substrate of *Ax*CuNiR without further purification. Phosphate buffer was prepared by mixing Na_2HPO_4 and NaH_2PO_4 (ultrapure, Merck) according to standard procedures. Milli-Q water (Millipore-Housing, 18.2 MΩ) was used throughout cleaning and solution preparations.

3.2 ASSEMBLY OF PROTEIN MONOLAYERS

The SAMs of various thiols were obtained by immersing freshly-quenched Au(111) electrodes in ethanol solutions containing the target thiols. After thorough rinsing with ethanol and Milli-Q water, the electrodes were transferred to protein solutions and incubated at 4°C for an appropriate period to form protein submonolayers or monolayers. The resulting electrodes were carefully rinsed with Milli-Q water and suitable buffer prior to measurements. Details are described in corresponding references for each protein: azurin (Chi, Zhang, Jensen, Christensen, and Ulstrup 2006; Chi, Zhang, Andersen, and Ulstrup 2001), *Ax*CuNiR (Zhang, Welinder, Hansen, Christensen, and Ulstrup 2003), cytochrome *c* (Jensen 2005), and cytochrome c_4 (Chi, Zhang, Christensen, and Ulstrup 2006).

3.3 INSTRUMENTAL METHODS

Electrochemical measurements were carried out at room temperature using an Autolab system (Eco Chemie, Netherlands) controlled by the general purpose electrochemical system software. A three-electrode system, consisting of a platinum coiled wire as counter electrode (CE), a reversible hydrogen electrode (RHE) as reference electrode (RE), and a Au(111)-based working electrode (WE), was used with the WE in a hanging-meniscus configuration. The RHE was checked versus a saturated calomel electrode (SCE) after each measurement. All electrode potentials are reported relative to SCE. Purified argon (Chrompack, 5 N) was applied to purge dioxygen from electrolyte solutions before the measurements, and the gas stream maintained over the solution during the measurements.

STM measurements were performed with a PicoSPM system (Molecular Imaging Co., USA) equipped with a bipotentiostat for potential control of both substrate and tip. Electrochemical control was conducted in a lab-designed cell with a three-electrode system as for regular electrochemical measurements. The tips were prepared from tungsten or Pt/Ir wires (ϕ25 mm) by electrochemical etching and insulated with apiezon wax to reduce or eliminate Faradaic currents.

4 EXPERIMENTAL OBSERVATIONS AND THEORETICAL SIMULATIONS

Appropriate assembly of proteins on solid electrodes is essential for device-like charge transfer studies. Protein molecules should be organized in such a way that favorable molecular orientation with full retention of their biological function and effective electronic coupling between the protein redox center and the addressable electrode is achieved. This requirement currently remains a daunting challenge for most proteins due to their structural complexity and conformational flexibility. Molecular wiring assembly has offered some solutions to coupling the biological redox center physically and electronically to the electrode surface. Molecular-wiring assembly generally refers to the use of small organic molecules to confine protein molecules on electrode surfaces for two-dimensional protein array formation. Small organic molecules, serving both as a linker and an ET-assisting bridge, should possess functional groups at both ends of the molecule. One group links to the electrode surface, while the other one interacts specifically with the protein. This strategy has achieved notable success for different types of redox proteins and enzymes. The four cases discussed below reflect the efficiency of the strategy.

4.1 CASE OBSERVATION I: CYTOCHROME *C*

Cytochromes *c* are among the most studied redox proteins in bioelectrochemistry and biochemistry (Armstrong and Wilson 2000; Bertini, Cavallaro, and Rasato 2006). Single-heme–containing cytochrome *c* is structurally relatively simple, but essential for electron flow in respiratory chain reactions (Bertini, Cavallaro, and Rasato 2006). For example, horse heart cytochrome *c* is composed of a single polypeptide chain containing 105 amino acids (Fig. 7.4c). A *c*-type heme prosthetic group is covalently attached to the polypeptide chain by two thioether bonds connecting the side chains of cysteine 14 and cysteine 17 to pyrrole rings. Electrostatic interaction is expected to play important roles in the interactions of cytochrome *c* with its natural redox partners under physiological conditions. This has inspired use of electrostatic immobilization of cyt *c* on electrode surfaces modified with anionic monolayers for *in vitro* ET investigations.

Electron transfer patterns of cyt *c* confined to various functionalized SAMs on gold and silver electrodes have been systematically studied by electrochemistry and different types of spectroscopies (Tarlov and Bowden 1991; Song, Clark, Bowden, and Tarlov 1993; Terrettaz, Cheng, and Miller 1996; Feng, Imabayashi, Kikiuchi, and Niki 1997; Kasmi, Wallace, Bowden, and Binet, and Linderman 1998; Avila, Gregory, Niki, and Cotton 2000; Murgida and Hildebrandt 2002; Niki, Pressler, Sprinkle, Li, and Margoliash 2002; Wei, Liu, Dick, Yamamoto, He, and Waldeck 2002; Niki, Hardy, Hill, Li, Sprinkle, Margoliash, Tanimura, Nakamura, Ohno, Richards, and Gray 2003; Wackerbarth and Hildebrandt 2003; Murgida and Hildebrandt 2004; Murgida, Hildebrandt, Wei, He, Liu, and Waldeck 2004; Imbayashi, Mita, andKakiuchi 2005). These investigations have revealed several interesting issues for protein ET features at interfaces, including (1) dynamics of interaction of the protein with the organic supporting layer, (2) distance-dependent ET features, (3) configuration-gated ET patterns, and (4) effects of electric field and electrostatic force.

One of the latest developments in the case of cytochrome *c* is the conjugation of metal or semiconductor nanoparticles to form biomolecule-inorganic hybrid systems. Novel ET properties are expected to emerge from such systems, which could facilitate their applications in biomolecular electronics. Recent efforts on direct adsorption of cytochrome *c* on gold (Aubin-Tam and Hamad-Schifferli 2005; Jiang, Jiang, Jin, Wang, and Dong 2005; Jiang, Shang, Wang, and Dong 2005) and silver (Liu, Zhong, Gan, Fan, Li, and Matsuda 2003; Tom and Pradeep, 2005) show some intriguing observations concerning the protein's conformational flexibility and biomolecule-metal interfacial interaction. We have demonstrated that significant enhancement of long-range interfacial ET of this protein can be achieved by using a gold nanoparticle as ET relay (Jensen, Chi, Grumsen, Abad, Horsewell, Schiffrin, and Ulstrup 2007). Water-soluble gold nanoparticles were synthesized and two-dimensionally arrayed on Au(111) surfaces (Fig. 7.5a and b). The gold nanoparticles act as nanoarrays for further immobilization of cytochrome *c*. Long-range interfacial ET across a large physical distance, detected by a well-defined voltammetric response (Fig. 7.5c), has been facilitated with remarkable enhancement of the kinetics up to more than an order of magnitude in comparison with that in the absence of gold nanoparticles. Although the detailed mechanism of enhancement is not fully understood, gold nanoparticles appear to serve as ET relays enhancing the electronic coupling between the protein redox center and the electrode surface.

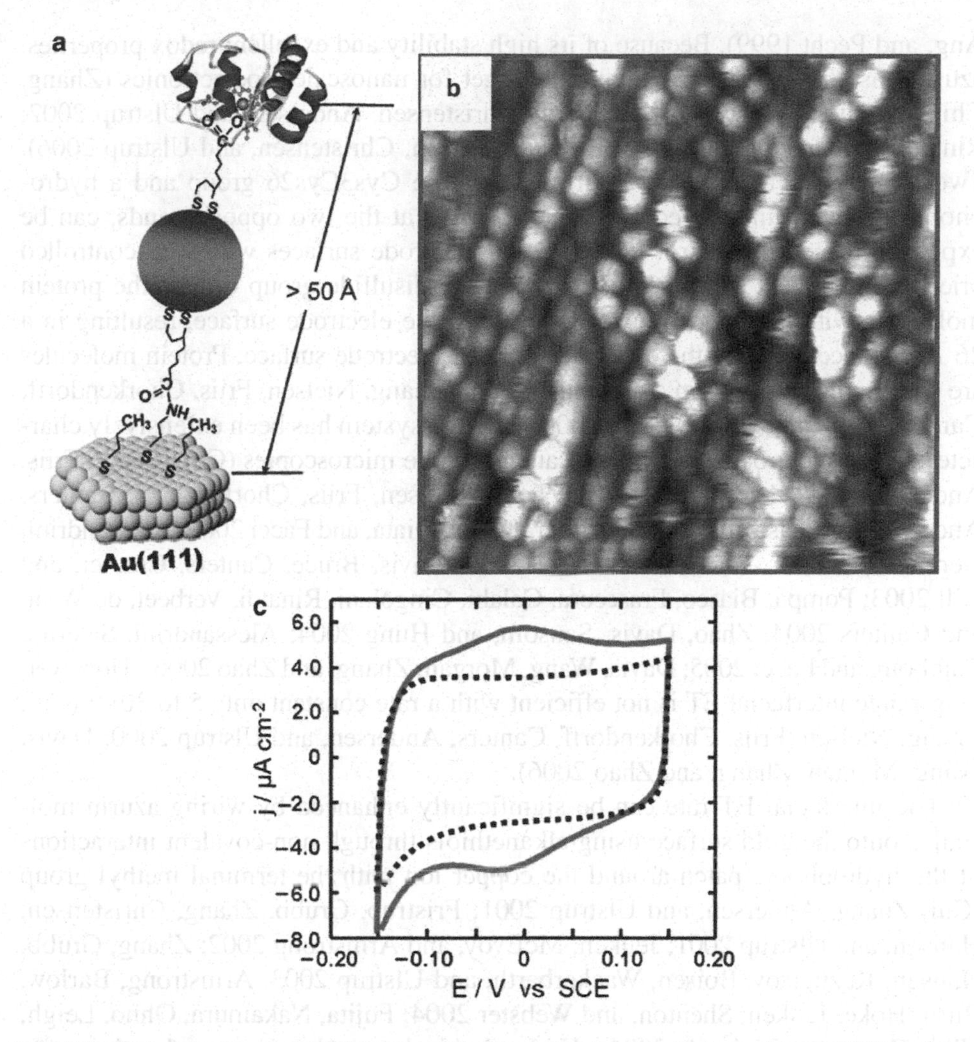

FIG. 7.5. Cytochrome *c* and gold nanoparticle bioconjugate assembled on the Au(111) surfaces. ***a.*** Schematic representation of the assembly. ***b.*** STM image of a nanoparticle array. ***c.*** Electrochemical cyclic voltammograms (CVs). Both STM image and CVs obtained in phosphate buffer (pH 7.2). Scan rate for CVs is 0.5 V s^{-1}.

4.2 CASE OBSERVATION II: AZURIN

Azurin is a blue single-copper protein that functions as an electron carrier physiologically associated with oxidative stress responses in many bacteria (Vijigenboom, Busch, and Canters 1997). This protein has been a long-standing model for exploring biological single-electron tunneling (Farver, Blatt, and Pecht 1982; Farver and Pecht 1989; Farver 1996; Gray and Winkler 1996; Farver, Li,

Ang, and Pecht 1999). Because of its high stability and excellent redox properties, azurin has also emerged as a favorite target for nanoscale bioelectronics (Zhang, Chi, Kuznetsov, Hansen, Wackerbarth, Christensen, Andersen, and Ulstrup 2002; Rinaldi and Cingolani 2004; Chi, Zhang, Jensen, Christensen, and Ulstrup 2006). Two structural features, i.e., a surface disulfide Cys3Cys26 group and a hydrophobic patch around the copper center located at the two opposite ends, can be exploited to confine azurin molecules on electrode surfaces with well-controlled orientations. Direct self-assembly through the disulfide group orients the protein molecules with the copper center opposite to the electrode surface, resulting in a 26 Å distance between the copper center and electrode surface. Protein molecules are well organized to form a monolayer (Chi, Zhang, Nielsen, Friis, Chorkendorff, Canters, Andersen, and Ulstrup 2000), and the system has been extensively characterized by electrochemistry and scanning probe microscopies (Chi, Zhang, Friis, Andersen, and Ulstrup 1999; Chi, Zhang, Nielsen, Friis, Chorkendorff, Canters, Andersen, and Ulstrup 2000; Schnyder, Kötz, Alliata, and Facci 2002; Alessandrini, Gerunda, Canters, Verbeet, and Facci 2003; Davis, Bruce, Canters, Crozier, and Hill 2003; Pompa, Biasco, Frasceera, Calabi, Cingolani, Rinaldi, Verbeet, de Waal, and Canters 2004; Zhao, Davis, Sansom, and Hung 2004; Alessandrini, Salerno, Frabboni, and Facci 2005; Davis, Wang, Morgan, Zhang, and Zhao 2006). However, long-range interfacial ET is not efficient with a rate constant only 5 to $30\,s^{-1}$ (Chi, Zhang, Nielsen, Friis, Chorkendorff, Canters, Andersen, and Ulstrup 2000; Davis, Wang, Morgan, Zhang, and Zhao 2006).

The interfacial ET rate can be significantly enhanced by wiring azurin molecules onto the gold surface using alkanethiols through non-covalent interactions of the hydrophobic patch around the copper ion with the terminal methyl group (Chi, Zhang, Andersen, and Ulstrup 2001; Fristrup, Grubb, Zhang, Christensen, Hansen, and Ulstrup 2001; Jeuken, McEvoy, and Armstrong 2002; Zhang, Grubb, Hansen, Kuznetsov, Boisen, Wackerbarth, and Ulstrup 2003; Armstrong, Barlow, Burn, Hoke, Jeuken, Shenton, and Webster 2004; Fujita, Nakamura, Ohno, Leigh, Niki, Gray, and Richards 2004). Hydrophobic interaction is proved to be sufficiently strong to wire azurin molecules with the copper center facing the electrode surface (Fig. 7.6a). Effective electronic coupling between the copper center and the electrode surface is thus established, leading to fast and reversible interfacial ET detected by voltammetric response (Fig. 7.6b). The alkanethiol in the present ET system serves both as a linker molecule and an ET-assisting bridge. The tunneling barrier can be adjusted by varying the bridge length, and the system can be used to study distance-dependent ET kinetics. The rate constant is virtually independent of the distance in the range of 4 to 8 methylene units, in which the ET kinetics is controlled mainly by conformational gating. Exponential distance decay is observed with longer bridges with a distance-decay factor (β) of 0.83 $Å^{-1}$ (Chi, Zhang, Andersen, and Ulstrup 2001; Chi, Zhang, Jensen, Christensen, and Ulstrup 2006). The decay factor is consistent with that for charge transfer through pure saturated hydrocarbons (Wold, Hagg, Rampi, and Frisbie 2002; Wang, Lee, and Reed 2003), indicative of efficient electronic coupling between the copper center and the gold electrode.

Topographic electronic structures of azurin monolayers are well resolved at the molecular level by in situ STM images (e.g. Fig. 7.7) acquired in buffer solutions. Each protein molecule is revealed as a contrasted spot with average diameter of ca 4 nm, close to expectations from the X-ray crystallographic dimensions. The distribution of molecules is quite uniform over the surface (Fig. 7.7a). Notably, high-resolution STM images where the ordered octanethiol is clearly seen underneath the azurin molecules (Fig. 7.7b), demonstrate a well-organized assembly in both organic and protein adlayers. This is crucial for observations of single-molecule protein ET.

Azurin sub-monolayer assembled on octanethiol-modified Au(111) surfaces is suitable for probing single-molecule protein ET (Chi, Farver, and Ulstrup 2005). STM observations were first performed with a larger scan area (e.g. 200 × 200 nm²) to obtain images of the azurin sub-monolayer with molecular resolution, and then focused on a few individual molecules by reducing the scan area. By keeping a constant bias voltage between the substrate and tip, STM imaging started with the substrate potential set at the equilibrium redox potential of azurin (i.e. zero overpotential). Imaging was continued toward either positive or negative overpotentials by adjusting the substrate and tip potentials in parallel (i.e., at constant bias voltage), and finally returned to the equilibrium potential. As a result, a series of STM images were acquired at various overpotentials (e.g. Fig. 7.8). The single-molecule contrast is clearly tuned by the redox state of azurin. The

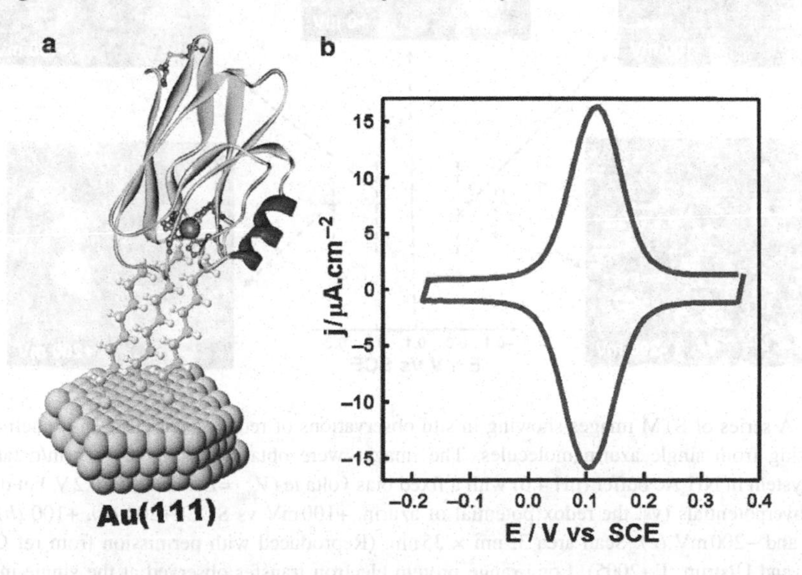

FIG. 7.6. Assembly and electrochemical response of Azurin on a Au(111) surface. ***a.*** Schematic representation of molecular wiring assembly of Azurin. ***b.*** Corresponding voltammetric response. The cyclic voltammogram (b) obtained in NH_4Ac buffer with the scan rate $2\,V\,s^{-1}$.

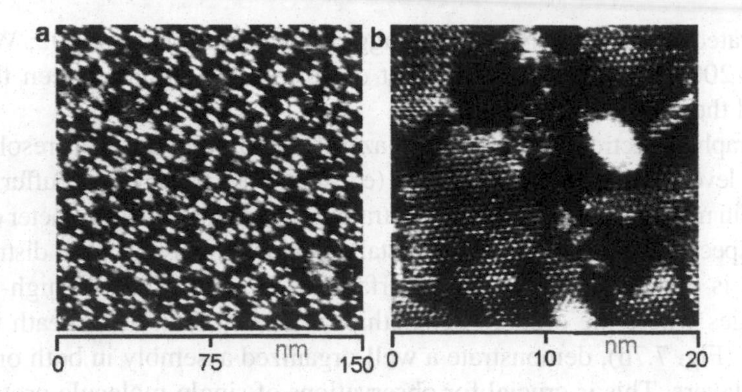

FIG. 7.7. STM images of azurin molecules wired on octanethiol-modified Au(111) surfaces. The images recorded in NH_4Ac buffer (pH 4.6) with a constant current mode: $I_t = 0.1$ nA, $V_b = -0.25$ V, $E_w = +0.20$ V (vs. SCE). Scan area 150 nm × 150 nm *(a)*, and 20 nm × 20 nm *(b)* (Reproduced with permission from ref Chi, Q., Zhang, J., Jensen, P.S., Christensen, H.E.M. and Ulstrup, J. (2006). Long-range interfacial electron of metalloproteins based on molecular wiring assemblies. Faraday Discussions 131, 181–195.)

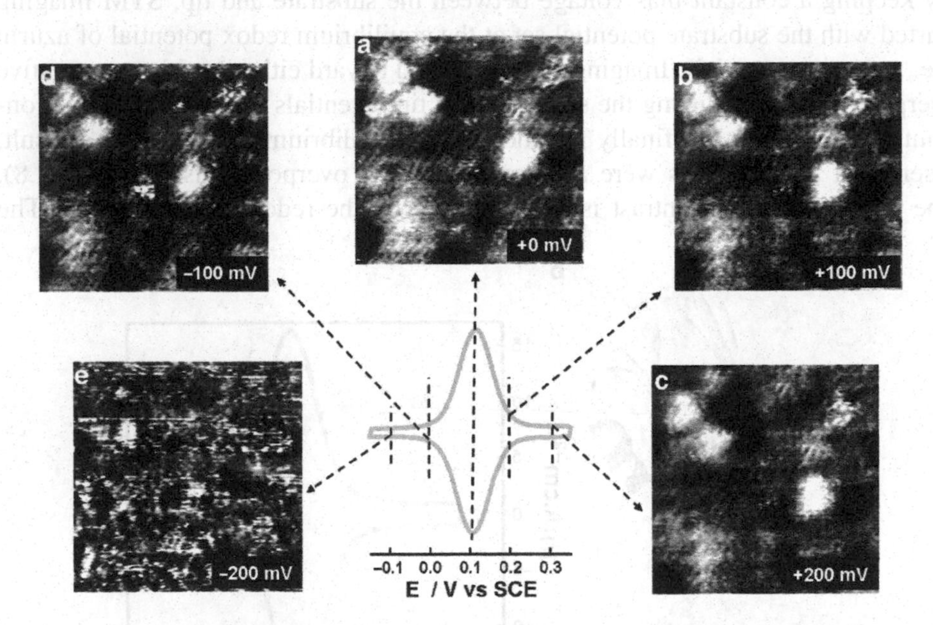

FIG. 7.8. A series of STM images showing in situ observations of redox-gated electron-tunneling resonance arising from single azurin molecules. The images were obtained using the azurin/octanethiol/Au(111) system in NH_4Ac buffer (pH 4.6) with a fixed bias voltage ($V_{bias} = E_T - E_S$) of −0.2 V, but different substrate overpotentials (vs. the redox potential of azurin, +100 mV vs SCE) +200 *(a)*, +100 *(b)*, 0 *(c)*, −100 *(d)*, and −200 mV *(e)*. Scan area 35 nm × 35 nm. (Reproduced with permission from ref Chi, Q., Farver, O. and Ulstrup, J. (2005). Long-range protein electron transfer observed at the single-molecule level: in situ mapping of redox-gated tunneling resonance. Proc. Natl. Acad. Sci. USA 102, 16203–16208 with modifications.)

current contrast-overpotential relation in Fig. 7.9a shows a maximum around the equilibrium redox potential. The current contrast decreases upon applying either positive or negative overpotentials, but the effect is asymmetric, the decay being faster at negative overpotentials. The maximum ON/OFF current ratio is ≈9, which is comparable with the cases of small redox molecules (Tao 1996; Gittins, Bethell, Schiffrin, and Nichols 2000; Haiss, Nichols, Higgins, Bethell, Hoobenreich, and Schiffrin 2003; Albrecht, Guckian, Ulstrup, and Vos 2005).

The asymmetric overpotential dependence can be represented formally by introducing an exponential overpotential-dependent pre-factor in the $\vec{k}^{o/r}$ expression form (Eq. 7) to obtain:

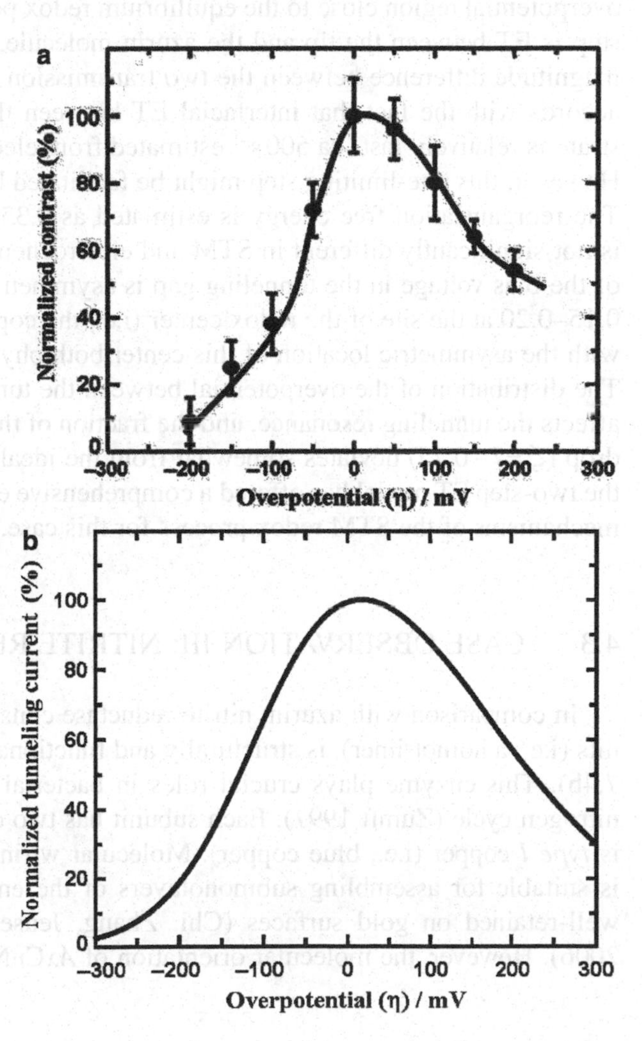

Fig. 7.9. Redox-gated electron tunneling resonance spectra of individual azurin molecules observed by ECSTM and calculated by Eqs. 4 to 7. ***a.*** Experimental results of relative electronic contrast vs. overpotential. ***b.*** Computed tunneling current vs. overpotential (Reproduced with permission from Chi, Q., Zhang, J., Jensen, P.S., Christensen, H.E.M. and Ulstrup, J. (2006). Long-range interfacial electron of metalloproteins based on molecular wiring assemblies. Faraday Discussions 131, 181–195.)

$$\vec{k}^{r/o} = \kappa_s \rho_s \frac{\omega_{eff}}{2\pi}\frac{2k_BT}{\alpha_s}\exp(e\theta\eta)\exp\left(-\frac{(E_r - eV_{bias} + e\xi\eta + e\gamma V_{bias})^2}{4E_rk_BT}\right) \tag{7}$$

θ can represent either overpotential dependence of the tunneling barrier or electrochemical double-layer effects, in the range of 3 to 6 eV^{-1} for the present case. Fig. 7.9b shows a calculated tunneling current-overpotential relation using Eqs 4 to 7, giving an asymmetric overpotential-dependence as in the experimental observations (Fig. 7.9a). We can thus obtain the main parameters that reflect the ET nature of this system in the ECSTM configuration. These parameters provide the following information: (a) the position of the resonance maximum is determined by all the factors; in the present case the maximum appears in a narrow positive overpotential region close to the equilibrium redox potential. (b) The rate-limiting step is ET between the tip and the azurin molecule, because at least an order of magnitude difference between the two transmission coefficients is required. This accords with the fact that interfacial ET between the azurin molecule and substrate is relatively fast (ca 500 s^{-1} estimated from electrochemical measurements). However, this rate-limiting step might be facilitated by multi-electron transfer. (c) The reorganization free energy is estimated as 0.35 to 0.45 eV, implying that E_r is not significantly different in STM and electrochemical ET. (d) The distribution of the bias voltage in the tunneling gap is asymmetric with optimal values of $\gamma \approx$ 0.15–0.20 at the site of the redox center (i.e., the copper ion in azurin), according with the asymmetric location of this center both physically and electronically. (e) The distribution of the overpotential between the tunneling gap also significantly affects the tunneling resonance, and the fraction of the substrate-solution potential drop ($\xi = 7 - 0.75$) deviates somewhat from the ideal expectation ($\xi = 1$). In short, the two-step ET model has offered a comprehensive explanation of the detailed ET mechanisms of the STM redox process for this case.

4.3 CASE OBSERVATION III: NITRITE REDUCTASE

In comparison with azurin, nitrite reductase consisting of three uniform subunits (i.e., a homotrimer), is structurally and functionally more composite (see Fig. 7.4b). This enzyme plays crucial roles in bacterial denitrification in the global nitrogen cycle (Zumft 1997). Each subunit has two copper centers, one of which is *type I* copper (i.e., blue copper). Molecular wiring using a π-conjugated thiol is suitable for assembling submonolayers of the enzyme with catalytic activity well-retained on gold surfaces (Chi, Zhang, Jensen, Christensen, and Ulstrup 2006). However, the molecular orientation of *Ax*CuNiR on 4-methylbenzenethiol

(MBT)-modified Au(111) surfaces is not fully known. Most likely the type I copper center is facing the electrode surface, as the local surface structure around the type I copper center resembles that of the azurin copper center (Fig. 7.10a).

The functional pathway of monolayers of this enzyme rests on both intramolecular and interfacial ET processes. The overall ET involves the following steps, completing an electrocatalytic cycle.

$$Cu_I^{2+} + e \longrightarrow Cu_I^{+} \qquad \text{(I)}$$

$$Cu_{II}^{2+} + Cu_I^{+} \longrightarrow Cu_{II}^{+} + Cu_I^{2+} \qquad \text{(II)}$$

$$\left\{ \begin{array}{l} NO_2^- + H_2O + e \longrightarrow NO + 2OH^- \\ Cu_{II}^{+} - e \longrightarrow Cu_{II}^{2+} \end{array} \right\} \qquad \text{(III)}$$

In the fully oxidized enzyme resting state, type I copper (T1Cu) is first reduced by an electron injected from the electrode. This is followed by intramolecular electron exchange between T1Cu and type II copper (T2Cu). T2Cu is reoxidized by nitrite, which is reduced to nitric oxide. In the natural protein ET processes, T1Cu serves as an entry site of electrons from the enzyme redox partner (e.g., cytochrome c_{551} or cupredoxins), whereas the T2Cu domain is the site for binding and reduction of the substrate nitrite. The tunneling pathway of intramolecular ET consists of eleven peptide bonds with a distance of 12.6 Å. Intramolecular ET is thus crucial in

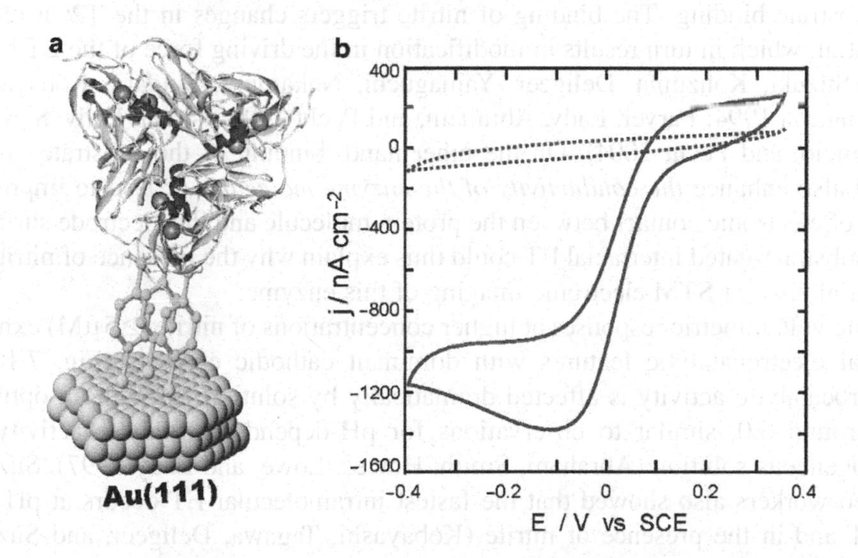

FIG. 7.10. ***a.*** Schematic representation of molecular wiring assembly nitrite reductase. ***b.*** Corresponding voltammetric response. The cyclic voltammograms obtained in 10 mM NaAc/HAc buffer (pH 6.0) containing 0 *(dotted line)* and 100 μM KNO_2 *(solid line)* with scan rate 10 mV s^{-1}

the catalytic cycle and could be a rate-limiting step in nature. In the electrochemical ET system, the electrode acts as an artificial redox partner to inject electrons to the T1Cu site, and intramolecular ET is a relay step in the overall ET process. With further consideration of noncovalent interactions between the enzyme and the wiring molecule, this system holds biomimetic features and ET should parallel biological ET. If diffusion effects of nitrite ions in solution are ignored, the rate-determining could be either interfacial ET between T1Cu and the electrode (step I) or intramolecular ET (step II), depending on the electronic coupling efficiency between the enzyme and the electrode through the wiring molecule. Because of the large size and multiple redox centers of this enzyme (compared to azurin), interfacial ET kinetics is proved to be very sensitive to the bridge length and other properties of the wiring molecule. A short π-conjugated thiol was employed to achieve the best possible effective electronic coupling (Chi, Zhang, Jensen, Christensen, and Ulstrup 2006).

In this case, no Faradaic response from the enzyme monolayer is detected in the absence of nitrite. A pair of redox peaks with a midpoint potential around 0.1 V (vs. SCE) is invoked by addition of nitrite at low concentrations ($\leq 1\,\mu M$). Anodic and cathodic peaks are largely symmetric, suggesting noncatalytic voltammetric character. Interfacial ET is thus only detected in the presence of nitrite, i.e., substrate-gated ET for this particular linker molecule. Studies of *Ax*CuNiR ET in homogeneous solution suggest that the rate of intramolecular ET is modified by substrate binding. The binding of nitrite triggers changes in the T2Cu redox potential, which in turn results in modification in the driving force of the ET reaction (Suzuki, Kohzuma, Deligeer, Yamaguchi, Nakamura, Shidara, Kobayashi, and Tagawa 1994; Farver, Eady, Abraham, and Pecht 1998; Farver, Eady, Sawers, Prudencio, and Pecht 2004). On the other hand, binding of the substrate nitrite could also enhance *the conductivity of the enzyme molecules* or lead to improvement of electronic contact between the protein molecule and the electrode surface. The substrate-gated interfacial ET could thus explain why the presence of nitrite is essential also for STM electronic imaging of this enzyme.

The voltammetric responses at higher concentrations of nitrite ($\geq 5\,\mu M$) exhibit typical electrocatalytic features with dominant cathodic currents (Fig. 7.10b). Electrocatalytic activity is affected dramatically by solution pH with an optimal pH around 6.0, similar to observations for pH-dependent catalytic activity in homogeneous solution (Abraham, Smith, Howes, Lowe, and Eady 1997). Suzuki and co-workers also showed that the fastest intramolecular ET occurs at pH 5.8 to 6.1 and in the presence of nitrite (Kobayashi, Tagawa, Deligeer, and Suzuki 1999). This suggests that the effects of solution pH on electrocatalytic activity arise mainly from intramolecular ET kinetics. The following approximate electrochemical version of the Michaelis-Menten equation applies for the relation between electrocatalytic current density (j) and the substrate concentration (S):

$$j = nF\Gamma\frac{k_{cat}[S]}{K_m + [S]} \quad (8)$$

where n is the number of electrons transferred, F is Faraday's constant, Γ is the surface population of the enzyme, k_{cat} is the apparent turnover rate constant, and K_m the apparent Michaelis-Menten constant.

The surface coverage can be estimated either from voltammetric measurements or from STM images as $7.6 \pm 0.8 \times 10^{-13}$ mol cm^{-2} (equivalent to ca 16% of a closely packed monolayer). Good agreement between both methods is observed, an indication that biological ET function of the enzyme is well retained in the assembly. K_m and k_{cat} were estimated as 44 μM and 30 s^{-1} by fitting of Eq. 8 to experimental data. K_m is close to the value for the free mobile enzyme (34 μM) (Abraham, Smith, Howes, Lowe, and Eady 1997), suggesting that there is no significant effect of immobilization on the enzyme-substrate affinity or structural conformations. k_{cat} is, however, clearly smaller than reported rate constants of intramolecular ET (185 – 1900 s^{-1}) observed by pulse radiolysis in homogeneous solution (Abraham, Smith, Howes, Lowe, and Eady 1997; Farver, Eady, Abraham, and Pecht 1998). This could be attributed to the differences in ET rate-determining steps and microenvironment. Although the catalytic rate in homogeneous solution is predominantly determined by intramolecular ET, the electrocatalytic efficiency is governed by concerted effects of intramolecular ET and interfacial ET between T1Cu and the electrode.

4.4. CASE OBSERVATION IV: CYTOCHROME C_4

Pseudomonas stutzeri cytochrome c_4 is a di-heme protein consisting of 190 amino acid residues (Fig. 7.4d). Two covalently bound heme groups are located in two separate cytochrome *c*-like domains connected by an extended polypeptide segment (Christensen 1994; Kadziola and Larsen 1997). The two heme groups are not coplanar but with an interplanar angle of ≈30°, and connected through hydrogen bonds with an Fe-Fe distance up to 19.1Å. The symmetry between the two domains is almost perfect near the axis, but gradually breaks down further away. This makes cytochrome c_4 a natural dipole under physiological conditions, with excess *negative* charges in the *N-terminal* domain and excess *positive* charges in the *C-terminal* domain. The two heme groups have thus intrinsically different redox properties, but with mutual interactions between the electronic states of the two centers.

Structural characteristics make this protein a good candidate for nanoscale bioelectronics, for example, for fabrication of a protein-based diode and a redox-gated biomolecular switch. Investigations exploiting these advantages are underway at our lab.

One of these efforts is to extract information of intramolecular ET from the electrochemical approach. The electrochemical approach to intramolecular ET of this protein in homogeneous solution is limited by weak signals and diffusion effects (Karlsson, Nielsen, Thuesen, and Ulstrup 1997). However, we have recently demonstrated that the protein can be well-oriented by engineering molecular assemblies at interfaces, leading to fast intramolecular ET. Details of intramolecular ET are thus obtained from single-crystal electrode-based electrochemical measurements and theoretical modeling (Chi, Zhang, Arslan, Christensen, Kuznetsov, and Ulstrup 2008).

5 CONCLUSIONS AND OUTLOOK

We have presented some observations of long-range interfacial ET and intramolecular ET of metalloproteins from a concerted effort of theoretical formulations and experimental developments. Molecular self-assembly at surfaces and interfaces provides a solid chemical basis for architecture of various metalloproteins with favorable molecular orientations. Detailed understanding of electronic structure and transport could be gained from electrochemistry and STM. For example, investigations of the azurin case have reached the nanoscale and single-molecule levels leading to comprehensive understanding of the ET mechanisms. Particularly, single-molecule ET tunneling resonance is directly observed. Experimental observations of other systems have also been achieved at a comparable level of resolution showing the feasibility of the method.

Several issues remain challenging and would be in focus for upcoming efforts: (1) development of *a general method* for controlled assembly of proteins; (2) improvement of *interfacial electronic coupling* between the large protein molecule (e.g. multicenter redox enzymes) and the electrode; (3) *solid state electrochemistry* of redox proteins and enzymes that could extend practical applications of protein-based electronic devices; and (4) *genetic engineering* of structures and functions of target proteins to enhance their mechanical strength and temperature-resisting stability suitable for room temperature bioelectronics.

ACKNOWLEDGMENTS

The authors acknowledge support of this work from the Danish Research Council for Technology and Production Sciences, the NanoScience Center at the University of Copenhagen, and Brødrene Hartmann's Foundation.

REFERENCES

Abraham, Z.H.L., Smith, B.E., Howes, B.D., Lowe, D.J., Eady, R.R. (1997). pH-dependence for binding a single nitrite ion to each type-2 copper center in the copper-containing nitrite reductase of *Alcaligenes xylosoxidans*. Biochem. J. 324, 511–516.

Albrecht, T., Guckian, A., Ulstrup, J. and Vos, J.G. (2005). Transistor-like behavior of transition metal complexes. Nano Lett. 5, 1451–1455.

Alessandrini, A., Gerunda, M., Canters, G.W., Verbeet, M.Ph. and Facci, P. (2003). Electron tunnelling through azurin is mediated by the active site Cu ion. Chem. Phys. Lett. 376, 625–630.

Alessandrini, A., Salerno, M., Frabboni, S. and Facci, P. (2005). Single-metalloprotein wet biotransistor. Appl. Phys. Lett. 86, 133902.

Armstrong, F.A., Barlow, N.L., Burn, P.L., Hoke, K.R., Jeuken, L.J.C., Shenton, C. and Webster, G.R. (2004). Fast, long-range electron-transfer reactions of a 'blue' copper protein coupled non-covalently to an electrode through a stilbenyl thiolate monolayer. Chem. Cummun. 316–317.

Armstrong, F.A. and Wilson, G.S. (2000). Recent developments in faradaic bioelectrochemistry. Electrochim. Acta 45, 2623–2645.

Aubin-Tam, M.E. and Hamad-Schifferli, K. (2005). Gold nanoparticle cytochrome c complexes: the effect of nanoparticle ligand charge on protein structure. Langmuir 21, 12080–12084.

Avila, A., Gregory, B.W., Niki, K. and Cotton, T.M. (2000). An electrochemical approach to investigate gated electron transfer using a physiological model system: cytochrome c immobilized on carboxylic acid-terminated alkanethiol self-assembled monolayers on gold electrodes J. Phys. Chem. B 104, 2759–2766.

Aviram, A. and Ratner, M.A. (1974). Molecular rectifiers. Chem. Phys. Lett. 29, 277–283.

Bertini, I., Cavallaro, G. and Rasato, A. (2006). Cytochrome c: occurrence and functions. Chem. Rev. 106, 90–115.

Besteman, K., Lee, J.O., Wiertz F.G.M. and Dekker, C. (2003). Enzyme-coated carbon nanotubes as single-molecule biosensors. Nano Lett. 3, 727–730.

Chi, Q., Farver, O. and Ulstrup, J. (2005). Long-range protein electron transfer observed at the single-molecule level: in situ mapping of redox-gated tunneling resonance. Proc. Natl. Acad. Sci. USA 102, 16203–16208.

Chi, Q., Zhang, J., Andersen, J.E.T. and Ulstrup, J. (2001). Ordered assembly and controlled electron transfer of the blue copper protein azurin at gold (111) single-crystal substrates. J. Phys. Chem. B 105, 4669–4679.

Chi, Q., Zhang, J., Arslan, T., Christensen, H.E.M., Kuznetrov, A.M. and Ulstrup, J. (2008). Electrochemical approach to fast intramolecular electron transfer of cytochrome c_4. in press.

Chi, Q., Zhang, J., Christensen, H.E.M. and Ulstrup, J. Unpublished results.

Chi, Q., Zhang, J., Friis, E.P., Andersen, J.E.T. and Ulstrup, J. (1999). Electrochemistry of self-assembled monolayers of the blue copper protein *Pseudomonas aeruginosa* azurin on Au(111). Electrochem. Commun. 1, 91–196.

Chi, Q., Zhang, J., Jensen, P.S., Christensen, H.E.M. and Ulstrup, J. (2006). Long-range interfacial electron of metalloproteins based on molecular wiring assemblies. Faraday Discussions 131, 181–195.

Chi, Q., Zhang, J., Nielsen, J.U., Friis, E.P., Chorkendorff, Ib., Canters, G.W., Andersen, J.E.T. and Ulstrup, J. (2000). Molecular monolayers and interfacial electron transfer of *Pseudomonas aeruginosa* azurin on Au(111). J. Am. Che. Soc. 122, 4047–4046.

Christensen, H.E.M. (1994). Cloning and characterization of the gene encoding cytochrome c(4) from *Pseudomonas-stutzeri*. Gene144, 139–144.

Conrad, L.S., Karlsson, J.J. and Ulstrup, J. (1995). Electron-transfer and spectral alpha-band properties of the di-heme protein cytochrome c(4). Eur. J. Biochem. 231, 133–140.

Davis, J.J., Bruce, D., Canters, G.W., Crozier, J. and Hill, H.A.O. (2003). Genetic modulation of metalloprotein electron transfer at bare gold. Chem. Commun., 576–577.

Davis, J.J., Morgan, C.L., Wrathmell, Axford, D.N., Zhao, J. and Wang, N. (2005). Molecular bioelectronics. J. Chem. Mater. 15, 2160–2172.

Davis, J.J., Wang, N., Morgan, A., Zhang, T. and Zhao, J. (2006). Metalloprotein tunnel junctions: compressional modulation of barrier height and transport mechanism. Faraday Discussions 131, 167–179.

Farver, O. (1996). In: Bendall, D. (Ed.), Protein Electron Transfer. BIOS Publishers, Oxford.

Farver, O., Blatt, Y. and Pecht, I. (1982). Resolution of two distinct electron transfer sites on azurin. Biochemistry 21, 3556–3561.

Farver, O., Eady, R.R., Abraham, H.L. and Pecht, I. (1998). The intramolecular electron transfer between copper sites of nitrite reductase: a comparison with ascorbate oxidase. FEBS Lett. 436, 239–242.

Farver, O., Eady, R.R., Sawers, G., Prudencio, M. and Pecht, I. (2004). Met144Ala mutation of the copper-containing nitrite reductase from *Alcaligenes xylosoxidans* reverses the intramolecular electron transfer. FEBS Lett. 561, 173–176.

Farver, O., Li, Y., Ang, M.C. and Pecht, I. (1999). Enhanced rate of intramolecular electron transfer in an engineered purple CUA azurin. Proc. Natl. Acad. Sci. U S A 96, 899–902.

Farver, O. and Pecht, I. (1989). Long-range intramolecular eletron-transfer in azurins. Proc. Natl. Acad. Sci. U S A 86, 6968–6972.

Feng, Z.Q., Imabayashi, T., Kikiuchi, T. and Niki, K. (1997). Long-range electron-transfer reaction rates to cytochrome c across long- and short-chain alkanethiol self-assembled monolayers: electroreflectance studies. J. Chem. Soc. Faraday Trans. 93, 1367–1370.

Flood, A.H., Stoddart, J.F.D., Steuerman, W. and Heath, J.R. (2004). Whence molecular electronics? Science 306, 2055–2056.

Friis, E.P., Andersen, J.E.T., Kharkats, Y.I., Kuznetsov, A.M., Nichols, R.J., Zhang, J. D., and Ulstrup, J. (1999). An approach to long-range electron transfer mechanisms in metalloproteins: in situ scanning tunneling microscopy with submolecular resolution. Proc. Natl. Acad. Sci. U S A 96, 1379–1384.

Friis, E.P., Kharkats, Y.I., Kuznetsov, A.M. and Ulstrup, J. (1998). In situ scanning tunneling microscopy of a redox molecule as a vibrationally coherent electronic three-level process. J. Phys. Chem. A 102, 7851–7859.

Fristrup, P., Grubb, M., Zhang, J., Christensen, H. E. M., Hansen, A. G. and Ulstrup, J. (2001). Voltammetry of native and recombinant *Pseudomonas aeruginosa* azurin on polycrystalline Au- and single-crystal Au(111)-surfaces modified by decanethiol monolayers. J. Electroanal. Chem. 511, 128–133.

Fujita, K., Nakamura, N., Ohno, H., Leigh, B.S., Niki, K., Gray, H.B. and Richards, J.H. (2004). Mimicking protein-protein electron transfer: voltammetry of *Pseudomonas aeruginosa* azurin and the *Thermus thermophilus* Cu-A domain at omega-derivatized self-assembled-monolayer gold electrodes. J. Am. Chem. Soc. 126, 13954–13961.

Gittins, D.I., Bethell, D., Schiffrin, D.J. and Nichols, R.J. (2000). A nanometre-scale electronic switch consisting of a metal cluster and redox-addressable groups. Nature 408, 67–69.

Gray, H.B. and Winkler, J.R. (1996). Electron transfer in proteins. Annu. Rev. Biochem. 65, 537–561.

Guo, M.L., Bhaskar, B., Li, H.Y., Barrows, T. and Poulos, T.L. (2004). Crystal structure and characterization of a cytochrome c peroxidase-cytochrome c site-specific cross-link. Proc. Natl. Acad. Sci. U S A 101, 5940–5945.

Haiss, W., van Zalinge, H., Higgins, S.J., Bethell, D., Höbenreich, H., Schiffrin, D.J. and Nichols, R.J. (2003). Redox state dependence of single molecule conductivity. J. Am. Chem. Soc. 125, 15294–15295.

Hamelin, A. (1996). Cyclic voltammetry at gold single-crystal surfaces. 1. Behaviour at low-index faces. J. Electroanal. Chem. 401, 1–16.

Ho, W.H., Ooi, B.L., Jøgensen, A.M., Borg, L., Jespersen, L.L. and Christensen, H.E.M. (2003). Cytoplasmic expression of the *Achromobacter xylosoxidans* blue copper nitrite reductase in *Escherichia coli* and characterisation of the recombinant protein. Protein Expr. Purif. 32, 288–292.

Hush, N.S. (2003). An overview of the first half-century of molecular electronics. Ann. NY Acad. Sci. 1006, 1–20.

Imabayashi, S., Mita, T. and Kakiuchi, T. (2005). Effect of the electrostatic interaction on the Redox reaction of positively charged cytochrome c adsorbed on the negatively charged surfaces of acid-terminated alkanethiol monolayers on a Au(111) electrode. Langmuir 21, 1470–1474.

Jensen, P.S. (2005). Master Thesis, Technical University of Denmark, Lyngby, Denmark.

Jensen, P.S., Chi, Q., Grumsen, F.B., Abad, J.M., Horsewell, A., Schiffrin, D.J. and Ulstrup, J. (2007). J. Phys. Chem. C. 111, 6124–6132. Gold nanoparticle assisted assembly of a here protein for enhancement of long-range interfacial electron transfer.

Jiang, X., Jiang, J., Jin, Y., Wang, E. and Dong, S. (2005). Effect of colloidal gold size on the conformational changes of adsorbed cytochrome c: probing by circular dichroism, UV-visible, and infrared spectroscopy. Biomacromolecules 6, 46–53.

Jiang, X., Shang, L, Wang, Y. and Dong, S. (2005). Cytochrome c superstructure biocomposite nucleated by gold nanoparticle: thermal stability and voltammetric behavior. Biomacromolecules 6, 3030–3036.

Jortner, J. and Ratner, M.A. (Eds.) (1997). Molecular Electronics. Blackwell, Oxford, UK.

Jeuken, L.J.C., McEvoy, J.P. and Armstrong, F.A. (2002). Insights into gated electron-transfer kinetics at the electrode-protein interface: a square wave voltammetry study of the blue copper protein azurin. J. Phys. Chem. B 106, 2304–2313.

Kadziola, A. and Larsen, S. (1997). Crystal structure of the dihaem cytochrome c(4) from *Pseudomonas stutzeri* determined at 2.2 angstrom resolution. Structure 15, 203–216.

Karlsson, J.J., Nielsen, M.F., Thuesen, M.H. and Ulstrup, J. (1997). Electrochemistry of cytochrome c(4) from *Pseudomonas stutzeri*. J. Phys. Chem. B 101, 2430–2436.

Karlsson, J.J., Rosstrup, T.E. and Ulstrup, J. (1996). pH and ionic strength effects on electron transfer rate constants and reduction potentials of the bacterial di-heme protein *Pseudomonas stutzeri* cytochrome c(4). Acta Chem. Scand. 50, 284–288.

Kasmi, E.I.A., Wallace, J.M., Bowden, E.F., Binet, S.M. and Linderman, R.J. (1998). Controlling interfacial electron-transfer kinetics of cytochrome c with mixed self-assembled monolayers. J. Am. Che. Soc. 120, 225–226.

Katz, E. and Willner, I. (2004a). Integrated nanoparticle-biomolecule hybrid systems: synthesis, properties, and applications. Angew. Chem. Int. Ed. 43, 6042–6108.

Katz, E. and Willner, I. (2004b). Biomolecule-functionalized carbon nanotubes: applications in nanobioelectronics. Chem. Phys. Chem. 5, 1085–1104.

Kobayashi, K., Tagawa, S., Deligeer, T. and Suzuki, S. (1999). The pH-dependent changes of intramolecular electron transfer on copper-containing nitrite reductase. J. Biochem. (Tokyo) 126, 408–412.

Kuznetov, A.M. and Ulstrup, J. (1999). Electron Transfer in Chemistry and Biology: An Introduction to the Theory. Wiley, Chichester, UK.

Kuznetsov, A.M., Sommer-Larsen, P. and Ulstrup, J. (1992). Resonance and environmental fluctuation effects in STM currents through large adsorbed molecules. Surf. Sci. 275, 52–64.

Kuznetsov, A.M. and Ulstrup, J. (2000). Mechanisms of in situ scanning tunnelling microscopy of organized redox molecular assemblies. J. Phys. Chem. A 104, 11531–11540. Errata: (2001). J. Phys. Chem. A 105, 7494.

Liu, T., Zhong, J., Gan, X., Fan, C., Li, G. and Matsuda, N. (2003). Wiring electrons of cytochrome c with silver nanoparticles in layered films. Chem. Phys. Chem. 4, 1364–1366.

Marcus, R. and Sutin, N. (1985). Electron transfers in chemistry and biology. Biochim. Biophys. Acta 811, 265–322.

Murgida, D.H. and Hildebrandt, P. (2002). Electrostatic-field dependent activation energies modulate electron transfer of cytochrome c. J. Phys. Chem. B 106, 12814–12819.

Murgida, D.H. and Hildebrandt, P. (2004). Electron-transfer processes of cytochrome c at interfaces: new insights by surface-enhanced resonance Raman spectroscopy. Acc. Chem. Res. 37, 854–861.

Murgida, D.H., Hildebrandt, P., Wei, J., He, Y.F., Liu, H.Y. and Waldeck, D.H. (2004). Surface-enhanced resonance Raman spectroscopic and electrochemical study of cytochrome c bound on electrodes through coordination with pyridinyl-terminated self-assembled monolayers. J. Phys. Chem. B 108, 2261–2269.

Niki, K., Pressler, K.R., Sprinkle, J.R., Li, H. and Margoliash, E. (2002). Russian J. Electrochem. 38, 63–67.

Niki, K., Hardy, W.R., Hill, M.G., Li, H., Sprinkle J.R., Margoliash, E., Fujita, K., Tanimura, R., Nakamura, N., Ohno, H., Richards, J.H. and Gray, H.B. (2003). Coupling to lysine-13 promotes electron tunneling through carboxylate-terminated alkanethiol self-assembled monolayers to cytochrome c. J. Phys. Chem. B 107, 9947–9949.

Nitzan, A. and Ratner, M.A. (2003). Science 300, 1384–1389.

Pompa, P.P., Biasco, A., Frascerra, V., Calabi, F., Cingolani, R., Rinaldi, R., Verbeet, M.Ph., de Waal, E. and Canters, G.W. (2004). Solid state protein monolayers: morphological, conformational, and functional properties. J. Chem. Phys. 121, 10325–10328.

Tom, R.T. and Pradeep, T. (2005). Interaction of azide ion with hemin and cytochrome c immobilized on Au and Ag nanoparticles. Langmuir 21, 11896–11902.

Rinaldi, R. and Cingolani, R. (2004). Electronic nanodevices based on self-assembled metalloproteins. Physica E 21, 45–60.

Schmickler, W. and Widrig, C. (1992). The investigation of redox reactions with a scanning tunneling microscope-experimental and theoretical aspects. J. Electroanal. Chem. 36, 213–221.

Schnyder, B., Kötz, R., Alliata, D. and Facci, P. (2002). Comparison of the self-chemisorption of azurin on gold and on functionalized oxide surfaces. Surf. Interface Anal. 34, 40–44.

Song, S., Clark, R.A., Bowden, E.F. and Tarlov, M.J. (1993). Characterization of cytochrome s alkanethiolate structures prepared self-assembled on gold. J. Phys. Chem. 97, 6564–6572.

Sumi, H. (1998). V-I characteristics of STM processes as a probe detecting vibronic interactions at a redox state in large molecular adsorbates such as electron-transfer metalloproteins. J. Phys. Chem. B 102, 1833–1844.

Suzuki, S., Kohzuma, T., Deligeer, Yamaguchi, K., Nakamura, N., Shidara, S., Kobayashi, K. and Tagawa, S. (1994). Pulse-radiolysis studies on nitrite reductase from achromobacter cycloclastes IAM 1013-Evidence for intramolecular electron transfer from type-1 to type-2 Cu. J. Am. Chem. Soc. 116, 11145–11146.

Tao, N.J. (1996). Probing potential-tuned resonant tunneling through redox molecules with scanning tunneling microscopy. Phys. Rev. Lett. 76, 4066–4069.

Tarlov, M.J. and Bowden, E.F. (1991). Electron-transfer reaction of cytochrome c adsorbed on carboxylic-acid terminated alkanethiol monolayer electrodes. J. Am. Chem. Soc. 113, 1847–1848.

Terrettaz, S., Cheng, J. and Miller, C.J. (1996). Kinetic parameters for cytochrome c via insulated electrode voltammetry. J. Am. Che. Soc. 118, 7857–7858.

Vijigenboom, E., Busch, J.E. and Canters, G.W. (1997). In vivo studies disprove an obligatory role of azurin in denitrification in *Pseudomonas aeruginosa* and show that azu expression is under control of RpoS and ANR. Microbiology 143, 2853–2863.

Wackerbarth, H. and Hildebrandt, P. (2003). Redox and conformational equilibria and dynamics of cytochrome c at high electric fields. ChemPhysChem 4, 714–724.

Wang, W., Lee, T. and Reed, M.A. (2003). Mechanism of electron conduction in self-assembled alkanethiol monolayer devices. Phys. Rev. B 68, 35416–35424.

Wei, J., Liu, H., Dick, A.R., Yamamoto, H., He, Y. and Waldeck, D.H. (2002). Direct wiring of cytochrome c's heme unit to an electrode: electrochemical studies. J. Am. Chem. Soc. 124, 9591–9599.

Wold, D.J., Hagg, R., Rampi, M.A. and Frisbie, C.D. (2002). Distance dependence of electron tunneling through self-assembled monolayers measured by conducting probe atomic force microscopy: unsaturated versus saturated molecular junctions. J. Phys. Chem. B 106, 2813–2816.

Zhang, J., Bilic, A., Reimers, J.R., Hush, N.S. and Ulstrup, J. (2005). Coexistence of multiple conformations in cysteamine monolayers on Au(111). J. Phys. Chem. B 109, 15355–5367.

Zhang, J., Chi, Q., Albrecht, T., Kuznetrov, A.M., Grubb, M., Hansen, A.G., Wackerbarth, Welinder, A.C. and Ulstrup, J. (2005). Electrochemistry and bioelectrochemistry towards the single-molecule level: theoretical notions and systems. Electrochim. Acta 50, 3143–3159.

Zhang, J., Chi, Q., Nielsen, J.U., Friis, E.P., Andersen, J.E.T. and Ulstrup, J. (2000). Two-dimensional cysteine and cystine cluster networks on Au(111) disclosed by voltammetry and in situ scanning tunneling microscopy. Langmuir 16, 7229–7237.

Zhang, J., Chi, Q., Kuznetsov, A.M., Hansen, A.G., Wackerbarth, H., Christensen, H.E.M., Andersen, J.E.T. and Ulstrup, J. (2002). Electronic properties of functional biomolecules at metal/aqueous solution interfaces. J. Phys. Chem. B 106, 1131–1152.

Zhang, J., Demetriou, A., Welinder, A.C., Albrecht, T., Nichols, R.J. and Ulstrup, J. (2005). Potential-induced structural transitions of DL-homocysteine monolayers on Au(111) electrode surfaces. Chem. Phys. 319, 210–221.

Zhang, J., Grubb, M., Hansen, A.G., Kuznetrov, A.M., Boisen, A., Wackerbarth, H. and Ulstrup, J. (2003). Electron transfer behaviour of biological macromolecules towards the single-molecule level. J. Phys. Condens. Mater. 15, S1873–S1890.

Zhang, J., Kuznetrov, A. M. and Ulstrup, J. (2003). In situ scanning tunnelling microscopy of redox molecules. Coherent electron transfer at large bias voltages. J. Electroanal. Chem. 541, 133–146.

Zhang, J., Welinder, A.C., Hansen, A.G., Christensen, H.E.M. and Ulstrup, J. (2003). Catalytic monolayer voltammetry and in situ scanning tunneling microscopy of copper nitrite reductase on cysteamine-modified Au(111) electrodes. J. Phys. Chem. B 107, 12480–12484.

Zhao, J.W., Davis, J.J., Sansom, M.S.P., and Hung, A. (2004). Exploring the electronic and mechanical properties of protein using conducting atomic force microscopy. J. Am. Chem. Soc. 126, 5601–5609.

Zumft, W.G. (1997). Cell biology and molecular basis of denitrification. Microbiol. Mol. Biol. Rev. 61, 533–593.

8

Ion Channels in Tethered Bilayer Lipid Membranes on Au Electrodes

Ingo Köper, Inga K. Vockenroth, and Wolfgang Knoll

1 INTRODUCTION

The natural biomembrane is a highly complex system to which nature has engineered a high specificity and a considerable amount of functions. In principle, a cell membrane consists of a lipid bilayer that forms a barrier between the inside and outside as well as between different compartments inside the cell. These barriers have to prevent any unwanted transport of, e.g., ions or small molecules across the membrane, but at the same time permit the selective transport of chosen components. This is achieved by the function of embedded membrane proteins. For example, selective ion channels govern the transport of specific types of ions across the bilayer to build and maintain concentration gradients and transmembrane potentials. This is essential since these potentials are one of the key components in signaling processes between cells. The use of these highly engineered biological systems in a biosensing unit would result in a fast and efficient device, for example, for drug screening or detection of toxins.

However, the biological membrane is a highly complex architecture that is almost impossible to use in practical applications. Approaches have been

A. Offenhäusser and R. Rinaldi (eds.), *Nanobioelectronics - for Electronics, Biology, and Medicine*, 211
DOI: 10.1007/978-0-387-09459-5_9, © Springer Science+Business Media, LLC 2009

described in which entire cells are placed on electronic devices such as field-effect transistors. However, these systems have severe drawbacks in terms of stability and reproducibility (Fromherz, Kiessling, Kottig, and Zeck 1999; Schöning and Poghossian 2002; Poghossian, Schultze, and Schöning 2003).

Another approach lies in the construction of a synthetic membrane architecture that mimics some of the features of a natural membrane, especially a high electrical sealing combined with the ability to incorporate membrane proteins in a functional form. The most common approaches to study for example protein–protein interactions use liposomes or vesicles as membrane models (Bernard, Guedeau-Boudeville, Sandre, Palacin, di Meglio, and Jullien 2000; Apel, Deamer, and Mautner 2002; Sofou and Thomas 2003). However, these models are rather limited for electrical measurements or practical applications. Measurements of ion channel translocation across the bilayer are commonly studied using patch clamp techniques using entire cells or using black lipid membranes (BLMs) as a model system. The latter system, however, suffers from an intrinsic instability. There are attempts to increase the stability of the BLM by using nanoporous substrates; practical applications have not yet been shown (Römer, Lam, Fischer, Watts, Fischer, Göring, Wehrspohn, Gösele, and Steinem 2004; Römer and Steinem 2004).

Solid supported membranes have been extensively used for the study of membrane proteins (Sackmann 1996; Abdelghani, Jacquin, Huber, Deutschmann and Sackmann 2001; Sackmann and Tanaka 2004). Incorporation of complex structures, the function of these units, and the specific binding of analytes could be shown. However, the study of proton- or ion-transport across a membrane has been shown in a few examples only, mainly due to the lack of sufficiently insulating membranes.

In this respect, a tBLM, as schematically shown in Fig. 8.1, is a promising system that mimics the principal features of a natural cell membrane (Cornell,

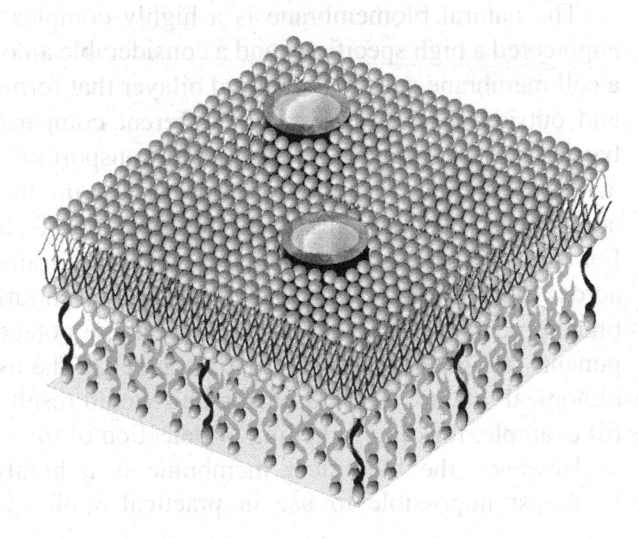

FIG. 8.1. Schematic picture of a tethered bilayer lipid membrane. The proximal bilayer leaflet is covalently attached to a solid support via a spacer group. The distal layer is typically formed by vesicle fusion and ion channels are subsequently incorporated.

Braach-Maksvytis, King, Osman, Raguse, Wieczorek, and Pace 1997; Knoll, Frank, Heibel, Naumann, Offenhäusser, Rühe, Schmidt, Shen, and Sinner 2000; Krishna, Schulte, Cornell, Pace, and Osman 2003; Naumann, Schiller, Giess, Grohe, Hartman, Kärcher, Köper, Lübben, Vasilev, and Knoll 2003; Atanasov, Knorr, Duran, Ingebrandt, Offenhäuser, Knoll, and Köper 2005; Terretaz and Vogel 2005). It provides a hydrophobic barrier between two compartments, the submembrane space and the bulk solution. Transport across the membrane is only possible due to the function of incorporated proteins, e.g., ion channels, transporters or proton pumping proteins. Additionally, the lipid bilayer of a tBLM is anchored via a spacer group to a solid support. This spacer serves as a stable anchoring moiety and provides an ion reservoir. The solid support also enhances the stability of the architecture compared with other model systems. For example, the classical membrane model system for the investigation of membrane proteins, the BLM, is very delicate to produce and handle and is, in the best-case scenario, stable for only a few hours. A typical tBLM, on the other hand, is easily assembled and allows for extended use over weeks.

Furthermore, the solid support offers the possibility of integrating the membrane architecture into (micro-)electronic readout systems. Scaling to an array format is a straightforward approach. Biosensing applications are possible where a tBLM acts as a synthetic interface between biology, e.g., membrane proteins as central sensing units, and electronics, e.g., electrodes or transistors (Cornell, Braach-Maksvytis, King, Osman, Raguse, Wieczorek, and Pace 1997; Knoll, Frank, Heibel, Naumann, Offenhäusser, Rühe, Schmidt, Shen, and Sinner 2000; Krishna, Schulte, Cornell, Pace, and Osman 2003; Naumann, Schiller, Giess, Grohe, Hartman, Kärcher, Köper, Lübben, Vasilev, and Knoll 2003; Atanasov, Knorr, Duran, Ingebrandt, Offenhäuser, Knoll, and Köper 2005).

During the last few years we have developed a new class of anchorlipids that are composed of three distinct parts (Schiller, Naumann, Lovejoy, Kunz, and Knoll 2003; Atanasov, Atanasova, Vockenroth, Knorr and Köper 2006; Köper, Schiller, Giess, Naumann, and Knoll 2006). The lipid head consists of diphytanylglycerol, a moiety well known to form stable, dense and fluid membranes even under harsh conditions. An oligo-ethyleneoxide spacer separates the lipid group from the support, allows for the incorporation of membrane proteins, and provides a certain ion reservoir. Finally, an anchor group covalently attaches the spacerlipid to the solid support. On gold electrodes this anchor consists of a thiol using the high Au-S affinity. Figure 8.2 summarizes the different molecules that have been synthesized. They possess two different anchor groups, an ethanthiol or a lipoic acid moiety and ethylenoxide-spacers with 4 to 14 units. Thus, the anchorlipids serve as a molecular "toolbox" that allows for an optimized assembly of the membrane architecture with respect to incorporation of proteins. We also modified the anchor group by adding silane groups to the molecule, which allows assembly of

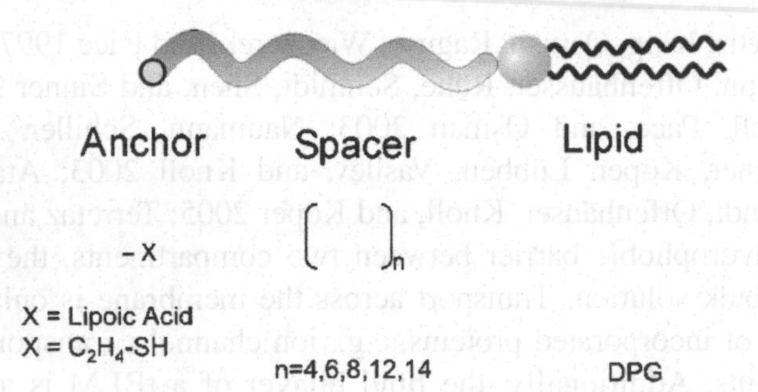

FIG. 8.2. Schematic depiction of the anchor lipids used in this study. The molecules consist of three distinct parts. The anchor group consists of a thiol, either as a single SH group or in form of a lipoic acid moiety, to allow for assembly on gold substrates. Oligo-ethylene oxide chains of different length form the spacer and the lipid head group consists of diphytanylglycerol (DPG).

the membrane for example on silicon oxide surfaces. We aim for the coupling of the tBLMs directly to a gate oxide of a field effect transistor (Atanasov, Knorr, Duran, Ingebrandt, Offenhäuser, Knoll, and Köper 2005; Atanasov, Atanasova, Vockenroth, Knorr, and Köper 2006).

On smooth surfaces, the anchorlipids form dense and homogeneous monolayers, either by self-assembly or by transfer of a pre-arranged film from an air–water interface. Typically, the layers provide a hydrophobic surface with static contact angles higher than 90°.

By fusion with small unilamellar vesicles or by incubation with lipids followed by rapid solvent exchange, an outer leaflet can be added to complete the bilayer.

The architecture can be characterized optically by surface plasmon resonance spectroscopy (SPR). The electrical parameters such as resistance and capacitance of the membrane can be investigated by using electrochemical impedance spectroscopy (EIS). Especially the latter gives important information about the bilayer quality. A high membrane resistance is essential for the study of incorporated proteins and their function; leakage currents across the bilayer have to be minimized.

The final experiment, however, remains the incorporation of membrane proteins and the study of their function. Soluble proteins can be simply added to a preformed bilayer and the hydrophobic interactions direct the protein into the lipid layer; however, most proteins cannot be solubilized. In this case, they can be incorporated into the tBLM by fusion of proteoliposomes with the self-assembled monolayer. The proper function of the incorporated proteins or peptides, which in the case of ion channels is the transport of ions across the bilayer, typically leads to increased membrane conductivity. This increase is easily detectable using EIS.

The following first describes the techniques used for the membrane characterization, then discusses the membrane assembly, and finally shows some examples for the functional incorporation of membrane peptides into different tBLM architecture.

2 MATERIALS AND METHODS

2.1 ELECTROCHEMICAL IMPEDANCE SPECTROSCOPY

EIS measurements were conducted using an AUTOLAB PGSTAT 12 impedance spectrometer. Three electrode measurements were performed in Teflon cells with the substrates as the working electrode, a coiled platinum wire as the counterelectrode, and DRIREF-2 reference the electrodes are from World Precision Instruments, Berlin, Germany. Spectra were recorded at a controlled potential with an ac modulation amplitude of 10 mV. Raw data were analyzed using the ZVIEW software package (Version 2.70, Scribner Associates). The home-built Teflon cells have a buffer volume of 1 ml and an electrochemically active area on the substrates of about 0.2 cm^2. The obtained data are fitted using an equivalent circuit of capacitors (C) and resistors (R). For typical experiments, we used R-(RC)-C circuits, in which the bilayer is represented by the central RC element. The spacer part and the electrical double layer at the gold electrode are described by a capacitor. The additional feed resistance describes effects of the electrolyte. For comparison, the final values were normalized to the electrode surface area.

2.2 SURFACE PLASMON RESONANCE SPECTROSCOPY

A lab-built setup in Kretschmann configuration with a 632 nm He/Ne laser was used (Knoll, Frank, Heibel, Naumann, Offenhäusser, Rühe, Schmidt, Shen, and Sinner 2000). In the scan mode, reflectivity changes are monitored as a function of the angle of incidence of the incoming laser beam. In the kinetic mode reflectivity changes occurring at a fixed angle are monitored as a function of time.

SPR spectra were analyzed using a three-layer model including the prism, gold, and thiolipid monolayer. After vesicle fusion a fourth layer corresponding to the outer leaflet of the bilayer was added. The refractive indices for the monolayer and bilayer were $n = 1.489$ and $n = 1.423$, respectively.

3 PROTEIN INCORPORATION

3.1 ASSEMBLY OF THE SYSTEM

The first step in the assembly of the system consists in the grafting of a lipid monolayer onto a substrate. This step is achieved by immersing the substrate (a gold electrode) into a diluted self-assembly solution (0.2 mg/ml in ethanol) of the anchor lipid. The process is typically finished within 24 hours, resulting in a hydrophobic and densely packed monolayer. In most cases a single anchor lipid is used. However, it is possible to introduce smaller spacer molecules in order to reduce the lateral density of the lipids in the proximal bilayer leaflet. Contact angles are typically higher than 90°. The latter empirically seems to be a crucial parameter for the second assembly step, the fusion of the monolayer with small unilamellar vesicles in order to form a homogeneous and electrically tight bilayer. The second important parameter that often determines the quality of the resulting bilayer is the roughness of the substrate. This factor becomes obvious when one considers a thickness of the spacer region of approximately 20 Å. Best results have been obtained on substrates with a mean-square roughness smaller than 5 Å. This is achieved in practice by using evaporation of a gold layer onto a template, e.g., an ultra-flat silicon wafer, and subsequently gluing the gold to a support. The support-glass assembly can be stripped from the template and the gold surface reveals the same roughness as the template.

The assembly process and the bilayer finally obtained can be characterized using divers surface analytical techniques such as surface plasmon resonance or quartz crystal microbalance.

As an example, Fig. 8.3 shows the fusion of DiPhyPC vesicles (50 nm by extrusion) with a DPTT monolayer as followed by SPR. The inset shows the angular shift of the plasmon curve, whereas the main graph shows the evolution of the reflectivity, here calculated as bilayer thickness measured at a fixed angle.

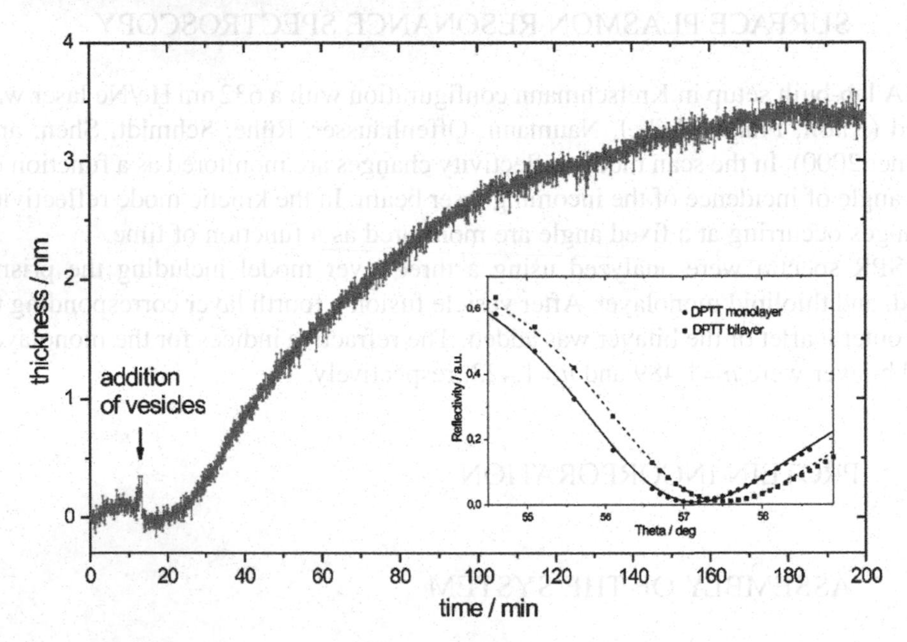

FIG. 8.3. SPR data showing the vesicle fusion process. The graph shows the evolution of the membrane thickness during the formation of the distal leaflet. The inset shows the corresponding angular SPR scans and the corresponding fits *(solid lines)*.

However, the final quality control remains the electrical characterization using electrochemical impedance spectroscopy. When investigating the ion transport through integrated ion channel proteins, a high resistance of the bilayer itself is a crucial parameter. Therefore, combined experiments in an electrochemical SPR cell give valuable information about the system. Figures 8.4 to 8.6 show the impedance spectra for bilayers built on different monolayer architectures. They all

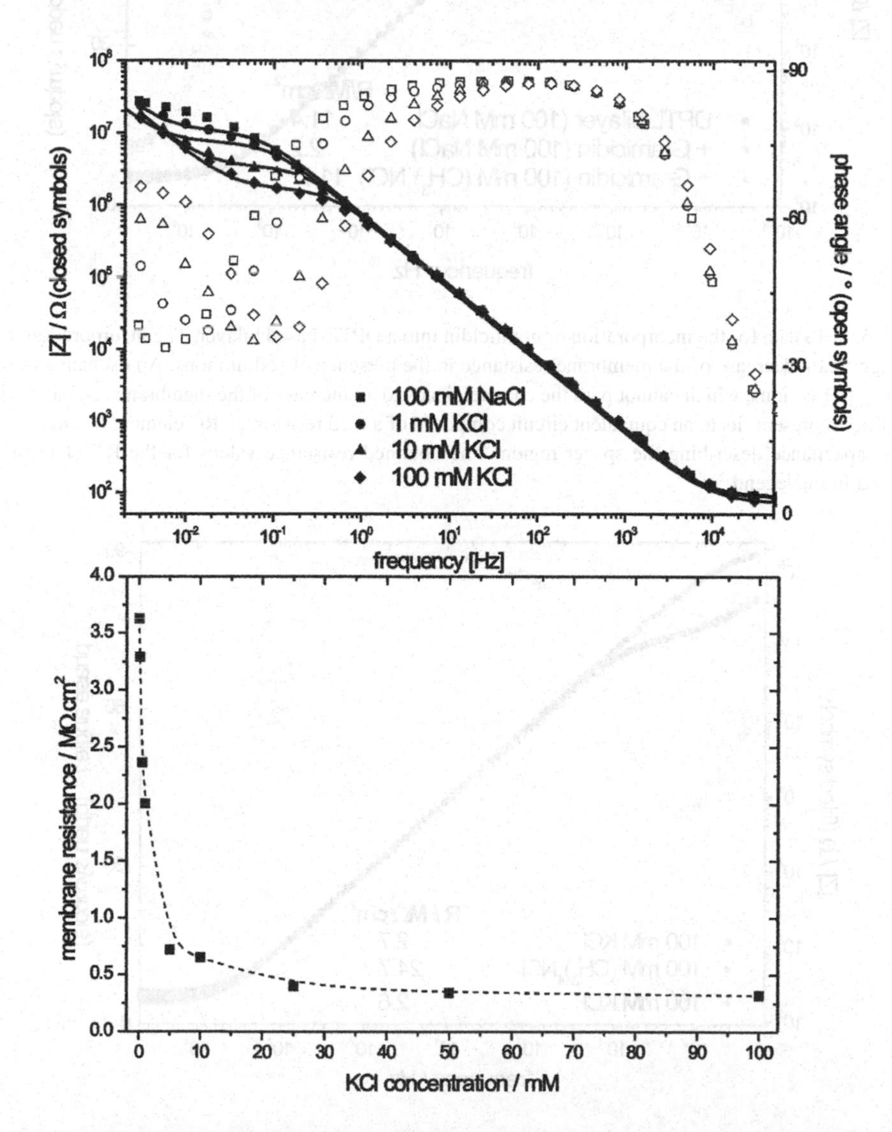

FIG. 8.4. Incorporation of valinomycin into a DPTT based tBLM. **Top:** Bode plot of the impedance and the phase shift as a function of frequency. Solid lines represent fits using an equivalent circuit. For better visibility, only the impedance fits are shown. **Bottom:** Bilayer resistance as a function of the potassium concentration in the electrolyte (ionic strength 100 mM). The dotted line is a guide to the eyes only.

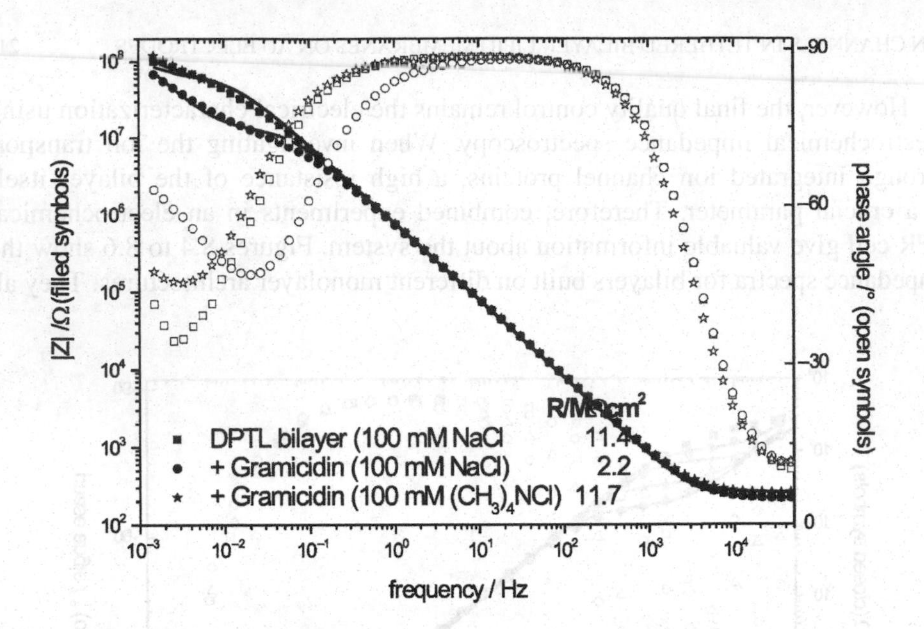

FIG. 8.5. EIS data for the incorporation of gramicidin into a DPTL-based bilayer. The incorporation leads to a significant decrease of the membrane resistance in the presence of sodium ions. An exchange against larger $(CH_3)_4N^+$ ions, which cannot pass the channel, leads to an increase of the membrane resistance. The solid lines represent fits to an equivalent circuit composed of a feed resistor, an RC element for the bilayer, and a capacitance describing the spacer region. The obtained resistance values for the RC element are enclosed in the legend.

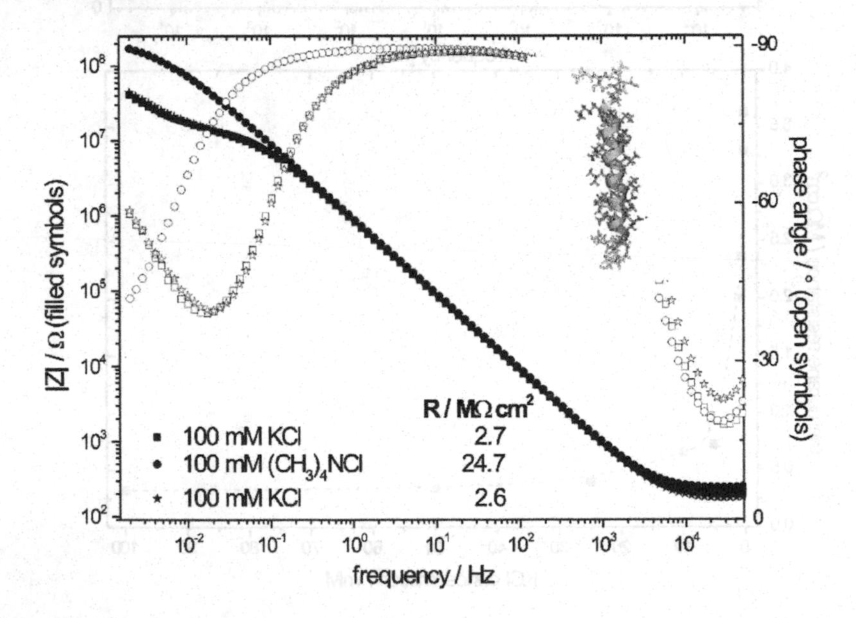

FIG. 8.6. Bode plot for the incorporation of M2 into a DPHT-based membrane. Upon exchange of the electrolyte to more bulky ions, the membrane resistance increases significantly. The inset shows the crystal structure of a M2 monomer.

show very high sealing resistances larger than 1 MΩ cm^2 and capacitances in the same order of magnitude as known from BLM and patch-clamp experiments.

3.2 VALINOMYCIN

Valinomycin is a small ion carrier peptide that is known to diffuse into the membrane (Totu, Josceanu, and Covington 2001; Naumann, Walz, Schiller, and Knoll 2003). It is highly selective for potassium ions, which it can transport from one side of the bilayer to the other. If the peptide is functionally incorporated, the resistance of the membrane should thus decrease with increasing potassium concentration in the electrolyte. In order to avoid any interference due to ionic strength variations of the electrolyte, 100 mM mixtures of KCl and NaCl were used with varying percentages of K^+ ions. Figure 8.4A shows the EIS data of a DPTT-based tBLM (4 EO units, thiol anchor) with incorporated valinomycin for four different electrolyte compositions. The experimental data can be fitted using the equivalent circuit described in the preceding. The obtained numerical values are plotted in Fig. 8.4B as a function of the potassium concentration. With increasing K^+ concentration the resistance decreases as expected, first rapidly then slower until it approaches asymptotically a small fraction of the original bilayer resistance for high potassium concentrations. At the same time, the membrane capacitance remains almost constant. After rinsing with 100 mM NaCl solution, the membrane resistance increases again to a value of 2.6 MΩ cm^2. The final resistance is slightly lower than the initial resistance value of the bilayer. This might be due to an incomplete removal of all potassium ions. It is most probable that some potassium remains in the submembrane reservoir or in the peptide itself.

3.3 GRAMICIDIN

Gramicidin is a small dimeric protein that forms bilayer spanning channels that allow for the passage of monovalent cations (Alonso-Romanowski, Gassa, and Vilche 1995; Jing, Wu, and Wang 1998; Andersen, Apell, Bamberg, Busath, Koeppe, Sigworth, Szabo, Urry, and Wooley 1999). The protein could be incorporated into a preformed bilayer based on the DPTL lipid (4 EO units, lipoic acid anchor). In Fig. 8.5, the EIS spectra of this experiment are shown. The experimental values are fitted using a simple equivalent circuit consisting of a series of a resistor, an RC element describing the bilayer, and a capacitor representing the spacer region. This model somewhat underestimates the complexity of the system, but still gives useful information about certain key properties of the membrane. Furthermore, more sophisticated SPICE modeling using dynamic parameters could confirm the overall trend of the equivalent circuit.

Gramicidin, 100 pg, dissolved in ethanol, was added to the bilayer. The protein spontaneously inserts into the membrane to form a channel, which is permeable for small monovalent ions. The pore formation results in a drop of the membrane resistance by a factor of 5. To prove that this effect is actually due to the function of the protein, larger molecules were used to replace the Na^+ ions. After exchange of the electrolyte solution (NaCl) against 100 mM $(CH_3)_4NCl$, the membrane resistance increases again to the original value. This effect is completely reversible. The different electrolyte solutions have no effect on the resistance of a pure bilayer. The capacitance values both for the bilayer as well as for the spacer region remain constant throughout the entire experiment.

3.4 M2δ

M2 is the major polypeptide component of the membrane pores responsible for the transport activities of ligand-gated ion channels (Oblatt-Montal, Bühler, Iwamoto, Tomich, and Montal 1993; Opella, Marassi, Gesell, Valente, Kim, Oblatt-Montal, and Montal 1999; Montal and Opella 2002). The nicotinic acetylcholine receptor (nAChR) has been one of the most extensively studied members of this superfamily. In the last few years details about the channel's structure and mechanism have been published. Opella and co-workers proposed that the pentameric helical bundle of M2 is the structural blueprint for the inner bundle that lines the pores of neurotransmitter-gated channels, such as acetylcholine and glutamate receptors (Opella, Marassi, Gesell, Valente, Kim, Oblatt-Montal, and Montal 1999; Montal and Opella 2002).

The nAChR is a gated pore composed of a pentameric assembly of subunits. Each subunit consists of the four membrane-spanning segments M1 to M4. The M2 segments (see inset of Fig. 8.6) face the inner side of the channel and thus shape the lumen of the aqueous pore. The straight α-helix formed by the peptides insert into the bilayer at an angle of 12°. The polar residues are found to line the inner side of the pore, whereas the nonpolar residues form the exterior of the bundle. Sévin-Landais and co-workers were able to immobilize the nAChR from *Torpedo californica* in a tethered membrane architecture using membrane patches and they could show binding of ligands and antibodies to the channel using optical techniques (Sévin-Landais, Rigler, Tzartos, Hucho, Hovius, and Vogel 2000). Furthermore, they probed the structural integrity of the receptor using monoclonal antibodies. However, no electrical characterization of the membrane or channel activities has been shown.

Figure 8.6 shows the impedance data for an M2-containing membrane. Peptide containing proteoliposomes were fused with a pre-assembled layer of DPHT (thiol anchor and six EO-units). Because of the formation of the membrane the bare membrane properties cannot be characterized. The peptide is incorporated into the membrane throughout the entire experiment. To verify the functional incorporation, an electrolyte exchange similar to the gramicidin experiment has been performed. The M2 channel is nonconductive for large ions such as $(CH_3)_4NCl$. Therefore, an exchange of the electrolyte from 100 mM KCl to 100 mM $(CH_3)_4NCl$ leads to a significant increase in the membrane resistance, as can be seen in Fig. 8.6. Upon rinsing with KCl the resistance drops back to the initial value.

4 CONCLUSION

Tethered bilayer lipid membranes play an important role in the growing field of biosensing applications. The use of biological components such as membrane proteins as sensing moieties allows one to profit from the high selectivity and specificity nature has engineered into such components. A stable platform is needed in order to assure proper function of the embedded proteins. It should provide a host system for the biological components, mimic natural properties to a maximum, and still ensure long-term stability.

We have developed a wide range of thiolipids that can form bilayer membranes on gold substrates. The membranes provide excellent stability and high electrical sealing properties. At the same time they are still flexible enough to host (small) membrane proteins in a functional form. We could show the functional incorporation of three different membrane proteins: valinomycin, gramicidin, and the M2 subunit of the acetylcholine receptor in three different membrane architectures. The toolbox-like approach for the lipid synthesis leaves further room for improvement. One can envision a molecular design of the membrane platform, e.g., also by using spacer molecules, that is adapted to the specific needs of a chosen membrane protein. Larger proteins, for example, might require extended space both between the anchor lipids and underneath the bilayer.

Ultimately, we would like to downscale the system from what are now macroscopic samples to miniaturized electrode arrays with the goal of even measuring of single-channel fluctuations of embedded membrane proteins. The electrical properties of the membrane should allow such experiments, although technical details of the measurements setup still need to be clarified.

ACKNOWLEDGMENTS

We gratefully thank Dr. J. Long for providing the M2 peptides. Partial financial support came from the Defense Advanced Research Projects Agency (DARPA) through the MOLDICE program. I.K. acknowledges support from the Laboratoire Européen Associé (LEA) on "Polymers in Confined Geometries."

REFERENCES

Abdelghani, A., Jacquin, C., Huber, M., Deutschmann, R. and Sackmann, E. (2001). Supported lipid membrane on semiconductor electrode. Mater. Chem. Phys. 70, 187–190.

Alonso-Romanowski, S., Gassa, L.M., and Vilche, J.R. (1995). An investigation by EIS of gramicidin channels in bilayer lipid membranes. Electrochim. Acta 40, 1561–1567.

Andersen, O.S., Apell, H.-J., Bamberg, E., Busath, D.D., Koeppe, R.E., Sigworth, F.J., Szabo, G., Urry, D.W. and Wooley, A. (1999). Gramicidin channel controversy—the structure in a lipid environment. Nat. Struct. Biol. 6, 609.

Apel, C.L., Deamer, D.W., and Mautner, M.N. (2002). Self-assembled vesicles of monocarboxylic acids and alcohols: conditions for stability and for the encapsulation of biopolymers. Biochim. Biophys. Acta 1559, 1–9.

Atanasov, V., Knorr, N., Duran, R.S., Ingebrandt, S., Offenhäuser, A., Knoll, W. and Köper, I. (2005). Membrane on a Chip. A functional tethered lipid bilayer membrane on silicon oxide surfaces. Biophys. J. 89, 1780–1788.

Atanasov, V., Atanasova, P., Vockenroth, I., Knorr, N. and Köper, I. (2006). Highly insulating tethered bilayer membranes. A generic approach for various substrates. Bioconj. Chem. 17, 631–637.

Bernard, A.-L., Guedeau-Boudeville, M.-A., Sandre, O., Palacin, S., di Meglio, L.-M. and Jullien, L. (2000). Permeation through lipid bilayers be adhesion of giant vesicles on decorated surfaces. Langmuir 16, 6801–6808.

Cornell, B.A., Braach-Maksvytis, V.L.B., King, L.G., Osman, P.D.J., Raguse, B., Wieczorek, L. and Pace, R.J. (1997). A biosensor that uses ion-channel switches. Nature, 387, 580–583.

Fromherz, P., Kiessling, V., Kottig, K. and Zeck, G. (1999). Membrane transistor with giant lipid vesicle touching a silicon chip. Appl. Phys. A 69, 571–576.

Jing, W., Wu, Z., and Wang, E. (1998). Electrochemical study of gramicidin D forming ion-permeable channels in the biolayer lipid membranes. Electrochim. Acta 44, 99–102.

Knoll, W., Frank, C.W., Heibel, C., Naumann, R., Offenhäusser, A., Rühe, J., Schmidt, E.K., Shen, W.W. and Sinner, A. (2000). Functional tethered lipid bilayers. Rev. Mol. Biotechnol.74, 137–158.

Krishna, G., Schulte, J., Cornell, B.A., Pace, R.J. and Osman, P.D. (2003). Tethered bilayer membranes containing ionic reservoirs: selectivity and conductance. Langmuir, 19, 2294–2305.

Köper, I., Schiller, S.M., Giess, F., Naumann, R. and Knoll, W. (2006). Functional tethered bimolecular lipid membranes (tBLMs). In: A. Ottova (Ed.), Advances in Planar Lipid Bilayer and Liposomes. Elsevier, Amsterdam, pp. 37–53.

Montal, M. and Opella, S.J. (2002). The structure of the M2 channel-lining segment from the nicotinic acetylcholine receptor. Biochim. Biophys. Acta 1565, 287–293.

Naumann, R., Schiller, S.M., Giess, F., Grohe, B., Hartman, K.B., Kärcher, I., Köper, I., Lübben, J., Vasilev, K. and Knoll, W. (2003). Tethered lipid bilayers on ultraflat gold surfaces. Langmuir 19, 5435–5443.

Naumann, R., Walz, D., Schiller, S.M. and Knoll, W. (2003). Kinetics of valinomycin-mediated K+ ion transport through tethered bilayer lipid membranes. J. Electroanal. Chem. 550–551, 241–252.

Oblatt-Montal, M., Bühler, L.K., Iwamoto, T., Tomich, J.M. and Montal, M. (1993). Synthetic peptides and four-helix bundle proteins as model systems for the pore-forming structure of channel-proteins. J. Biol. Chem. 268, 14601–14607.

Opella, S.J., Marassi, F.M., Gesell, J.J., Valente, A.P., Kim, Y., Oblatt-Montal, M. and Montal, M. (1999). Structures from the M2 channel-lining segments from nicotinic acetylcholine and NMDA receptors by NMR spectroscopy. Nat. Struct. Biol. 6, 374.

Poghossian, A., Schultze, J.W. and Schöning, M.J. (2003). Application of a (bio-)chemical sensor (ISFET) for the detection of physical parameters in liquids. Electrochim. Acta 48, 3289–3297.

Römer, W. and Steinem, C. (2004). Impedance analysis and single-channel recordings on nano-black lipid membranes based on porous alumina. Biophys. J. 86, 955–965.

Römer, W., Lam, Y.H., Fischer, D., Watts, A., Fischer, W.B., Göring, P., Wehrspohn, R.B., Gösele, U. and Steinem, C. (2004). Channel activity of a viral transmembrane peptide in micro-BLMs: Vpu1-32 from HIV-1. J. Am. Chem. Soc. 126,16267–16274.

Sackmann, E. (1996). Supported membranes: scientific and practical applications. Science 271, 43–48.

Sackmann, E. and Tanaka, M. (2000). Supported membranes on soft polymer cushions: fabrication, characterization and applications. Trends Biotechnol. 18, 58–64.

Schiller, S.M., Naumann, R., Lovejoy, K., Kunz, H. and Knoll, W. (2003). Archaea analogue thiolipids for tethered bilayer lipid membranes on ultrasmooth gold surfaces. Angewandte Chemie 42, 208–211.

Schöning, M.J. and Poghossian, A. (2002). Recent advances in biologically sensitive field-effect transistors (BioFETs). Analyst 127, 1137–1151.

Sévin-Landais, A., Rigler, P., Tzartos, S., Hucho, F., Hovius, R. and Vogel, H. (2000). Functional immobilisation of the nicotinic acetylcholine receptor in tethered lipid membranes. Biophys. Chem. 85, 141–152.

Sofou, S. and Thomas, J.L. (2003). Stable adhesion of phospholipid vesicles to modified gold surfaces. Biosens. Bioelectr. 18, 445–455.

Terretaz, S. and Vogel, H. (2005). Investigating the function of ion channels in tethered lipid membranes by impedance spectroscopy. MRS Bull. 30, 207–210.

Totu, E., Josceanu, A.M., and Covington, A.K. (2001). Improved potassium-selective membrane using valinomycin as ionophore for ion-selective microdevices. Mater. Sci. Eng. C 18, 87–91.

9
Fluorescent Nanocrystals and Proteins

Pier Paolo Pompa, Teresa Pellegrino, and Liberato Manna

Colloidal semiconductor nanocrystals (NCs) have recently gained recognition as versatile bioprobes for advanced molecular and cellular imaging techniques. Fluorescent NCs may in fact overcome some of the typical limitations presented by organic fluorophores, owing to high photobleaching threshold, good chemical stability, broad absorption spectrum, along with the peculiar possibility of accurately tuning their spectral properties (thanks to their size-dependent emission spectrum). The fluorescence of NCs can be precisely tuned in the whole visible spectrum or even the near-infrared region, thus covering a very wide spectral range (from 400 nm to 2.2 μm). The broad absorption spectrum permits efficient excitation at any wavelength shorter than the emission peak, while maintaining the same characteristic narrow, symmetric emission spectrum, regardless of the excitation wavelength. Such spectral feature allows excitation of different sizes of NCs by means of a single wavelength of light, resulting in many fluorescent colors that may be detected simultaneously. Interestingly, a number of reliable procedures to transfer semiconductor nanocrystals from hydrophobic solutions to aqueous environment have been recently developed, in order to exploit NCs as fluorophores in biological processes, which obviously occur in aqueous environments. This chapter illustrates several demonstrations of protein–NCs hybrid systems that have been recently developed by several groups. The implementation of different

A. Offenhäusser and R. Rinaldi (eds.), *Nanobioelectronics - for Electronics, Biology, and Medicine*, 225
DOI: 10.1007/978-0-387-09459-5_10, © Springer Science+Business Media, LLC 2009

conjugation strategies of luminescent semiconductor NCs to proteins is discussed, along with an analysis of the photophysical features of the bio-inorganic nanocomposites. Novel NCs-based sensing schemes and optical processes involving FRET or BRET mechanisms are also examined.

1 COLLOIDAL NANOCRYSTALS AS VERSATILE FLUORESCENT BIOPROBES

Fluorescence is a widely used technique in biology, biochemistry, and biophysics. Labeling of biomolecules using fluorescent markers (e.g., to investigate inter- or intramolecular interactions) is a widespread and very powerful approach. Generally, small fluorescent molecules (e.g., organic dyes) are exploited by the researchers for a large number of biological applications, and several detection strategies have been developed lately. Fluorescent labeling has been proved to be effective for both in vivo cellular imaging and in vitro assay detection (Miyawaki 2003). Moreover, such technique has been successfully employed, not only in single detection schemes, but also for "multiplexing" (i.e., simultaneous detection of multiple signals) (Schrock, du Manoir, Veldman, Schoell, Wienberg, Ferguson-Smith, Ning, Ledbetter, Bar-Am, Soenksen, Garini, and Ried 1996; Roederer, DeRosa, Gerstein, Anderson, Bigos, Stovel, Nozaki, Parks, Herzenberg, and Herzenberg 1997), even though, in the latter case quite complex instrumentation and processing are usually required. For instance, the simultaneous measurement of 10 parameters on cellular antigens was demonstrated in a flow cytometry study, based on a three-laser system and eight-color–marking scheme (Roederer, DeRosa, Gerstein, Anderson, Bigos, Stovel, Nozaki, Parks, Herzenberg, and Herzenberg 1997).

Tagging of biological molecules with organic fluorophores; however, may have significant limitations. Organic and genetically encoded fluorophores are usually characterized by narrow excitation spectra and broad emission bands, often exhibiting extended red tails (Miyawaki 2003). Such spectral features make simultaneous excitation of different dyes difficult in most cases. Also, due to the possible occurrence of non-negligible spectral overlaps between different fluorophores (which may introduce spectral cross-talk between the different detection channels), simultaneous analyses of several probes present in the same sample (e.g., to measure multiple biological indicators simultaneously) are extremely challenging. Hence, ideal fluorescent markers should have a narrow, symmetric emission spectrum, and a broad excitation profile, so that multiple probes can all display useful emission upon excitation at one single wavelength. In addition, organic dyes generally exhibit low photobleaching thresholds, and this may strongly affect their effectiveness in long-term imaging experiments.

In this frame, colloidal semiconductor nanocrystals (NCs), also known as quantum dots (QDs), have recently gained recognition as versatile bioprobes for advanced molecular and cellular imaging techniques. Fluorescent QDs may in fact overcome some of the typical limitations presented by organic fluorophores, owing to the high photobleaching threshold, good chemical stability, broad absorption spectrum, along with the peculiar possibility of accurately tuning their spectral properties (thanks to the size-dependent emission spectrum of QDs) (Bruchez, Moronne, Gin, Weiss, and Alivisatos 1998; Mattoussi, Mauro, Goldman, Anderson, Sundar, Mikulec, and Bawendi 2000) (Figs. 9.1 and 9.2).

In semiconductor nanocrystals, the size dependence of the optical (and electronic) properties, such as absorption, photoluminescence (PL), and electroluminescence, arises from quantum-size confinement effects (Colvin, Schlamp, and Alivisatos 1994; Alivisatos 1996). This results in fluorescence emission maximum, which progressively shifts to longer wavelengths with increasing particle size (Dabboussi, Bawendi, Onitsuka, and Rubner 1995; Hines and Guyot-Sionnest 1996). The broad

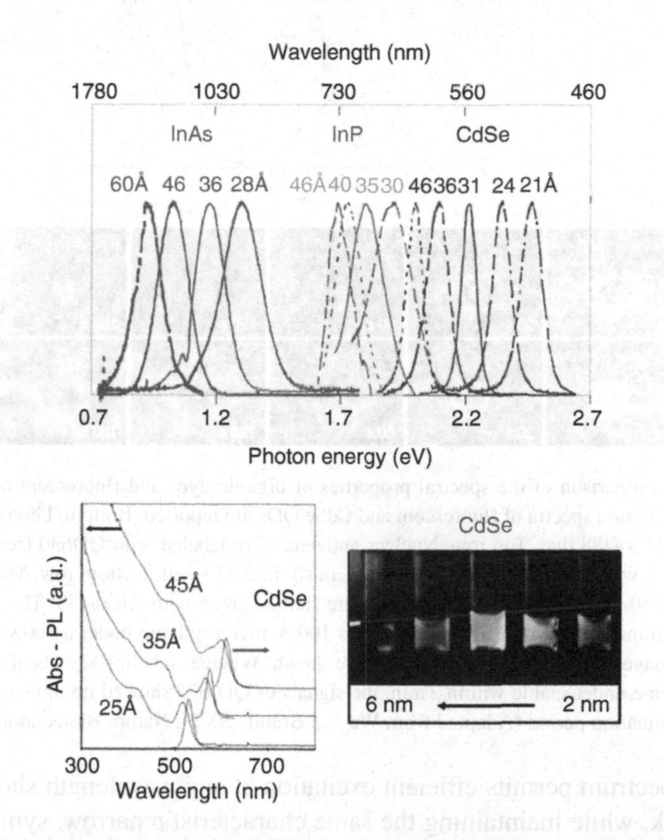

FIG. 9.1. Size-dependent optical properties of QDs. Top. A very wide spectral range (from visible to infrared) can be covered by exploiting different materials and dimensions. Bottom. CdSe nanocrystals. The fluorescence emission progressively shifts to longer wavelengths with increasing particle size. The broad absorption spectrum allows efficient excitation at any wavelength shorter than the emission peak. This also allows simultaneous excitation of different sizes of QDs (i.e., different fluorescence colors) by means of a single wavelength (Adapted from Bruchez, Moronne, Gin, Weiss and Alivisatos (1998). Science 281, 2013.)

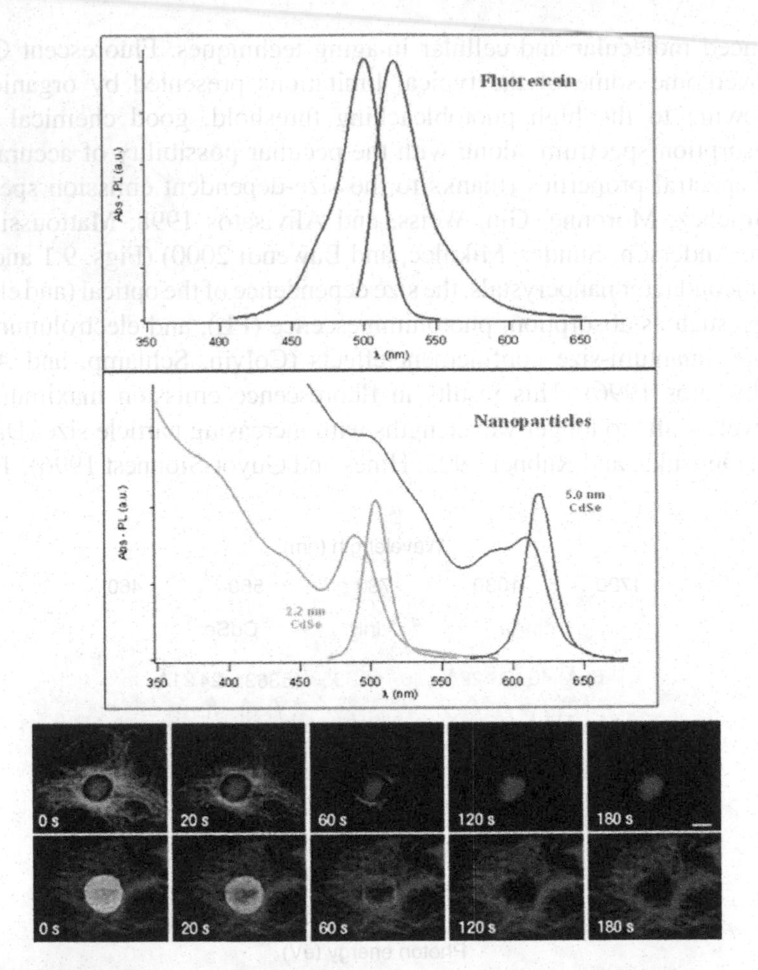

FIG. 9.2. **Top.** Comparison of the spectral properties of organic dyes and fluorescent nanocrystals: the absorption and emission spectra of fluorescein and CdSe QDs are reported. Bottom. Photostability of QDs as compared to Alexa488 dye. Top row. Nuclear antigens were labeled with QD630 (red), and microtubules were labeled with Alexa488 (green) simultaneously in a 3T3 cell. Bottom row. Microtubules were labeled with QD630 (red), and nuclear antigens were stained green with Alexa488. The specimens were continuously illuminated for 3 min with light from a 100 W mercury lamp under a 100X 1.30 oil-immersion objective. Images at 0, 20, 60, 120, and 180 s are shown. Whereas labeling signals of Alexa 488 faded quickly and became undetectable within 2 min, the signals of QD 630 showed no obvious change for the entire 3 min illumination period (Adapted from Wu and Brand (2002). Nature Biotechnol. 21, 41.)

absorption spectrum permits efficient excitation at any wavelength shorter than the emission peak, while maintaining the same characteristic narrow, symmetric emission spectrum (full-width at half-maximum ~25–40 nm, for green–red luminescence), regardless of the excitation wavelength. Moreover, such spectral feature allows excitation of different sizes of QDs by means of a single wavelength of light, resulting in many fluorescent colors that may be detected simultaneously.

Additional QD properties of significant interest for biological applications include the elevated quantum yield in aqueous solutions, the high molar extinction coefficients (typically >10^5 M^{-1} cm^{-1}, namely 10 ÷ 100-fold greater as compared to that of organic dyes), and large effective Stokes shifts (as the excitation may occur at a wavelength far away (>100 nm) from the correspondent QD emission) (Leatherdale et al., 2002 ; Murphy, 2002; Niemeyer, 2001; Dabboussi, Bawendi, Onitsuka, and Rubner 1997; Alivisatos 2004). The fluorescence of QDs can be precisely tuned in the whole visible spectrum (e.g., by using ZnSe, CdS, CdSe, and CdTe nanocrystals) (Steigerwald, Alivisatos, Gibson, Harris, Kortan, Muller, Thayer, Duncan, Douglass, and Brus 1988; Murray, Norris, and Bawendi 1993; Hines and Guyot-Sionnest 1996), or even in the near-infrared region (e.g., by exploiting different materials, such as InAs, InP, and PbSe) (Guzelian, Banin, Kadavanich, Peng, and Alivisatos 1996; Wundke, Potting, Auxier, Schulzgen, Peyghambarian, and Borrelli 2000), thus covering a very wide spectral range (from 400 nm to 2.2 μm). The most employed QDs, for both research and applicative purposes, are typically spherical nanocrystals, with core diameters in the 15- to 120-Å range (Murray, Norris, and Bawendi 1993; Mattoussi, Cumming, Murray, Bawendi, and Ober 1996, 1998), and relatively narrow size distributions (5%) (Murray, Norris, and Bawendi 1993; Hines and Guyot-Sionnest 1996). This latter feature is very important, since the monodispersity results in narrow photoluminescence spectra (Dabboussi, Bawendi, Onitsuka, and Rubner 1997).

Overcoating of the nanocrystal core (e.g., CdSe) with a wider band gap semiconducting material (ZnS or CdS) allows passivity of surface states and induces a decrease of the leakage of excitons outside the core (Hines and Guyot-Sionnest 1996; Dabboussi, Bawendi, Onitsuka, and Rubner 1997; Peng et al. 1997). This procedure, because of the strong enhancement of the luminescent quantum yield, along with the improvement of photochemical stability and resistance to photobleaching (without any perturbation of the other spectral features of QDs emission) led to the development of core-shell nanocrystals. All these properties make colloidal semiconductor nanoparticles ideal probes for labeling biomolecules for fluorescent tagging applications.

2 SYNTHESIS OF SEMICONDUCTOR NANOCRYSTALS

Semiconductor nanocrystals have a strong dependence of properties on size. It is therefore desirable to develop synthetic approaches to nanocrystals that yield particles with a distribution of sizes that is as narrow as possible. More specifically, a good synthetic approach should produce nanoparticles that are highly crystalline, well dispersed in a solvent, with a narrow size distribution and possibly with the control of their shape. Today, thanks to various refined synthetic approaches, colloidal inorganic nanocrystals can be grown of different materials, including metals, semiconductors, and insulators, all with the desired requisites as outlined. In one of the most successful approaches, based on the thermal decomposition of precursors in hot surfactants (see a recent review on the topic; Donega, Liljeroth, and Vanmaekelbergh 2005), the growing medium is a liquid mixture of surfactants and the whole synthesis process is carried out at high temperatures under inert atmosphere.

Surfactant molecules are composed of a polar head group and of one or more hydrocarbon chains, which form the hydrophobic part of the molecule. Surfactants commonly used in a colloidal NCs synthesis include polymers, alkyl thiols, amines, carboxylic and phosphonic acids, phosphines, phosphine oxides, phosphates, phosphonates, and various coordinating solvents.

In this "hot injection scheme," each of the atomic species that will be incorporated in the nanocrystal is introduced in the reactor as a molecular precursor. Once the precursors are injected in the reaction flask, at the reaction temperature they decompose and release the atomic species that will be responsible for the nucleation and for the growth of the nanocrystals. Surfactants play an important role in the growth and stability of nanocrystals. Each nanocrystal is in fact coated by a monolayer of surfactants, which are bound to the nanocrystal surface via their polar head groups, while they expose their hydrophobic tails to the outer environment. This binding, however, is dynamic. During growth, surfactants absorb and desorb continuously from the surface of nanocrystals, allowing for the controlled addition or removal of atomic species from their surface, and thus the growth or the shrinkage of nanocrystals.

The growth of nanocrystals depends on a number of parameters, such as the nanocrystal surface energy, the concentration of free species in solution, and nanocrystal size, and can be tuned by varying the relative kinetic and thermodynamic parameters of the reaction. The choice of high temperatures and availability a wide variety of surfactants and precursor molecules expands the range of materials that is possible to synthesize with respect to other techniques. One can in fact identify a proper combination of surfactants and a reaction temperature that is suitable for growing a given material. In addition, at high temperatures, various defects in the crystal lattice of the nanoparticle, which can form during the synthesis, are annealed out. Therefore, the combination of specific surfactants and high temperatures facilitates the formation of NCs with narrower size distributions, fewer internal defects, and a more uniform surface reconstruction, hence having well-defined physical properties (e.g., strong luminescence in a relatively narrow interval of energies from semiconductor nanocrystals). In addition, when the synthesis is stopped by lowering the reaction temperature, the surfactant coating around the NCs remains tightly bound to their surface and guarantees their full solubility in a variety of solvents (from nonpolar solvents to moderately polar solvents).

3 WATER SOLUBILIZATION STRATEGIES

In order to exploit semiconductor nanocrystals as fluorophores in biological processes, which obviously occur in aqueous environments, a hydrophilic nanocrystal surface is desired. Unfortunately, as most semiconductor nanocrystals are synthesized by the hot injection method described in the preceding, their surface is hydrophobic. Therefore, several procedures to transfer nanoparticles from hydrophobic solutions to an aqueous environment have been developed so far (Table 9.1 and Fig. 9.3).

QDs as well as amphiphilic molecules can be employed to transfer surfactant-capped QDs in water. Amphiphilic molecules have a hydrophobic portion that can strongly interact with hydrocarbon chains of the surfactants and a hydrophilic portion pointing to the aqueous phase, which ensures the water solubility of the QDs. Also in this case, a cross-linking step can be performed to better stabilize the shell.

In the first reported works on biological applications of QDs, the nanocrystals employed were transferred in water by means of stabilizing molecules. This strategy, which was rapidly extended to other types of nanocrystals, was based on the replacement of the hydrophobic surfactant's molecules, which coated the surface of the as-grown nanocrystals with molecules that had a functional group capable of binding directly to the nanocrystal surface and a hydrophilic portion that bears

TABLE 9.1. An overview of the various water solubilization procedures of QDs.

Ligand exchange	Ligand exchange and cross-linking	Amphiphilic coating
Mercaptoalchylcarboxy derivatives		
Mercaptoacetic acid (MAA) (Mitchell et al. 1999)	Silanization (Parak et al. 2002)	Phospholipids micelles (Dubertret et al. 2002)
Mercaptopropionic acid (MPA) (Wielard et al. 2001)		
Mercaptoundecanoic acid (MUA) (Aldana et al. 2001)		
16-Mercaptohexadecanoic acid (MHDA) (Aldana et al. 2001)		
Dithiol ligands	Oligomeric ligands (Kim et al. 2003)	Triblock copolymers (Wuister et al. 2003; Gao et al. 2004)
Dihydrolipoic acid (DHLA) (Mattoussi et al. 2000)		
Ditiothreitol (DTT) (Pathak et al. 2001)		
Adesive domain peptides (Pinaud et al. 2004)	Dendrimers bridging (Guo et al. 2003)	Amphyphilic polymers (Larson et al. 2003; Pellegrino et al. 2004)

Categorized by the three main different approaches: ligand exchange, ligand exchange followed by cross-linking, and the use of amphiphilic molecules.

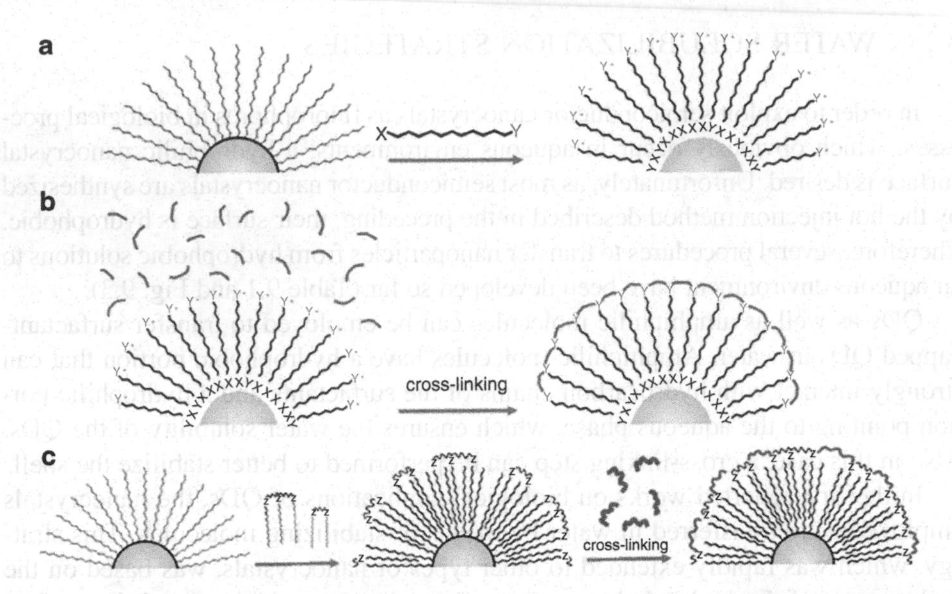

FIG. 9.3. Scheme showing the general approaches developed so far for water solubilization of colloidal QDs. A. Ligand exchange procedure. The surfactant molecules on the QDs surface can be displaced by bi-functional molecules that have at one end a functional group that can strongly bind the QDs' surface and at the other end a group that confers charge to the nanoparticles surface and thus stabilizes the nanoparticles in water by electrostatic repulsions. B. Ligand exchange and cross-linking procedure. To improve the stability of the shell obtained after ligand exchange, a cross-linking step is required. Cross-linker molecules are properly chosen to be reactive to more than one functional group present on the outmost shell of the QDs' surface; thus they can bridge the ligand molecules on the surface of the nanoparticles. C. Amphiphilic molecule-coating procedure.

a charge (e.g., a carboxylate group) (Chen, Ji, and Rosenzweig 2003; Mayya and Caruso 2003). These molecules made the nanocrystals soluble in water by means of electrostatic repulsion, as nanoparticles of equal charge repel each other, thus preventing their aggregation. This so-called ligand or surfactant exchange approach is rather straightforward and easy. A practical example of application of this method is the water solubilization of CdSe/ZnS nanocrystals coated by a layer of trioctylphosphine oxide (TOPO), which is the most common surfactant used in the synthesis of this type of nanocrystal. The layer of TOPO molecules is responsible for the hydrophobicity of these particles, which are soluble in solvents such as toluene or chloroform (Chan and Nie 1998). The TOPO layer is exchanged with a layer of a mecaptocarboxylic acid, such as mercaptoacetic acid (Willard, Carillo, Jung, and Van Orden 2001), mercaptopropionic acid (Mitchell, Mirkin, and Letsinger 1999), or mercaptodecanoic acid (Aldana, Wang, and Peng 2001). Through this process, the mercapto groups are anchored to the QD's surface. The carboxylic groups, on the other hand, are already negatively charged in water at neutral pH; therefore, nanocrystals repel each other electrostatically.

Although the surfactant exchange procedure as described is simple and direct, some drawbacks are associated with it. In electrolytic solutions, charges are in fact shielded by counter ions and the electrostatic repulsions can be weakened considerably. A sufficiently high salt concentration can shield electrostatic repulsion so effectively that it can lead to interparticle aggregation. One additional disadvantage of this method is that the charge of the stabilizing molecules depends on the pH of the solution. The carboxylic groups (COOH), for instance, are deprotonated at pH values above 5 to 6, and so they are negatively charged, whereas for pH values below 5 to 6 they are deprotonated, and so they are neutral. Nanoparticles stabilized by carboxylates are therefore stable only at neutral or alkaline pH conditions. A pH dependence of the particle stability in solution is clearly not desirable, as it makes any further processing step (e.g., bioconjugation to a variety of molecules) less straightforward. Furthermore, the surfactants are directly bound to the surface of the nanocrystals; hence the functional group for these stabilizing molecules must be tailored to have a high affinity for the specific material of which the nanocrystal is made. Therefore, appropriate stabilizing molecules have to be chosen individually for every particle material. For example, thiols are known to bind strongly to gold nanocrystals but less to CdSe QDs, whereas amines bind relatively strongly to cobalt nanocrystals (Brust, Walker, Bethell, Schiffrin, and Whyman 1994; Kobayashi, Horie, Konno, Rodriguez-Gonzalez, and Liz-Marzan 2003). As a matter of fact, no general procedure exists that can work for all types of colloidal nanocrystals, because there is no universal binding group. In addition, in several cases these stabilizing molecules can detach easily from the nanocrystal surface, and they can even undergo degradation processes. Thiol (-SH) functionalized molecules, commonly used for binding to the surface of fluorescent CdSe/ZnS nanoparticles suffer from several drawbacks. The binding affinity of thiols to ZnS surfaces is only moderate; thus these nanoparticles are not stable in aqueous solution for long periods (Aldana, Wang, and Peng 2001). Eventually, the thiol-ZnS bonds break and the stabilizing molecules come off from the particle surface. This occurs because either the thiol group is hydrolized or photo-oxidized.

To improve the ligands' affinity for the surface of nanocrystals, stabilizing molecules have been employed that have more than one site available for coordination to the particle surface, such as dithiol-functionalized ligands (e.g., dihydrolipoic acid and ditiothreitol) (Mattoussi, Mauro, Goldman, Anderson, Sundar, Mikulec, and Bawendi 2000; Pathak, Choi, Arnheim, and Thompson 2001). A more elaborate example of such a strategy is the use of engineered proteins that have a positively charged polylysine chain to solubilize negatively charged nanocrystals. This approach has been proposed by Mattoussi and co-workers (Mattoussi, Mauro, Goldman, Anderson, Sundar, Mikulec, and Bawendi 2000). The proteins decorate the nanocrystal surface via electrostatic interactions with the charged molecules already present on the nanocrystal surface. As a matter of fact, the overall method suffers from the same constraints of the "electrostatic" ligand exchange procedure as described. In a more elegant approach, Pinaud and co-workers (Pinaud, King, Moore, and Weiss 2004) solubilized TOPO-coated CdSe/ZnS nanocrystals in water by coating their surface with phytochelatin-related peptides. These peptides were characterized by short amphiphilic sequences. In their approach, the adhesive domain of the peptide stuck to the nanocrystals' surface and guaranteed their solubilization in the organic phase, whereas the hydrophilic domain ensured the solubility in water. This domain can be modified to include functional groups for bioconjugation. The sticky domain is usually a sequence of cysteine residues flanked by hydrophobic derivates alanine residues, which are able to bind to the nanocrystals' surface by means of cysteinyl thiolates. In order to increase the binding efficiency of the peptide and better fit the surface curvature of nanoparticles of different sizes, the length of the adhesive domain can be changed by varying the number of cysteine residues present in the adhesive domain (Parak, Gerion, Zanchet, Woerz, Pellegrino, Micheel, Williams, Seitz, Bruehl, Bryant, Bustamante, Bertozzi, and Alivisatos 2002).

A stabilization mechanism that is alternative to electrostatic repulsion is based on steric repulsion. In this case, the nanocrystals are coated by ligand molecules that form some sort of hydrophilic brushes around the nanoparticles' surface. These brushes prevent inorganic particle cores from touching each other. Typical molecules used for this purpose are polyethylene glycol (PEG) (Wilhelm, Billotey, Roger, Pons, Bacri, and Gazeau 2003) and dextrane (Alejandro-Arellano, Ung, Blanco, Mulvaney, and Liz-Marzan 2000). Such molecules also replace the surfactant molecules at surface of the nanoparticles.

More advanced water solubilization procedures, still based on the ligand exchange scheme, have been developed recently. In these approaches, ligand exchange is carried out by molecules that can be cross-linked to each other once they have been anchored to the surface of the nanocrystals; therefore, the new layer of stabilizing molecules forms a cross-linked, organic shell around the original nanocrystal. One remarkable advantage of this approach is that even if the bond between one stabilizing molecule and the atoms at the nanocrystal surface breaks, this molecule is still kept in place by the network formed with the neighboring molecules. Therefore, the overall stability of the nanoparticles is greatly improved. Also in these cases, several priming molecules have been tested to realize the initial exchange of the surfactant that coats the nanocrystals and achieve an efficient cross-linkage.

Surface silanization is a typical example of surfactant exchange followed by cross-linking and is a powerful strategy for preparing water-soluble nanocrystals. It consists mainly in the growth of a glass shell around the nanocrystals, and was first developed for CdSe/ZnS nanoparticles. The silanization procedure is now a well-established technique, and recent works have extended this type of coating to different classes of semiconductor nanocrystals as well as to metal nanoparticles(Holzinger et al. 2006 (Correa-Duarte, Giersig, and Liz-Marzan 1998; Alejandro-Arellano, Ung, Blanco, Mulvaney, and Liz-Marzan 2000). In the first step of the protocol, the surface of hydrophobic nanoparticles, such as TOPO-coated CdSe/ZnS nanocrystals, is functionalized (primed) with a mercaptosilane molecule via the formation of a covalent bond (e.g., a thiol-Zn bond). The mercapto groups bind to the ZnS surface and displace the original surfactant molecules (TOPO). Heating promotes the hydrolyzation of the trimethoxisilane groups, followed by their cross-linking. The cross-linkage stabilizes further the silane molecules on the surface of the nanocrystals as it leads to the formation of a priming layer made of a lattice of siloxane bonds that encases the nanocrystals. In the second step, the nanocrystals are exposed to additional "hydrophilic" trimetoxisilane molecules, which can carry charged (e.g., phosphonates) or neutral groups (e.g., PEG) or various other functional groups (e.g., thiols, amines, or carboxylic acids). These new trimethoxisilane molecules are then cross-linked with the priming layer, again through the formation of siloxane bonds, and also with each other, forming a thin silica shell around the nanocrystal. This shell exposes to the external environment the functional groups of the "hydrophilic" trimethoxisilane molecules that were used in the second step, as well as unreacted silanol groups. Because of the presence of these unreacted groups, another step is required to stop the growth of the shell. This is done by reacting the free silanol groups with a quenching reagent (trimethylchlorosilane). By varying the composition of the hydrophilic trimethoxysilanes, it is possible to synthesize water-soluble nanocrystals with a specific net charge, which can range from positive to neutral to negative (Parak, Gerion, Zanchet, Woerz, Pellegrino, Micheel, Williams, Seitz, Bruehl, Bryant, Bustamante, Bertozzi, and Alivisatos 2002) and with the desired functional groups on the surface, which are useful for performing further surface functionalization reactions (i.e., with biomolecules).

Based on a similar cross-linking scheme, also oligomeric polydentate phosphine ligands have been proposed to cage TOPO-coated CdSe QDs (Kim and Bawendi 2003). The shell was designed to realize a case around the QDs consisting of an inner phosphine layer, a cross-linking layer, and a outer functionalized layer. Initially, the TOPO on the QDs surface was exchanged with a monomeric alkyl phosphine (tryhydroxy propylphosphine). A following oligomerization step was carried out by using diisocyanatohexane as cross-linker, which led to the formation of a thin shell that encloses the QDs. The choice of oligomeric phosphines that bear also carboxylic groups enabled the shell to be processed further, for instance, by undergoing bioconjugation reactions via the carboxylic groups.

One major problem of functionalized nanoparticles is that over time they can suffer from degradative processes, such as photo-oxidation. In order to limit or delay such processes, it would be desirable to have a thick and densely packed ligand layer at the nanocrystal surface. For this purpose, hyperbranched organic dendron ligands have been exploited as a valid alternative to the various functionalization methods described in the preceding (Wang, Li, Chen, and Peng 2002; Guo, Li, Wang, and Peng 2003). These dendrons are branched organic molecules equipped with thiol moieties that serve as anchoring groups to the nanocrystal surface. The multiple hydroxyl terminal groups of each OH-dendron make it feasible to further cross-link the dendrons molecules with each other by means of simple cross-linking chemistry. To this aim, a two-phase approach has been employed. First, hydrophobic QDs and dendrons were dissolved together in an organic solvent. An aqueous phase was then added to the organic phase and after incubation the QD–dendron conjugates were transferred in the aqueous phase. At this point, the terminal hydroxyl groups on the dendrons were first activated and then reacted with a dendrimer bridging molecule, which formed a compact outer shell that was rich in amino groups. These amino groups could be converted to a wide variety of activated molecules (e.g., NHS ester, EDC, Schiff base, anhydride groups) that could be processed further for performing standard coupling reactions to biomolecules. In analogy with the silanization scheme, also this dendron-based approach offers a toolkit of differently functionalized nanoparticles, which are activated with versatile linker molecules, and as such they are ready for bioconjugation reactions.

Water-soluble nanoparticles prepared by ligand exchange followed by cross-linking, due to the highly cross-linked shell, have the indisputable advantage of remaining stable in a large range of pH and salt concentrations. In fact, stability in physiological conditions has been demonstrated in several protocols. Moreover, nonspecific binding to cells can be controlled by manipulating the surface charge of the nanocrystals. Finally, good control of the thickness of the cross-linked shell can be achieved by choosing the appropriate primer ligand and cross-linker molecules. This results in more stable optical properties of water-soluble QDs.

Based on what has been discussed so far, it is clear that simple ligand exchange of the stabilizing molecules, although still frequently used, is not the optimal method to water-solubilization of nanoparticles; ligand exchange followed by cross-linking is definitely more advantageous. Yet, the first step in all these procedures relies in the binding of specific molecules to the surface atoms of the nanocrystals. As such it varies by material; therefore, no general procedure can be devised. To overcome these problems in part, strategies have been developed that do not involve ligand exchange. These methods are based on the addition of an extra layer around the original hydrophobic coating of surfactant molecules. The added layer consists of amphiphilic molecules that can intercalate the first hydrophobic surfactant layer with their hydrophobic portion, and that ensure water-solubility of the nanocrystal with their hydrophilic groups. The amphiphilic molecules employed to make CdSe/ZnS nanocrystals soluble in water for instance include different classes of compounds, such as phospholipids, amphiphilic polymers, and tri-block copolymers. Some of the main strategies are briefly described in the following.

Dubertret and co-workers (Dubertret, Skourides, Norris, Noireaux, Brivanlou, and Libchaber 2002) showed that it is possible to encapsulate individual QDs along with their hydrophobic ligands in the core of block-copolymer micelles. The spherical micelles employed were made of a mixture of n-poly(ethylene glycol) phophatidylethanolamine and phosphatidylcholine. The nanometer sizes of the resulting micelles were such that each of them could encapsulate a single nanocrystal. The procedure is simple and straightforward, as it is based on a single reaction step. The QDs and phospholipids were first dissolved together in an organic solvent. The slow removal of the solvent led to the formation of micelles, each encapsulating a single QD. Afterward, the addition of water led to the solubilization of QD–micelle structures. In addition, the use of amino-functionalized PEG molecules allowed the introduction of functional groups, so that it was possible to bind biomolecules to these QD–micelle structures. It is worth noting that in this approach no cross-linking step was required. Moreover, the resulting QD–micelles were stable and the QDs retained their original photophysical properties.

An alternative route to high-quality water-soluble QDs nanocrystals has been developed recently and is based on an amphiphilic polymer as a coating for hydrophobic CdSe/ZnS nanocrystals (Larson, Zipfel, Williams, Clark, Bruchez, Wise, and Webb 2003). In this procedure, the hydrophobic tails of the polymer intercalate with the surfactant molecules of the nanocrystal and form an additional shell around the nanocrystals. The water solubility of the polymer-coated nanocrystals is ensured by the hydrophilic groups that are located at the outer region of the polymer shell. Finally, the polymer shell is stabilized by cross-linking. More specifically, octylamine-modified polyacrylic acid was chosen as polymer. It was dissolved in organic phase together with the hydrophobic TOPO-capped QDs. After solvent evaporation, the dried film was re-dissolved in water and purified from the excess of polymer by gel filtration. Afterward, the surface coating was cross-linked by PEG-modified lysine using EDC chemistry. Based on a similar scheme, Pellegrino and co-workers (Pellegrino et al. 2004) developed a general strategy for decorating hydrophobic nanocrystals of various materials, including $CoPt_3$, Au, CdSe/ZnS, and Fe_2O_3 with a hydrophilic polymer shell, by exploiting the nonspecific hydrophobic interactions between the alkyl chains of poly(maleic anhydride alt-1-tetradecene) and nanocrystal surfactant molecules. Addition of bis(6aminohexyl)amine resulted in the cross-linking of the polymer chains around each nanoparticle. The nanocrystals became soluble in water upon hydrolyzation of the unreacted anhydride groups (which effectively leads to an amphiphilic polymer shell) and could be further processed according to a universal protocol that relies solely on the chemistry of the outer polymer shell. The remarkable advantage of this procedure is that, since it does not involve ligand exchange, it can be applied to almost any hydrophobically capped nanocrystal, regardless of the type of inorganic core.

One last example of amphiphilic molecules as a way to solubilize QDs in water can be found in recently reported work on new QD probes suitable for in vivo targeting, in which QDs were encapsulated in an amphiphilic tri-block copolymer (Gao, Cui, Levenson, Chung, and Nie 2004). The tri-block copolymer had two hydrophobic sections (polybutylacrylate and polymethylacrylate), a hydrophilic section (polymethilacrylic acid), and hydrocarbon side chains. Because of the strong hydrophobic interaction between the alkyl chains of the TOPO bound to the nanoparticles' surface and the side chains of the tri-block copolymer, a spontaneous assembly of the polymer around each QD was achieved. The resulting water soluble nanoparticles showed extremely stable fluorescence in a wide range of pH and salt conditions.

In conclusion, several reliable methods have been developed to stabilize nanocrystals in aqueous solution. Although at this point there is no optimal protocol available that includes all the advantages of the individual procedures, state-of-the-art nanocrystals have reached a degree of performance for what concerns stability in aqueous solutions that is sufficient for most biological experiments. One conceptual advantage of fluorescent colloidal nanocrystals to organic fluorophores is evident. Organic fluorophores with different colors of fluorescence are typically just different molecules; thus each of them requires tailored linking strategies to molecules/biomolecules. On the contrary, the optical properties of colloidal nanocrystals are defined by the inorganic semiconductor core, whereas their surface chemistry is defined by their outer coating, which can be identical for different types of nanocrystals. Therefore, general functionalization strategies can be developed for them.

4 PROTEIN–QD HYBRID SYSTEMS

One of the first demonstrations of protein conjugation to luminescent semiconductor QDs was reported by Chan and Nie (1998). The authors showed that nanometer-sized QDs can be used as efficient fluorescent probes, whereas attached biomolecules recognize specific analytes, such as proteins, DNA, or viruses. Remarkably, the nanoconjugates were shown to be biocompatible and suitable for use in cell biology and immunoassay. Mercaptoacetic acid was used for QDs solubilization and covalent protein attachment. As explained, when reacted with ZnS-capped CdSe QDs in chloroform, the mercapto group binds to a Zn atom, and the polar carboxylic acid group renders the QDs water soluble. The free carboxyl group is also available for covalent coupling to various biomolecules (e.g., proteins, peptides, and nucleic acids) by cross-linking to reactive amine groups. TEM analysis showed that the solubilization and cross-linking steps did not result in aggregation, and the QD bioconjugates were primarily single particles. Moreover, the authors observed that the optical properties of QDs were not significantly perturbed upon solubilization and conjugation. (The emission spectra and efficiencies of the water-soluble QDs were substantially unchanged with respect to the original sample, in chloroform.) The photophysical properties of the QD conjugates were demonstrated to be superior as compared with common organic dyes; the QDs' fluorescence spectrum was narrow and symmetric (no red tails), and their emission was nearly 100-fold more stable against photobleaching with respect to rhodamine 6G.

The authors also demonstrated the possibility of obtaining fluorescence images from cultured HeLa cells that had been incubated with transferrin–QD bioconjugates (mercapto–QDs served as control). Interestingly, when transferrin was present, receptor-mediated endocytosis occurred, and the luminescent QDs were transported into the cell. (In contrast, in the absence of transferrin, no QDs were observed inside the cell.) Therefore, the attached transferrin molecules were still active and were recognized by the receptors on the cells' surface. (The QD labels did not significantly interfere with ligand-receptor binding or endocytosis.) Furthermore, QD labels also were found to be suitable for sensitive immunoassays. QD-immunoglobulin G (IgG) conjugates were incubated with bovine serum albumin (BSA) and with specific polyclonal antibody. Fluorescence imaging revealed that the polyclonal antibody could recognize the immunoglobulin, leading to extensive aggregation of the QDs, whereas well-dispersed and primarily single QDs were observed in the presence of BSA. This means that the attached immunomolecules can recognize specific antibodies or antigens.

A direct method for conjugating protein molecules to luminescent CdSe/ZnS core/shell QDs was later developed by Mattoussi and co-workers (Mattoussi, Mauro, Goldman, Anderson, Sundar, Mikulec, and Bawendi 2000), with the aim of using the hybrid assemblies as bioactive fluorescent probes in sensing and other diagnostic applications. A chimeric fusion protein was designed by the authors in order to electrostatically bind to the oppositely charged surface of lipoic acid-capped QDs. Protein–QD conjugates retained a high quantum yield, along with biological activity, and no significant particle aggregation was observed. Remarkably, this approach combines the advantages of lipoic acid-capped QDs (photochemical stability and water solubility along with the usual interesting spectral features of QDs, such as the possibility of size-tuning the emission wavelengths) with the undemanding electrostatic conjugation of bioactive proteins to QDs.

The conjugation strategy was based on self-assembly, driven by electrostatic attractions, between the negatively charged lipoic acid capped CdSe-ZnS nanocrystals and engineered bifunctional recombinant proteins consisting of positively charged attachment domains (containing a leucine zipper) genetically fused with desired biologically relevant domains. The alkyl-COOH terminated capping groups of QDs allowed dispersion of the nanocrystals in aqueous solutions at basic pH, and also provided a surface charge distribution that can promote direct self-assembly with other molecules that have a net positive charge. (Importantly, this approach can be easily extended to a variety of different core-shell nanocrystals.) The coding DNA sequence for the two-domain model maltose binding protein-basic zipper fusion protein (MBP-zb) was constructed using standard gene assembly and cloning techniques. Conjugation of lipoic acid capped nanoparticles with MBP-zb protein was carried out by simply mixing dissolved fusion protein with QDs in buffered media. This procedure yielded stable self-assembled QD–protein conjugates (which retained the spectroscopic properties of the starting QDs) in a few minutes.

Interestingly, the authors observed that, upon conjugation with MBP-zb, QDs displayed a significant enhancement of their fluorescence emission. The fluorescence quantum yield increased from ~10% for unconjugated CdSe-ZnS QDs to ~20% to 30% for QD/MBP-zb bioconjugates. In addition, the bioinorganic conjugates were shown to preserve the functional properties of MBP, namely, the binding to amylose affinity resin and displacement from the resin by soluble maltose. The increase of QDs' fluorescence emission was found to be dependent on the protein:QD molar ratio, revealing that each QD can bind multiple MBP-zb molecules. A saturation behavior was observed at a molar ratio of ca. 15 to 20 proteins per QD, approximately corresponding to a maximum enhancement in the emission of about 2.5-fold (as compared with unconjugated QDs). Examination of the ionic strength dependence of the protein–QD interaction as a function of increasing NaCl concentration revealed a poor salt dependence up to at least a 1 M concentration. The enhancement of fluorescence efficiency observed in the experiments was explained by the authors in terms of a surface charge neutralization of QDs, induced by the self-assembly process. This is likely to modify the gross electrostatic/polar environment of the inorganic core, and affect the efficiency of core electron–hole recombination (thus leading to the observed effect on the overall luminescence intensity). The maximum number of MBP-zb proteins than can be attached to a single QD was theoretically estimated via steric considerations (Mattoussi, Mauro, Goldman, Anderson, Sundar, Mikulec, and Bawendi 2000). This value, which depends on the relative sizes of the QD compared with the protein, was calculated to be ~19 proteins per QD. Remarkably, the predicted value was in very good agreement with the experimental data derived from fluorescence measurements.

The superior spectral features of luminescent QDs have also been exploited for several FRET-based investigations, due to the possibility of overcoming some of the limitations arising from the use of organic dyes as the donor-acceptor pairs. In one of the first studies, a biotin-streptavidin binding assay was demonstrated, by using CdSe/ZnS core/shell QDs as the FRET donors (Willard, Carillo, Jung, and Van Orden 2001). Specific binding of streptavidin, labeled with tetramethylrhodamine (SAv-TMR), to biotinylated bovine serum albumin (bBSA), conjugated to water-soluble QDs (QD-bBSA), was shown by the authors (Fig. 9.4); the binding process was assessed by monitoring the enhancement of the TMR fluorescence induced by QD → TMR resonance energy transfer (Fig. 9.4).

Importantly, the size of the QDs was chosen in order to maximize the spectral overlap between the donor emission and acceptor absorption spectra, while avoiding significant overlap between the donor and acceptor emission (Fig. 9.4, bottom).

Moreover, knowledge of the photophysical properties of the QDs and TMR species, along with calculation of their spectral overlap, led to a Forster radius (R_0)

of 54 Å for this donor-acceptor pair, which is comparable to the highest R_0 values obtained for the most commonly used organic dye pairs in FRET applications (Van Der Meer, Coker III, and Chen 1994; Wu and Brand 1994).

A 400-nm excitation was used for all the samples. At this wavelength, QDs excitation occurs with high efficiency, whereas direct TMR excitation is minimal. Upon specific binding of SAv-TMR to QD-bBSA, a strong enhancement in the TMR fluorescence was observed, owing to an efficient QD → TMR FRET process. Interestingly, the fluorescence enhancement was found to be dependent on the SAv-TMR to QD-bBSA molar ratio (up to approx a fourfold increase). However, the authors observed that, although QD → TMR FRET is clearly responsible for the enhanced TMR fluorescence, other energy transfer processes (in addition to FRET) also contributed to the quenching of the QD emission. This was particularly evident at low SAv-TMR concentrations, where only 10% of the QD emission was found to be quenched via FRET mechanisms. (This value, however, increased up to ~50% at higher SAv-TMR concentrations.) The additional quenching of the QD emission was explained by the authors in terms of QD self-quenching (due to the formation of QD aggregates, possibly induced by the binding of multiple biotin moieties to a single SAv-TMR molecule) and nonspecific interactions between the TMR molecules and the QD surfaces (e.g., QD to TMR electron transfer). In any case, apart for the intricate interaction mechanisms responsible for QD fluo-

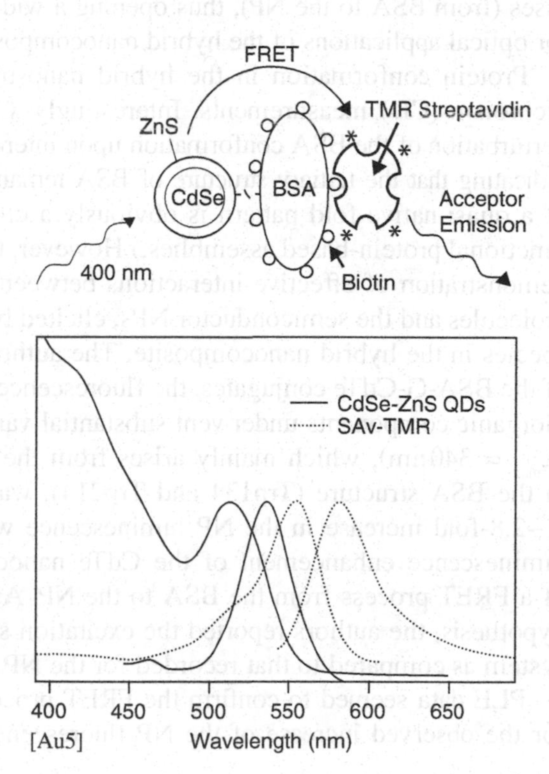

FIG. 9.4. **Top.** Schematic of the FRET-based binding assay. Bottom. Normalized absorption and emission spectra of CdSe/ZnS QDs conjugated to bBSA and SAv-TMR (Adapted from Willard, Carillo, Jung, and Van Orden 2001.)

rescence quenching, the possibility of probing the specific binding of different proteins via FRET was demonstrated in this study. An important feature of the proposed strategy was that the binding resulted in a strong enhancement of the dye fluorescence, which was well resolved from the donor (QD) emission spectrum. This method has a wide applicability and can be easily extended to the design of novel binding assays of great interest, such as antibody-antigen binding, DNA hybridization, and enzyme–substrate interactions.

The possibility of realizing an optical coupling between a semiconductor nanoparticle and a protein, upon conjugation of the two species, was shown by Mamedova and co-workers (Mamedova, Kotov, Rogach, and Studer 2001). A fluorescent, amino acid–coated, CdTe nanocrystal was conjugated with bovine serum albumin (BSA), a typical plasma protein, with the perspective of using the hybrid nanostructure for both the implementation of various supramolecular assemblies and the study of BSA interactions with live blood cells. Conjugates were prepared by applying glutaric dialdehyde (G) linkers to the proteins, and by using hydrophilic stabilizer L-cysteine to coat (stabilize) the nanoparticles. The authors found that BSA:CdTe conjugates could be produced in a controlled way (as probed by gel-electrophoresis analyses), with dimers (1:1) as the most abundant species (although a weak amount of some 2:1 BSA:CdTe assemblies was also present). Moreover, they showed that the protein and the NP could interact via fluorescence resonance energy transfer (FRET) processes (from BSA to the NP), thus opening a wide range of interesting possibilities for optical applications of the hybrid nanocomposite.

Protein conformation in the hybrid nanostructure was assessed by circular dichroism (CD) measurements. Interestingly, CD spectra revealed only a small perturbation of the BSA conformation upon interaction with the CdTe nanoparticle, indicating that the tertiary structure of BSA remains mostly intact. (The preservation of a quasi-native fold pattern is obviously a crucial point for the preparation of functional protein-based assemblies.) However, the key issue of this study was the demonstration of effective interactions between the excited states of the protein molecules and the semiconductor NPs, elicited by the spatial proximity of the two species in the hybrid nanocomposite. The authors observed that, upon formation of the BSA-G-CdTe conjugates, the fluorescence intensities of both biological and inorganic components underwent substantial variations. The protein fluorescence (λ_{max} = 340 nm), which mainly arises from the two tryptophan residues present in the BSA structure (Trp134 and Trp214), was completely quenched, whereas a ~2.8-fold increase in the NP luminescence was simultaneously detected. The luminescence enhancement of the CdTe nanocrystals was interpreted in terms of a FRET process from the BSA to the NP. As an additional evidence for such hypothesis, the authors reported the excitation spectrum (PLE) of the conjugated system as compared to that recorded for the NP alone (Fig. 9.5).

PLE data seemed to confirm the FRET process as the mechanism responsible for the observed increase of the NP fluorescence, even though the line-shape of

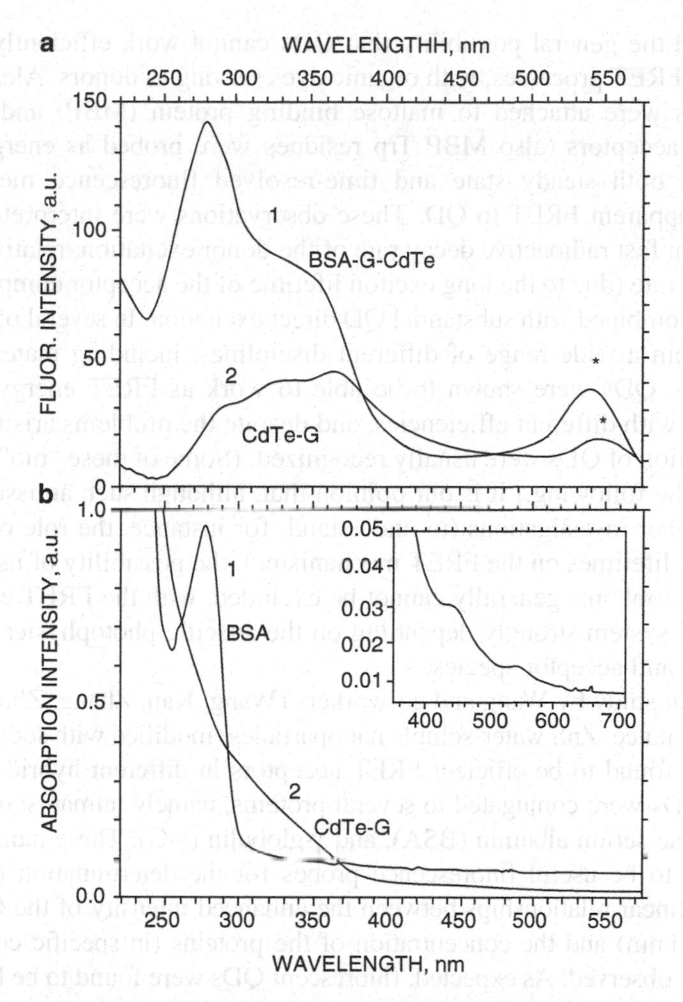

FIG. 9.5. A. Excitation spectra of *(1)* BSA-G-CdTe conjugate and *(2)* CdTe-G (λ_{em} = 580 nm). Peaks marked with a star correspond to the excitonic transition in CdTe nanoparticles. B. UV-vis absorption spectra of *(1)* BSA and *(2)* CdTe-G. The inset shows the UV-vis absorption spectrum of BSA-G-CdTe conjugate in the visible region of the spectrum (Taken from Mamedova, Kotov, Rogach, and Studer 2001.)

the PLE spectrum of the CdTe NP in the unconjugated system was not completely convincing. (The reported PLE is very different from the corresponding absorption spectrum.)

In principle, this latter discrepancy might affect the luminescence enhancement factor found by the authors, despite the highly controlled formation of the BSA:CdTe conjugates documented in this study. Actually, an analogous approach was carried out later by another group (Clapp, Medintz, Fisher, Anderson, and Mattoussi 2005), and no detectable increases of QD emission were observed upon conjugation. Indeed, on the basis of these experiments, Clapp and co-workers hypothesized the general possibility that QDs cannot work efficiently as energy acceptors in FRET processes, with organic dyes serving as donors. AlexaFluor488 or Cy3 dyes were attached to maltose binding protein (MBP) and used with various QD acceptors (also MBP Trp residues were probed as energy donors). Remarkably, both steady state and time-resolved fluorescence measurements showed no apparent FRET to QD. These observations were interpreted in terms of a dominant fast radioactive decay rate of the donor excitation, relative to a slow FRET decay rate (due to the long exciton lifetime of the acceptor compared to that of the dye), combined with substantial QD direct excitation. In several other studies, however (from a wide range of different disciplines, including material science and biology), QDs were shown to be able to work as FRET energy acceptors, even though with different efficiencies, and despite the problems arising from the direct excitation of QDs were usually recognized. (Some of these "bio" studies are reported in the following.) It is our opinion that, although such an issue certainly deserves further investigations (to understand, for instance, the role of the donor and acceptor lifetimes on the FRET mechanisms), the possibility of using QDs as acceptor fluorophores generally cannot be excluded, with the FRET efficiency in the designed system strongly depending on the specific photophysical properties of the donor and acceptor species.

In a recent study by Wang and co-workers (Wang, Kan, Zhang, Zhu, and Wang 2002), for instance, ZnS water-soluble nanoparticles (modified with sodium thioglycolate) were found to be efficient FRET acceptors in different hybrid systems, in which the QDs were conjugated to several proteins, namely human serum albumin (HSA), bovine serum albumin (BSA), and γ-globulin (γ-G). These nanoassemblies were shown to be useful fluorescence probes for the determination of the three proteins, as linear relationships between the enhanced intensity of the QD fluorescence (at 441 nm) and the concentration of the proteins (in specific concentration ranges) were observed. As expected, fluorescent QDs were found to be brighter and more stable against photobleaching, as compared with organic fluorophores. The proposed method was applied to the analysis of proteins in human serum samples, and the results were consistent with those obtained by conventional techniques, thus indicating reliability of the NPs-based approach for practical applications.

The functionalized nanoparticles were reported to have a long fluorescence lifetime and good fluorescence quantum yields at room temperature (0.23). Remarkably, the authors observed that the fluorescence of the functionalized NPs was enhanced by the interaction with proteins, and that the extent of the luminescence increase is proportional to the concentration of the proteins. (The excitation wavelength was 288 nm and the emission wavelength 441 nm.) The influence of NaCl content on this assay was examined, revealing that the ionic strength had a minor role in the binding mechanism. This suggested that the interaction of the proteins and QDs was mainly the result of nonelectrostatic binding. (In the opposite case, with high NaCl concentration, the effect of the electrostatic shielding of charges should reduce the binding of the QDs to protein, thus resulting in a decreased fluorescence signal.) Importantly, the authors demonstrated also that the QDs-based assay was weakly affected by the presence of some coexisting substances (e.g., arginine, lysine, glucose, citric acid, and isoleucine) and some metal ions.

The photophysical mechanism underlying the NPs fluorescence enhancement was discussed by the authors in terms of FRET processes from the proteins to the ZnS NPs. They observed that the excitation peak of the hybrid system, at 288 nm, well coincided with the UV-vis absorption spectra of proteins, such as BSA and HSA, thus suggesting that the excitation energy is nonradiatively transferred from the proteins to the excitonic state of the nanoparticles. (Such energy transfer resulted in NP fluorescence enhancement.)

Another very interesting scheme, based on QDs as FRET donors, was developed by Medintz and co-workers (Medintz, Clapp, Mattoussi, Goldman, Fisher, and Mauro 2003). In this study, QD–protein conjugates were designed to work as optical sensors, and were demonstrated to function as sugar receptors. The hybrid assemblies were formed by conjugating maltose-binding protein (MBP) (from *Escherichia coli*) to QDs via a C-terminal oligohistidine segment. CdSe/ZnS core/shell nanocrystals (maximum emission wavelength, λ_{max} = 530 or 560 nm) were made water-soluble by an organic monolayer of dihydrolipoic acid attached to their ZnS shell through thiol coordination. Two different strategies were proposed (Fig. 9.6):

1. In one configuration, the binding of β-cyclodextrin-QSY9 dark quencher conjugate in the MBP saccharide binding site resulted in FRET quenching of QD photoluminescence. Upon maltose addition, the β-cyclodextrin-QSY9 was displaced, leading to a significant increase of QD photoluminescence.
2. The second approach consisted of a two-step FRET mechanism. QDs were coupled with Cy3-labeled MBP, bound to β-cyclodextrin-Cy3.5. Upon β-cyclodextrin-Cy3.5 displacement by maltose, Cy3 emission was detected.

Both schemes exploited covalent β-cyclodextrin-acceptor dye conjugates, capable of binding within the saccharide-binding pocket of MBP and competing with maltose (the preferred substrate of MBP). A variant of MBP expressed with a pentahistidine segment at its C-terminus was used for coordinative conjugation of the protein to QD surfaces. The binding of strongly positively charged protein domains with the negatively charged QDs resulted in a significant enhancement of the nanocrystals luminescence, in close agreement with previous results (Mattoussi Mauro, Goldman, Anderson, Sundar, Mikulec, and Bawendi 2000). Such increase was used by the authors to assess protein–QD interactions, revealing an average coverage of ~15 proteins per single QD (560 nm emission maximum).

The simpler of the two QD–MBP sugar-sensing nanoassemblies consisted of 560-nm donor QDs (initial quantum yield ~15%), each conjugated with several

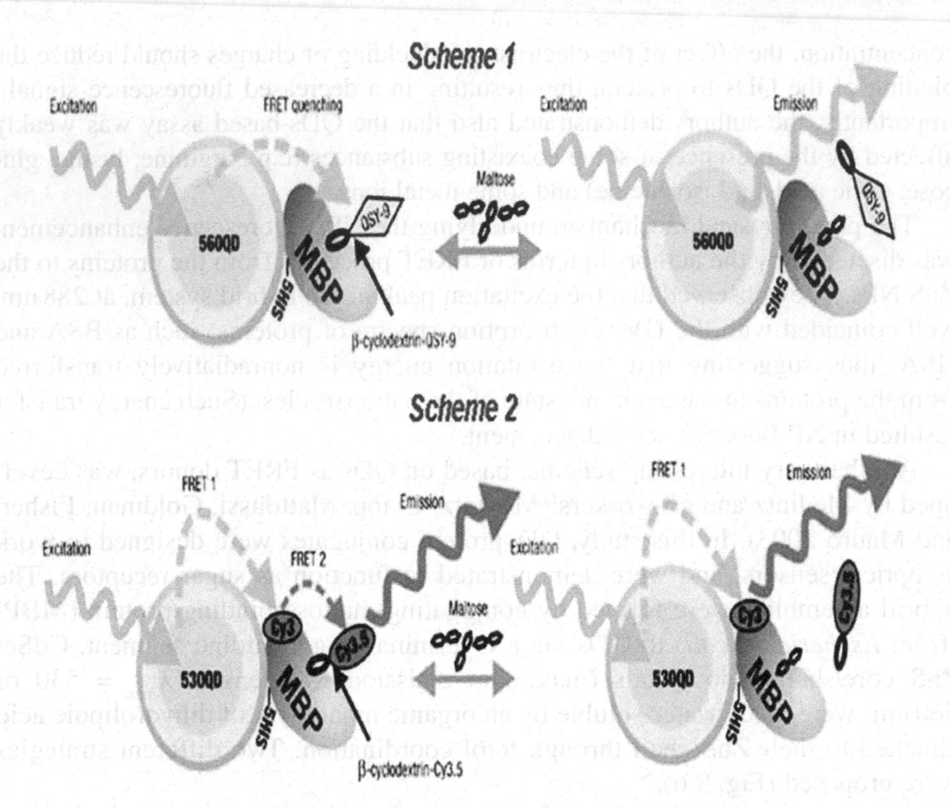

FIG. 9.6. **Top. Scheme 1:** QD-MBP nanosensor. Each 560-nm–emitting QD is surrounded by several protein molecules (a single MBP is shown for simplicity). Formation of QD-MBP-β-CD-QSY9 (absorption maximum ~565 nm) results in quenching of QD emission. Added maltose displaces β-CD-QSY9 from the sensor assembly, resulting in an increase of QD emission. **Bottom. Scheme 2:** QD-MBP-Cy3-β-CD-Cy3.5 maltose sensor assembly. A 530-nm QD is surrounded by several MBPs (only one shown for clarity), each monolabeled with Cy3 at cysteine 95 (absorption maximum ~556 nm, emission maximum ~570 nm). Specifically bound β-CD-Cy3.5 (absorption maximum ~575 nm, emission maximum ~595 nm) completes the sensor complex. Excitation of the QD results in FRET excitation of the MBP-Cy3, which in turn FRET excites the β-CD-Cy3.5. Added Maltose Displaces β-CD-Cy3.5, leading to increased Cy3 emission (Adapted from Medintz, Clapp, Mattoussi, Goldman, Fisher, and Mauro 2003.)

MBP proteins (quantum yield ~39%), with their saccharide binding sites occupied by the displaceable β-CD-QSY9 energy acceptor dye conjugate, which provided signal transduction (Fig. 9.6, top). The excellent overlap between QD emission and QSY9 absorption (R_0 ~ 5.5 nm) allowed β-CD-QSY9 bound within the MBP sugar-binding pocket to quench QD emission by ~50%. In addition, since each conjugate was characterized by ~10 to 15 MBP proteins per QD, the quenching efficiency was further improved. Added maltose readily competed with and displaced the β-CD-QSY9 quencher, resulting in a concentration-dependent increase in QD luminescence, namely, quantitative maltose sensing (sugar binding is transduced into an optical signal). These results were also confirmed by time-resolved measurements. The average lifetime (τ) for the QD-10MBP of ~8 ns decreased to 2.7 ns when β-CD-QSY9 occupied MBP sugar-binding sites. Importantly, addition of excess maltose, with resulting displacement of β-CD-QSY9, elicited a nearly full recovery of the QD donor lifetime (τ ~ 7.3 ns).

In the two-step FRET assembly, each 530-nm QD donor (initial quantum yield ~10.6%, with 10 MBPs ~15%) was conjugated with 10 Cy3-labeled MBP molecules. The protein-bound Cy3 served as a bridging acceptor/donor for ultimate energy transfer to maltose-displaceable β-CD-Cy3.5 bound at MBP saccharide binding sites (Fig. 9.6, bottom). For nanosensor homogeneity, a single-cysteine MBP variant that was site-specifically labeled at the cysteine with thiol-reactive Cy3 dye was used. In this case, efficient quenching of QD emission by Cy3-labeled MBP was observed (~95% quenching at 10:1 dye/QD ratio, with simultaneous increase in Cy3 emission). In analogy with the first scheme, the presence of several MBP-Cy3 acceptors (~10) associated with each QD contributed significantly to the high overall FRET efficiency. With the QD-MBP-Cy3 donor/bridging dye assembly in place, and after formation of the complete nanosensor assembly by occupation of MBP sugar-binding sites with displaceable β-CD-Cy3.5, sensitivity to added maltose was demonstrated. Also in this latter approach, time-resolved data were consistent with the steady state luminescence intensity measurements.

One year later, a FRET-based modeling technique was reported in an interesting study to determine the orientation of MBP proteins upon coordination to the QD surface (Medintz, Konnert, Clapp, Stanish, Twigg, Mattoussi, Mauro, and Deschamps 2004). Six different single-cysteine MBP mutants, spatially distributed on the protein surface, were specifically labeled with Rhodamine red (RR) dye. The authors prepared individual QD-RR-labeled MBP protein assemblies, in which QDs and RR dye were designed to work as the FRET donor/acceptor pairs. The distance from each of the six different RR-acceptor MBP locations to the center of the energy-donating QD was then derived from FRET efficiency data. The orientation of the bound MBP with respect to the QD surface was modeled and refined from the donor–acceptor distances in conjunction with the crystallographic coordinates of MBP, by using a method analogous to a nanoscale global positioning system determination. The final orientation suggested that MBP prefers a certain configuration relative to the QD, and that the MBP-binding site in this configuration is accessible to analytes, which is consistent with previous findings (Medintz, Clapp, Mattoussi, Goldman, Fisher, and Mauro 2003). Such an approach may provide a general strategy for determining the orientation of a protein on a QD. Importantly, information from such a model can be used to specifically enhance desirable properties in other constructs. In addition, the derived orientation can be further exploited through homology modeling to produce entirely new constructs from closely related proteins.

Another hybrid system was realized by conjugating a fluorescent semiconductor nanocrystal and the metalloprotein azurin (Az) (Pompa, Chiuri, Manna, Pellegrino, del Mercato, Parak, Calabi, Cingolani, and Rinaldi 2006). The assembly was designed so that the photophysical properties of the two species were exploited to elicit FRET mechanisms from the biomolecule to the QD (Fig. 9.7).

CdSe/ZnS core/shell water soluble silanized QDs were functionalized with surface exposed thiol groups (–SH) (Parak, Gerion, Zanchet, Woerz, Pellegrino, Micheel, Williams, Seitz, Bruehl, Bryant, Bustamante, Bertozzi, and Alivisatos 2002), in order to bind the protein via the accessible cysteine residues (Cys3–Cys26) (see Fig. 9.7).

In this way, a specific conjugation process between the two inorganic/biological species was achieved; also, a "directional" binding was obtained (the Az molecule is oriented with respect to the QD), thus providing a fixed distance in the FRET donor–acceptor pairs. In this system the average radius of the silanized coated QD was 3 nm (although inhomogeneities in the silanized shell could not be excluded), and the donor–acceptor distance was about 5 nm (from the QD center to the Trp48 amino acid residue inside the protein).

The conjugation of the metalloprotein to the semiconductor nanoparticle was studied by analyzing the behavior of the QD fluorescence intensity as a function of Az/QD molar ratio (MR). An increase of the QD emission as a function of the molar ratio was observed, suggesting that part of the excitation energy absorbed by the protein was non-radiatively transferred to the CdSe/ZnS nanocrystal. This also suggested that a single QD could bind to several Az molecules. The experimental conditions of maximum conjugation were characterized by an average number of 10 to 15 proteins per each nanoparticle (further Az–QD binding was rather unlikely due to steric hindrance limitations). A theoretical estimation of the number of Az proteins that can be packed around the spherical CdSe/ZnS nanocrystal was derived from steric considerations (Mattoussi, Mauro, Goldman, Anderson, Sundar, Mikulec, and Bawendi 2000), and the calculated value was consistent with experimental data.

The maximum enhancement of the fluorescence intensity at the asymptotic plateau (MR ~15) was rather limited with respect to the unconjugated QDs (about 6–7% larger). Such weak PL increase, however, was interpreted as due to the intrinsic spectral features of the two species, and basically explained in terms of the absorption properties of the biomolecule relative to the semiconductor nanocrystals. The authors pointed out that the molar extinction coefficient (ε) of the QD largely exceeded the corresponding value of the metalloprotein ($\varepsilon_{QD} \sim 100\varepsilon_{Az}$ at $\lambda = 280$ nm), so that the fraction of the excitation energy that was absorbed by azurin (and then partially transferred to the QD via FRET) was very low when compared with the excitation directly absorbed by the QD. PLE experiments were also performed in order to further characterize the optical properties of the hybrid system. Notably, such investigation supported the hypothesis of (Trp48 → QD) FRET process as the mechanism responsible for the observed increase in the QD

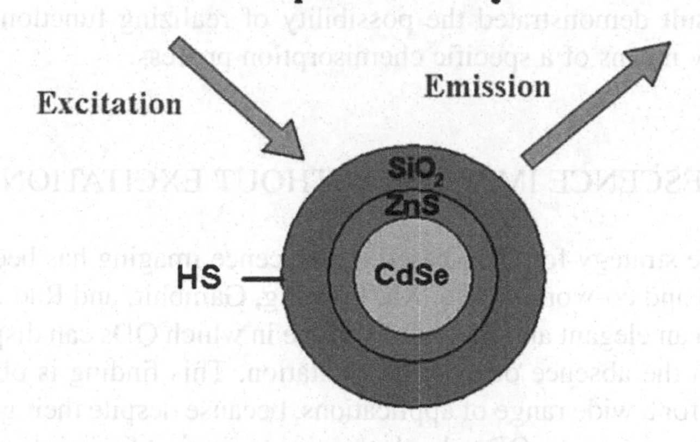

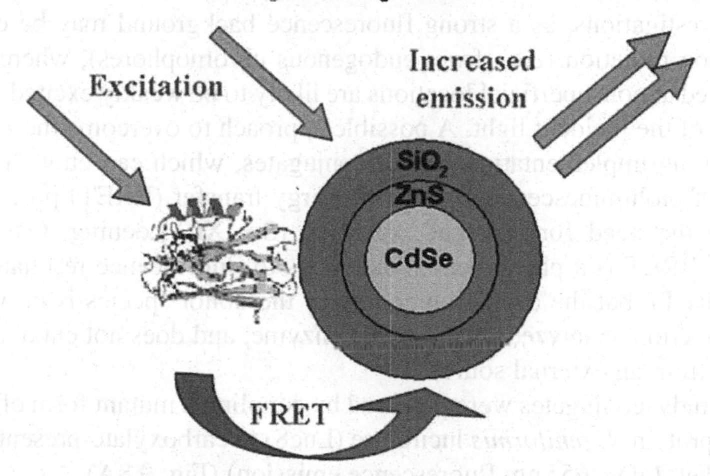

FIG. 9.7. Scheme of Az-QD hybrid system. **Top.** Nonconjugated species: Fluorescence emission *(green)* is due to the nanoparticle only. **Bottom.** Az-QD system: The increase in the fluorescence intensity is elicited by an energy transfer process (FRET) from the protein to the NP (Taken from Pompa, Chiuri, Manna, Pellegrino, del Mercato, Parak, Calabi, Cingolani, and Rinaldi 2006.)

luminescence. The raise of the fluorescence excitation mostly took place in the region around 270–280 nm, where the maximum of the Az absorption band was located (due to protein aromatic residues). The overall excitation spectrum of the hybrid system was due to the convolution of the PLE spectrum of the QD with the PLE spectrum of the protein.

The conformational state of the azurin molecules conjugated to the QDs was analyzed by means of intrinsic fluorescence spectroscopy experiments. The conjugated proteins exhibited a photoluminescence spectrum that was identical to that observed for the native state, revealing that the covalent binding of the biomolecule to the functionalized QD surface did not significantly affect the overall fold pattern. This result demonstrated the possibility of realizing functional Az–QD hybrid system by means of a specific chemisorption process.

5 FLUORESCENCE IMAGING WITHOUT EXCITATION

An innovative strategy for QDs-based fluorescence imaging has been recently developed by So and co-workers (So, Xu, Loening, Gambhir, and Rao 2006). The authors designed an elegant and powerful scheme in which QDs can display useful emission even in the absence of external excitation. This finding is obviously of extreme interest for a wide range of applications, because despite their great potential for imaging experiments, QDs clearly require excitation from external sources to fluoresce. This latter "limitation" is not trivial, especially in the case of in vivo imaging investigations, as a strong fluorescence background may be elicited by the excitation radiation (e.g., from endogenous chromophores), whereas fluorophores placed at nonsuperficial locations are likely to be weakly excited, due to the attenuation of the incident light. A possible approach to overcome these problems is based on the implementation of QD conjugates, which can emit fluorescence by means of bioluminescence resonance energy transfer (BRET) processes, thus eliminating the need for external excitation (So, Xu, Loening, Gambhir, and Rao 2006). BRET is a phenomenon analogous to fluorescence resonance energy transfer (FRET), but the excitation energy of the donor species is provided by a chemical reaction, catalyzed by the donor enzyme, and does not entail absorption of photons from an external source.

In this study, conjugates were prepared by coupling a mutant form of the bioluminescent protein *R. reniformis* luciferase (Luc8) to carboxylate-presenting CdSe/ZnS core/shell QDs (655 nm fluorescence emission) (Fig. 9.8A).

Upon addition of its substrate coelenterazine, Luc8 displays blue emission (λ_{max} = 480 nm). Importantly, this blue-emitting protein was chosen because of the high extinction coefficient characterizing QDs in such spectral range with respect to longer-wavelength regions (Fig. 9.8B).

Conjugation was assessed by gel electrophoresis analyses, and each hybrid structure was estimated to contain, on average, six Luc8 proteins per QD. An efficient BRET mechanism from Luc8 to the QD was demonstrated by the authors, as a strong QD emission (at 655 nm) was observed in the conjugates upon addition of co-elenterazine, despite the absence of any exciting radiation. Also, a clear dependence of the efficiency of the Luc8 $\rightarrow$ QD energy transfer process on the donor–acceptor distance was found in the experiments. (When the mean distance was increased by ~2–3 nm, the BRET efficiency significantly decreased.)

Beyond the remarkable interest for such hybrid system from a photophysical point of view (in this study the QDs seem to work very efficiently as energy accep-

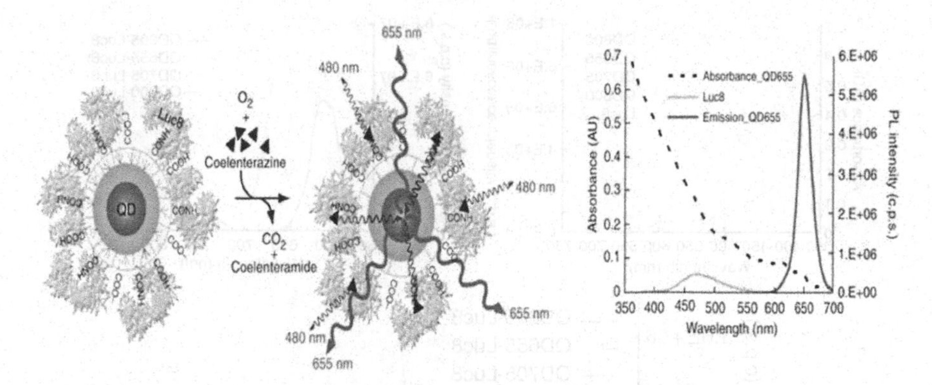

FIG. 9.8. Design of the QD conjugates based on BRET. **A.** A schematic of a quantum dot covalently coupled to a BRET donor, Luc8. The bioluminescence energy of Luc8-catalyzed oxidation of coelenterazine is transferred to the QDs, resulting in quantum dot emission. **B.** Absorption and emission spectra of the QD (λ_{exc} = 480 nm), and bioluminescence spectrum of Luc8 (Adapted from So, Xu, Loening, Gambhir, and Rao 2006.)

tors, an issue which is currently under intensive investigation, see the preceding), the broad applicative potential of these "self-illuminating QDs," especially in terms of "deep-tissue imaging" and multiplexing approaches, was clearly disclosed by the authors. Notably, they demonstrated readily detection of QDs emission (via BRET) in conjugates injected in deep tissue (located at a depth of ~3 mm), at variance with QDs luminescence induced by external excitation (which requires light to pass efficiently through tissues). Because of the substantial absorption and scattering of the short-wavelength excitation photons in tissues, fluorescence emission by BRET is more efficient than by external illumination. Moreover, the absence of the excitation light avoids preferential emission from surface fluorophores, thus circumventing the problems arising from background fluorescence. On the other hand, the multiplexing capabilities of the hybrids conjugates arise from the broad absorption spectrum of QDs, which allows the blue emission of Luc8 to be non-radiatively transferred efficiently to QDs of different emission colours. Actually, Luc8 was demonstrated to be an efficient BRET donor also in the case of nanocrystals fluorescing at 605, 705, and 800 nm (both CdSe/ZnS and CdTe/ZnS nanocrystals were used), although the BRET efficiencies were found to be dependent on the specific spectral features of each QDs (namely, extinction coefficients and quantum yields) (Fig. 9.9).

Because of the spectrally separated emission spectra of the four conjugates, bioluminescence-based multiplexed imaging was proven both in vitro and in vivo. Furthermore, BRET conjugates were used by the authors to label cells, and to monitor such cells in animals; experimental evidence indicated that the QDs conjugates were functional and could produce BRET emission even after uptake into cells.

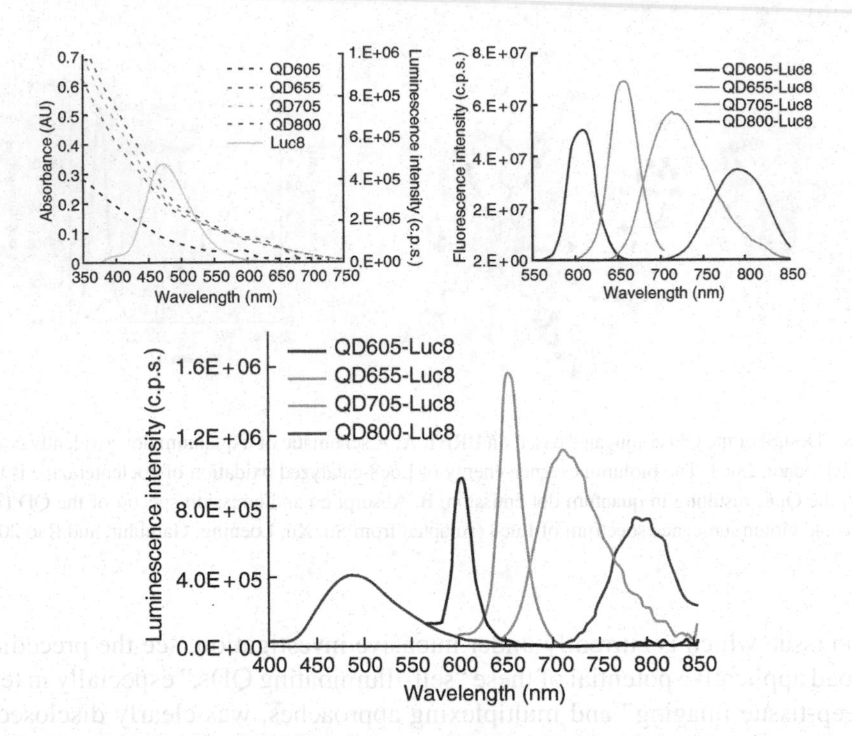

FIG. 9.9. **Top, left.** Overlap of the bioluminescence emission of Luc8 with the absorption spectra of different QDs (emitting at 605, 655, 705, and 800 nm). **Top, right.** Fluorescence emission spectra (λ_{exc} = 480 nm) of indicated conjugates. **Bottom.** Bioluminescence emission spectra of indicated conjugates (no external excitation) (Adapted from So, Xu, Loening, Gambhir, and Rao, 2006.)

REFERENCES

Aldana, J., Wang, Y.A. and Peng, X.G. (2001). Photochemical instability of CdSe nanocrystals coated by hydrophilic thiols. J. Am. Chem. Soc. 123, 8844.

Alejandro-Arellano, M., Ung, T., Blanco, A., Mulvaney, P. and Liz-Marzan, L.M. (2000). Silica-coated metals and semiconductors, stabilization and nanostructuring. Pure Appl. Chem. 72, 257.

Alivisatos, A.P. (1996). Perspectives on the physical chemistry of semiconductor nanocrystals. J. Phys. Chem. 100, 13226.

Alivisatos, A.P. (2004). The use of nanocrystals in biological detection. Nat. Biotechnol. 22, 47.

Bruchez, M., Moronne, M., Gin, P., Weiss, S. and Alivisatos, A.P. (1998). Semiconductor nanocrystals as fluorescent biological labels. Science 281, 2013.

Brust, M., Walker, M., Bethell, D., Schiffrin, D.J. and Whyman, R. (1994). Synthesis of thiol derivatised gold nanoparticles in a two-phase liquid/liquid system. J. Chem. Soc., Chem. Commun. 7, 801.

Chan, W.C.W. and Nie, S. (1998). Quantum dot bioconjugates for ultrasensitive nonisotopic detection. Science 281, 2016.

Chen, Y.F., Ji, T.H. and Rosenzweig, Z. (2003). Synthesis of glyconanospheres containing luminescent CdSe–ZnS quantum dots. Nano Lett. 3, 581.

Clapp, A.R., Medintz, I.L., Fisher, B.R., Anderson, G.P. and Mattoussi, H. (2005). Can luminescent quantum dots be efficient energy acceptors with organic dye donors? J. Am. Chem. Soc. 127, 1242.

Colvin, V.L., Schlamp, M.C. and Alivisatos, A.P. (1994). Light-emitting diodes made from cadmium selenide nanocrystals and a semiconducting polymer. Nature 370, 354.

Correa-Duarte, M.A., Giersig, M. and Liz-Marzan, L.M. (1998). Stabilization of CdS semiconductor nanoparticles against photodegradation by a silica coating procedure. Chem. Phys. Lett. 286, 497.

Dabbousi, B.O., Bawendi, M.G., Onitsuka, O. and Rubner, M.F. (1995). Electroluminescence from CdSe quantum-dot/polymer composites. Appl. Phys. Lett. 66, 1316.

Dabboussi, B.O., Rodriguez-Viejo, J., Mikulec, F.V., Heine, J.R., Mattoussi, H., Ober, R., Jensen, K.F. and Bawendi, M.G. (1997). (CdSe)ZnS core-shell quantum dots: synthesis and characterization of a size series of highly luminescent nanocrystallites. J. Phys. Chem. B 101, 9463.

Donega, C.D., Liljeroth, P. and Vanmaekelbergh, D. (2005). Physicochemical evaluation of the hot-injection method, a synthesis route for monodisperse nanocrystals. Small 1, 1152.

Dubertret, B., Skourides, P., Norris, D.J., Noireaux, V., Brivanlou, A.H. and Libchaber, A. (2002). In vivo imaging of quantum dots encapsulated in phospholipid micelles. Science 298, 1759.

Gao, X.H., Cui, Y.Y., Levenson, R.M., Chung, L.W.K. and Nie, S.M. (2004). In vivo cancer targeting and imaging with semiconductor quantum dots. Nat. Biotechnol. 22, 969.

Guo, W.Z., Li, J.J., Wang, Y.A. and Peng, X.G. (2003). Conjugation chemistry and bioapplications of semiconductor box nanocrystals prepared via dendrimer bridging. Chem. Mater. 15, 3125.

Guzelian, A.A., Banin, U., Kadavanich, A.V., Peng, X.G. and Alivisatos, A.P. (1996). Colloidal chemical synthesis and characterization of InAs nanocrystal quantum dots. Appl. Phys. Lett. 69, 1432.

Hines, M.A. and Guyot Sionnest, P. (1996). Synthesis and characterization of strongly luminescing ZnS-capped CdSe nanocrystals. J. Phys. Chem. 100, 468.

Holzinger, D., Liz-Marzan, L.M. and Kickelbick, G. (2006). Synthesis of core-shell nanoparticles using organic surface-functionalized gold and Au@SiO_2 nanoparticles as multifunctional initiators in atom transfer radical polymerization. J. Nanosci. Nanotechnol. 6, 445.

Kim, S. and Bawendi, M.G. (2003). Oligomeric ligands for luminescent and stable nanocrystal quantum dots. J. Am. Chem. Soc. 125, 14652.

Kobayashi, Y., Horie, M., Konno, M., Rodriguez-Gonzalez, B. and Liz-Marzan, L.M. (2003). Preparation and properties of silica-coated cobalt nanoparticles. J. Phys. Chem. B 107, 7420.

Larson, D.R., Zipfel, W.R., Williams, R.M., Clark, S.W., Bruchez, M.P., Wise, F.W. and Webb, W.W. (2003). Water-soluble quantum dots for multiphoton fluorescence imaging in vivo. Science 300, 1434.

Leatherdale, C.A., Woo, W.K., Mikulec, F.V. and Bawendi, M.G. (2002). On the absorption cross section of CdSe nanocrystal quantum dots. J. Phys. Chem. B 106, 7619.

Mamedova, N.N., Kotov, N.A., Rogach, A.L. and Studer, J. (2001). Albumin–CdTe nanoparticle bioconjugates: preparation, structure, and interunit energy transfer with antenna effect. Nano Lett. 1, 281.

Mattoussi, H., Cumming, A.W., Murray, C.B., Bawendi, M.G. and Ober, R. (1996). Characterization of CdSe nanocrystallite dispersions by small angle x-ray scattering. J. Chem. Phys. 105, 9890.

Mattoussi, H., Cumming, A.W., Murray, C.B., Bawendi, M.G. and Ober, R. (1998). Properties of CdSe nanocrystal dispersions in the dilute regime: structure and interparticle interactions. Phys. Rev. B 58, 7850.

Mattoussi, H., Mauro, J.M., Goldman, E.R., Anderson, G.P., Sundar, V.C., Mikulec, F.V. and Bawendi, M.G. (2000). Self-assembly of CdSe–ZnS quantum dot bioconjugates using an engineered recombinant protein. J. Am. Chem. Soc. 122, 12142.

Mayya, K.S. and Caruso, F. (2003). Phase transfer of surface-modified gold nanoparticles by hydrophobization with alkylamines. Langmuir 19, 6987.

Medintz, I.L., Clapp, A.R., Mattoussi, H., Goldman, E.R., Fisher, B. and Mauro, J.M. (2003). Self-assembled nanoscale biosensors based on quantum dot FRET donors. Nat. Mater. 2, 630.

Medintz, I.L., Konnert, J.H., Clapp, A.R., Stanish, I., Twigg, M.E., Mattoussi, H., Mauro, J.M. and Deschamps, J.R. (2004). A fluorescence resonance energy transfer-derived structure of a quantum dot-protein bioconjugate nanoassembly. Proc. Natl Acad. Sci. USA 101, 9612.

Mitchell, G.P., Mirkin, C.A. and Letsinger, R.L. (1999). Programmed assembly of DNA functionalized quantum dots. J. Am. Chem. Soc. 121, 8122.

Miyawaki, A. (2003). Visualization of the spatial and temporal dynamics of intracellular signaling. Dev. Cell 4, 295.

Murphy, C.J. (2002). Optical sensing with quantum dots. Anal. Chem. 74, 520A.

Murray, C.B., Norris, D.J. and Bawendi, M.G. (1993). Synthesis and characterization of nearly monodisperse CdE (E = S, Se, Te) semiconductor nanocrystallites. J. Am. Chem. Soc. 115, 8706.

Niemeyer, C.N. (2001). Nanoparticles, proteins and nucleic acids: biotechnology meets materials science. Angew. Chem. Int. Ed. Engl. 40, 4128.

Parak, W.J., Gerion, D., Pellegrino, T., Zanchet, D., Micheel, C., Williams, S.C., Boudreau, R., M.A Le Gros, Larabell, C.A. and Alivisatos, A.P. (2003). Biological applications of colloidal nanocrystals. Nanotechnology 14, R15.

Parak, W.J., Gerion, D., Zanchet, D., Woerz, A.S., Pellegrino, T., Micheel, C., Williams, S.C., Seitz, M., Bruehl, R.E., Bryant, Z., Bustamante, C., Bertozzi, C.R. and Alivisatos, A.P. (2002). Conjugation of DNA to silanized colloidal semiconductor nanocrystalline quantum dots. Chem. Mater. 14, 2113.

Pathak, S., Choi, S.K., Arnheim, N. and Thompson, M.E. (2001). Hydroxylated quantum dots as luminescent probes for in situ hybridization. J. Am. Chem. Soc. 123, 4103.

Pellegrino, T., Manna, L., Kudera, S., Liedl, T., Koktysh, D., Rogach, A.L., Keller, S., Rädler, J., Natile, G. and Parak, W.J. (2004). Hydrophobic nanocrystals coated with an amphiphilic polymer shell: a general route to water soluble nanocrystals. Nano Lett. 4, 703.

Peng, X.G., Schlamp, M.C., Kadavanich, A.V. and Alivisatos, A.P. (1997). Epitaxial growth of highly luminescent CdSe/CdS core/shell nanocrystals with photostability and electronic accessibility. J. Am. Chem. Soc. 119, 7019.

Pinaud, F., King, D., Moore, H.P. and Weiss, S. (2004). Bioactivation and cell targeting of semiconductor CdSe/ZnS nanocrystals with phytochelatin-related peptides. J. Am. Chem. Soc. 126, 6115.

Pompa, P.P., Chiuri, R., Manna, L., Pellegrino, T., del Mercato, L.L., Parak, W.J., Calabi, F., Cingolani, R. and Rinaldi, R. (2006). Fluorescence resonance energy transfer induced by conjugation of metalloproteins to nanoparticles. Chem. Phys. Lett. 417, 351.

Roederer, M., DeRosa, S., Gerstein, R., Anderson, M., Bigos, M., Stovel, R., Nozaki, T., Parks, D., Herzenberg, L. and Herzenberg, L. (1997). Nine color eleven parameter immunophenotyping using three laser flow cytometry. Cytometry 29, 328.

Schrock, E., du Manoir, S., Veldman, T., Schoell, B., Wienberg, J., Ferguson-Smith, M.A., Ning, Y., Ledbetter, D.H., Bar-Am, I., Soenksen, D., Garini, Y. and Ried, T. (1996). Multicolor spectral karyotyping of human chromosomes. Science 273, 494.

So, M.K., Xu, C., Loening, A.M., Gambhir, S.S. and Rao, J. (2006). Self-assembled nanoscale biosensors based on quantum dot FRET donors. Nat. Biotechnol. 24, 339.

Steigerwald, M.L., Alivisatos, A.P., Gibson, J.M., Harris, T.D., Kortan, R., Muller, A.J., Thayer, A.M., Duncan, T.M., Douglass, D.C. and Brus, L.E. (1988). Surface derivatization and isolation of semiconductor cluster molecules. J. Am. Chem. Soc. 110, 3046.

Van Der Meer, B.W., Coker III, G. and Chen, S.-Y.S. (1994). Resonance Energy Transfer: Theory and Data. VCH, New York.

Wang, L.-Y., Kan, X.-W., Zhang, M.-C., Zhu, C.-Q. and Wang, L. (2002). Fluorescence for the determination of protein with functionalized nano-ZnS. Analyst 127, 1531.

Wang, Y.A., Li, J.J., Chen, H.Y. and Peng, X.G. (2002). Stabilization of inorganic nanocrystals by organic dendrons. J. Am. Chem. Soc. 124, 2293.

Wilhelm, C., Billotey, C., Roger, J., Pons, J.N., Bacri, J.C. and Gazeau, F. (2003). Intracellular uptake of anionic superparamagnetic nanoparticles as a function of their surface coating. Biomaterials 24, 1001.

Willard, D.M., Carillo, L.L., Jung, J. and Van Orden, A. (2001). CdSe–ZnS quantum dots as resonance energy transfer donors in a model protein–protein binding assay. Nano Lett. 1, 469.

Wu, P. and Brand, L. (1994). Resonance energy transfer: methods and applications. Anal. Biochem. 218, 1.

Wuister, S.F., Swart, I., van Driel, F., Hickey, S.G. and Donega, C.D. (2003). Highly luminescent water-soluble CdTe quantum dots. Nano Lett. 3, 503.

Wundke, K., Potting, S., Auxier, J., Schulzgen, A., Peyghambarian, N. and Borrelli, N.F. (2000). PbS quantum-dot-doped glasses for ultrashort-pulse generation. Appl. Phys. Lett. 76, 10.

Part C
Cell-Based Nanobioelectronics

The cell is the smallest structural and functional unit of all known living organisms, and is sometimes called the building block of life. Some organisms, such as most bacteria, consist of a single cell, while other organisms are multicellular. All living cells sense and respond to their environment by a set of mechanisms known as cell signaling – part of a complex system of communication that governs basic cellular activities and coordinates cell actions. The ability of cells to perceive and correctly respond to their microenvironment is the basis of development, tissue repair, and immunity as well as normal tissue homeostasis. Errors in cellular information processing are responsible for diseases such as cancer, autoimmunity, and diabetes. Understanding cell signaling is required to treat these diseases effectively and, potentially, to build artificial tissues.

Cells receive information from their environment through a class of proteins known as receptors. The information is then processed through signaling pathways and decoded in the nucleus and other areas of the cell. To understand cell signaling, the spatial and temporal dynamics of both receptors and the components of signaling pathways need to be unraveled. Nowadays it has become possible to understand the complex processes occurring inside a single neuron and in a network of neurons that eventually produces the intellectual behavior, cognition, emotion and physiological responses. The nervous system is composed of the electrically exciting neurons and other supportive cells (such as glial cells). Neurons are able to process and transmit information and form functional circuits. Neurons are usually comprised of a cell body (or soma) with neurites (dendrites and an axon). The majority of neurons receive input on the cell body and dendritic tree, and transmit output via the axon. However, there is great heterogeneity throughout the nervous system and the animal kingdom, in the size, shape and function of neurons. Neurons communicate via chemical and electrical synapses, in a process known as synaptic transmission. The fundamental process that triggers synaptic transmission is the action potential, a propagating electrical signal that is generated by exploiting the electrically excitable membrane of the neuron. This is also known as a wave of depolarization.

However, there is still the need to understand the underlying structure of signaling networks and how changes in these networks can affect the transmission of information. To accomplish this task, new technologies and approaches are needed. On the other hand researchers would like to combine the high specificity of biochemical reactors with universal microelectronics to develop selective measurement techniques for diagnostics, drug research, and the detection of poisons. Cell-based bioelectronics and in particular neuroelectronics deals with interfacing the inorganic electronic systems with (nerve) cells. Challenges are the interface between (nerve) cell and electronic device and the patterning of (neuronal) cells.

"Spontaneous and synchronous firing activity in solitary micro-cultures of cortical neurons on chemically patterned multi-electrode arrays" is the title of the first chapter of the third section. Here, the authors report on the use of microelectronic devices as tools to investigate bio-electrical neuronal network dynamics or development of cell-based biosensors for pharmaceutical drug screening and/or neurotoxic detection devices. Compared to the intracellular recording of biosignals with patch clamp techniques, the disadvantage of less stable neuron-electronic contacts on micro-electronic devices is considered to be compensated by the advantage to be able to record over many channels simultaneously and over longer periods of time as neuronal membranes is not intentionally damaged. In particular, the authors are interested in the study of patterned neuronal networks on multi-electrode arrays to create a functional and preferably more stable contact between the electrodes of microdevices and neuronal tissue. Here, they study the adhesion, growth, and spontaneous bio-electrical activity of cortical neurons in microcultures (< 100 neurons) with different numbers of neurons forming a small neuronal network. Subsequently, the growth and bioelectrical activity is followed over a time period of 8 (16) days.

"Nanomaterials for Neural Interfaces; Emerging New Function and Potential Applications" is the title of the second chapter of this section. The authors focus on controlling the connectivity of the neurons as a prime requirement for the successful building of neural nets in culture. Microfabrication and microcontact printing techniques are successfully to control the adhesion and outgrowth of neurons and the authors discuss whether nanofabrication will produce still more useful results. Finally, fabrication methods, biological effects, biological problems and solutions, importance of order and symmetry in nanofeatures are discussed.

The next chapter "Interfacing Neurons and Silicon-based Devices" deals again with interfacing the inorganic electronic systems with nerve cells. Challenges are the two-way interface between nerve cell and electronic device and the patterning of neurons. In order to establish a two-way interface between a neuron and an electronic device, several approaches have been followed in the past. One can use field-effect transistors (FET) and metal microelectrode arrays (MEA) to detect extracellular neuronal signals. In order to study the signal transfer between cell and sensor spot, signals from different cell types with non-metallized FETs or MEAs were recorded and compared with classical patch-clamp measurements. The obtained signals can be described in a first approximation by a simplified coupling model (Point Contact Model). However, in some special cases this model is not precise enough to fully describe the recorded signal shapes. At the moment a more comprehensive cell-sensor coupling model for explanation of extracellularly recorded signals is developed. On the other side one can use either the MEAs or floating-gate FET structures for extracellular stimulation of excitable cells. First successful stimulations were possible. The aim of both lines is a large-scale integration of two-way cell-sensor interfaces on-chip.

The last chapter of this section is entitled "Hybrid Nanoparticles for Cellular Applications". Artifical nanoparticles of a size and complexity approaching those of several supramolecular structures of living systems are being increasingly used as tools for the analysis and manipulation of cells and organisms. This paper focuses on hybrid organic/inorganic nanoparticles of the three most widely used classes, based on their inorganic core: seminconductors, metals and oxides. Their essential properties most relevant to cellular applications are summarised. Current strategies to bring nanoparticles to interact with living cells are described, and recent results, both *in vitro* and *in vivo,* are illustrated.

Neuron-Based Information Processing

10

Spontaneous and Synchronous Firing Activity in Solitary Microcultures of Cortical Neurons on Chemically Patterned Multielectrode Arrays

T.G. Ruardij, W.L.C. Rutten, G. van Staveren, and B.H. Roelofsen

1 INTRODUCTION

The growing interest in the development and study of patterned neuronal networks on multielectrode arrays (MEAs) (Corey, Wheeler, and Brewer 1996; Branch, Wheeler, Brewer, and Leckband 2000; Chang, Brewer, and Wheeler 2000, 2002; Blau, Weinl, Mack, Kienle, Jung and Ziegler 2001) and field effect transistors (Corey, Wheeler, and Brewer 1996; Yeung, Lauer, Offenhauser, and Knoll, 2001) is, among other factors, driven by the necessity to create a functional and preferably more stable contact between the electrodes of microdevices and neuronal tissue. Compared with the intracellular recording of biosignals with patch clamp techniques, the disadvantage of less stable neuron-electronic contacts on microelectronic devices is considered to be compensated for by the advantage of

A. Offenhäusser and R. Rinaldi (eds.), *Nanobioelectronics - for Electronics, Biology, and Medicine*,
DOI: 10.1007/978-0-387-09459-5_11, © Springer Science+Business Media, LLC 2009

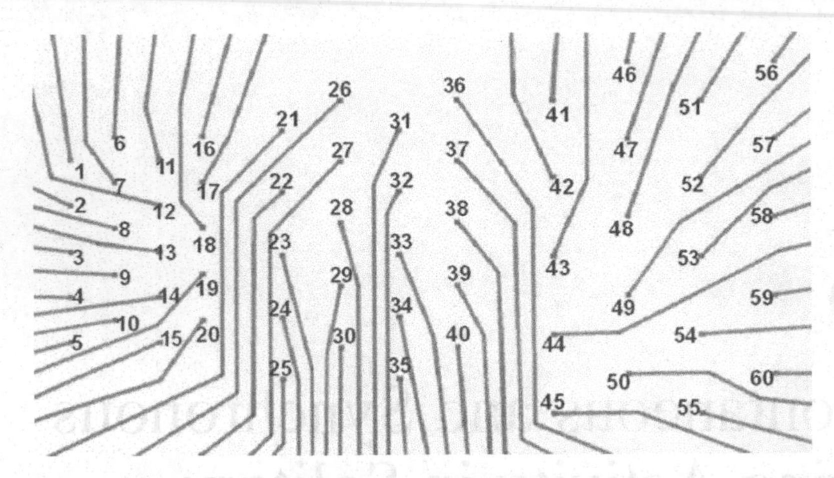

FIG. 10.1. Layout of the multielectrode array with three subsections of different distances between electrodes. Center to center distances between electrodes are 140 μm (electrode 1–20), 190 μm (electrode 21–40), and 240 μm (electrodes 41–60).

being able to record over many channels simultaneously and over longer periods of time, as neuronal membranes are not intentionally damaged. Many authors are interested in the use of these microelectronic devices as tools to investigate bioelectrical neuronal network dynamics or development of cell-based biosensors for pharmaceutical drug screening and/or neurotoxic detection devices.

Our group is interested in the development of a cultured neuron probe with a living in vitro cultured neuronal interface (i.e., an island of precultured neurons) on top of the microelectrodes of a MEA device.

The cultured probe may be used either as a stimulatory or recording device. For stimulation, the method is to attract single axons (sprouts) toward the living interface and finally stimulate these axons directly from the electrodes. Therefore, in this case an island of precultured cells serves solely as a sprout-friendly host; also, antidromic stimulation of sprouts by natural synaptic connections between axonal sprouts and network cells is impossible. In the stimulatory case, consequently, spontaneous activity of the hosting culture is not the principal source of disturbance.

However, in the recording application of a cultured probe, spontaneous activity of the intermediate islands gives unwanted activity. Therefore, we are specifically interested in the occurrence (and absence!) of spontaneous bioelectrical activity in isolated disconnected networks of neurons containing only a limited number of neurons (<100). Solitary viable microcultures showing no spontaneous bioelectrical activity could be optimally suited for a precultured neuroprosthetic stimulation device as interference between stimulation signals and spontaneously formed signals is avoided.

The number of strategies to fabricate connected patterns of neurons on solid surfaces is numerous, and all depend on procedures leading to chemical patterns with a neurophobic and neurophilic part. In most cases, neurophilic chemical patterns were fabricated on neurophobic background surfaces using photoresist liftoff procedures, or microcontact printing procedures with polydimethylsiloxane stamps (Ruardij, Goedbloed, and Rutten 2000; Ruardij, van den Boogaart, and Rutten 2002). A very critical step is the choice of the background surface, which should not only prevent the initial adhesion of neurons a few hours after seeding, but also prevent the overgrowth of neurites/axons over preferably longer culturing periods of approximately several weeks or even months. Branch and co-workers (Branch, Wheeler, Brewer, and Leckband 2000, 2001) showed that a polyethylenglycol (PEG) coating, covalently coupled to hydrophilic silicon oxide, is an excellent neurophobic substrate because adsorption of cell-adhesive proteins is effectively inhibited. However, hydrophobic substrates such as fluorocarbon-coatings (Jansen, Gardeniers, Elders, Tilmans, and Elwenspoekare 1994) also considered to be suitable as background surfaces as the most abundant protein in culturing media, e.g., albumin is irreversibly adsorbed onto the surface first and subsequently blocks the adsorption of other cell-adhesive proteins that play a specific role in the adhesion process. An additional advantage of hydrophobic substrates is the possibility to create excellent neurophobic coatings over a period of approximately 1 month (Amiji and Park 1992; Li and Caldwell 1996; Lee, Ju, and Kim 2000) with adsorbed PPO-PEO-PPO triblock-copolymers layers. Therefore, multielectrode arrays equipped with a standard hydrophilic siliconoxide surface, were treated with a hydrophobic layer of dichlorodimethylsilane (DDS) (Ruardij, Goedbloed, and Rutten 2000).

The type of interaction between neurons and the adhesive part of a chemical pattern can be nonspecific, specific, or both. The specific approaches with biomolecules (Yeung, Lauer, Offenhauser, and Knoll 2001), polypeptides (Branch, Wheeler, Brewer, and Leckband 2000), or epitopes (Ruardij, Goedbloed, and Rutten 2000) are familiar neuron-adhesive choices, but a common drawback is the hydrolysis of amide linkages in polypeptide backbones. The nonspecific electrostatic interaction between synthetic amide-linkage free compounds such as polyethylenimine (PEI) and negatively charged neuron-membranes is more promising from that perspective and demonstrated its applicability in neuronal patterning studies using photoresist liftoff technology (Ruardij, Goedbloed, and Rutten 2000; Ruardij, van den Boogaart, and Rutten 2002). Application of the microcontact printing technique in combination with PEI as the printing substance is even more appealing as it opens the possibility of printing relatively thick layers of PEI with multiple internal electrostatic attachment points.

The aim of this chapter is to study the adhesion, growth, and spontaneous bioelectrical activity of cortical neurons in microcultures (<100 neurons) with different numbers of neurons forming a small neuronal network. To this end, a chemical pattern with circles of neuron-adhesive polyethylenimine (PEI) with diameters of 50, 100, and 150 μm are microprinted onto the embedded electrodes of a DDS-coated multielectrode array. Subsequently, the growth and bioelectrical activity is followed over a time period of 8 days (Van Pelt, Wolters, Corner, Rutten, and Ramakers 2004).

2 METHODS

2.1 CORTICAL NEURON ISOLATION AND PROCEDURES

Cerebral cortex from 1-day-old newborn rats was dissected out under sterile circumstances and cut into pieces of approximately 1 mm^3. After collection, the tissue was trypsinized (0.25% Trypsin/EDTA, Gibco, Breda, The Netherlands) for 45 minutes in an incubator at 37°C at 5% CO_2 and subsequently treated with soybean trypsin inhibitor (STI, 1 mg/ml) and deoxyribonuclease I (DNAse I, 1.1 U/ml). The dissociated tissue was spun down at 1200 rpm during 5 minutes and resuspended in chemically defined medium R12 (DMEM/Ham's F12, Gibco, Breda, The Netherlands) without serum (Romein, van Huizen, and Wolters 1984). Trypan blue stain (0.4%) was used to discriminate and count living neurons in a Bürker chamber, prior to the sedimentation of the neurons onto the surfaces. Neurons were seeded onto the patterned structures with a plating density of 5000 living cells/mm^2. Cells were allowed to adhere onto the surfaces during a time period of 4 hours. Samples were rinsed with a 0.9% NaCl solution to remove nonadherent cells.

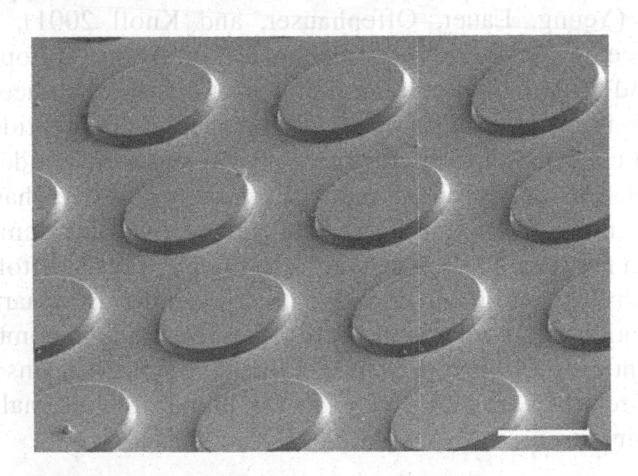

FIG. 10.2. Scanning electron microscopy image of a PDMS stamp. Scaling bar = 100 μm.

2.2 PREPARATION OF PDMS MICROSTAMPS

Sylgard 184 silicone (Mavom bv, Alphen, The Netherlands) was mixed with the curing agent in a 10:1 ratio. Air bubbles were removed from the mixture by evacuation with a water jet pump. Collapse of air bubbles was promoted by following a cycle of evacuation and pressure release six times. A metal ring with an inner diameter of 4.3 mm (height 0.8 mm) was placed around the central area of a polyimide mold containing three different regions of 20 microwells (12-µm deep) with diameters of 50, 100, and 150 µm at the bottom. The spacing distance between the wells was fixed at 90 µm for all three regions. The ring was filled with the mixed silicone, covered with a 76-mm microscope slide, and cross-linked at room temperature in the mold for 4 days. Finally, stamps were carefully removed from the mold and stored in plastic tubes until use.

2.3 FABRICATION OF MULTIELECTRODE ARRAYS

Multielectrode areas (MEAs) were fabricated from 5 × 5 cm glass plates with gold deposited wires leading to 60 hexagonal ordered electrodes. MEAs were isolated with a sandwich layer of SiO_2-Si_3N_4-SiO_2 (ONO) using a PECVD process. Electrode tips were deinsulated with a SF6 reactive ion etching (RIE) technique and platinized to reduce the electrode impedance down to 200 kΩ at 1 kHz. Three different subsections of electrode configurations were assembled on the MEAs. The center-to-center distance between electrodes (10-µm diameter) was 140 µm (electrodes 1–20; Section 1), 190 µm (electrodes 21–40, Section 2), and 240 (electrodes 41–60; Section 3). The ONO sandwich layer was made hydrophobic by treatment with 0.1% (v/v) dichlorodimethylsilane (Merck, Darmstadt, Germany) dissolved in xylene for 10 min.

2.4 MICROPRINTING OF POLYETHYLENIMINE ON MULTIELECTRODE ARRAYS

PDMS microstamps were assembled onto a transparent Plexiglas cylinder, which was connected to the holder of a micromanipulator through a Plexiglas table and a metal rod. Polyethylenimine (PEI; molecular weight is 6*105–106, Fluka Zwijndrecht, The Netherlands) was applied as the printing substance (foreground) while several background substrates were tested. Stamps were made hydrophilic by exposing the PDMS to a gas flame shortly (Lucas, Czisny, and Gross 1986). Then stamps were wetted with PEI (10 mg/ml) by manually pressing the stamp for 15 seconds into lint-free tissue (Hanotex BV, Joure, The Netherlands) moistened with PEI.

The alignment and pressing of the PEI-wetted PDMS stamp upon the multielectrode array was monitored continuously by visual inspection of the alignment and stamping procedure on a phase contrast microscope (Nikon Diaphot, Tokyo, Japan). A micromanipulator enabled precise control of the alignment of the stamp to the electrodes of the multielectrode array in the horizontal x-y plane of focus. Thereafter, stamps with PEI were pressed onto the MEA by manipulation of the microstamp in the vertical z-direction.

2.5 MORPHOLOGICAL ASSESSMENT OF NEURONAL TISSUE

Microphotographs using a 35-mm photocamera (Nikon-FE, Tokyo, Japan) were taken on the three separate subsections of the multielectrode arrays after 1, 4, 8, and 15 days. Each time lapse procedure was done on six different samples. Adhesion was evaluated after 1 day in vitro with a manual counting procedure of individually identifiable neurons. Furthermore, the development of fasciculated neurites across the neurophobic DDS-coated ONO was quantified by calculation of the average number of surrounding electrodes, connected to a single PEI-coated circle NC through neurite fascicles. Electrodes at the edges of the subsections were disregarded to keep number of surrounding electrodes at a constant number of six. Neurite development was studied after 8 and 15 days.

2.6 BIOELECTRICAL RECORDING

The presence of neurons in postnatal tissue was demonstrated through extracellular recording of spontaneous bioelectrical activity in postnatal cultures (P1s) using a home-built measurement system. In order to measure bioelectrical activity, each MEA was placed in an aluminum mini-incubator providing contact between the gold leads and a selector. The selector connected 30 or 31 electrodes to 16 channels of an amplifier, allowing switching between different electrode configurations.

Electrode signals were amplified, filtered between 0.3 and 6 kHz (first-order) and captured by a 12-bit National Instruments PCI-6023E Data Acquisition PC card. The input range as well as the sampling frequency was software controlled by a Labview program. The real-time data processing software reduced the data stream by rejection of data that did not contain bioelectrical activity. Artefact rejection was severe: If activity was measured at the same time in different channels, the waveforms were rejected. In each channel, the rms noise level was constantly monitored and determined the setting of a level detector to detect spike activity. The threshold was set at six times the level of the noise (typically 7 μV rms). Each time bin of 10 ms with recorded activity was stored and analyzed with Matlab computer software. For wave shape classification, three spike features were used:

- The peak-to-peak amplitude Vpp
- The width of the main peak
- The area under the peak

A fourth feature, called peak balance, distinguished different shapes. It calculated the difference between the waveform area before and after the peak value.

3 RESULTS

Table 10.1 presents the manually counted average number of adhering cells on the PEI-microprinted circles after 1 day together with data on the percentage of covered electrodes with neuronal tissue $P_{COV\text{-}EL}$, percentage of electrically active electrodes $P_{AC\text{-}EL}$, and the neuronal connectivity N_C between electrodes after 8 and 15 days. The different sizes of the PEI-printed areas lead to significant differences in the average number of adhering cells per PEI circle. Approximately 16, 46, and 81 neurons adhered on PEI-circles with diameters of 50, 100, and 150 µm, respectively. As a result, the percentage of neuronal tissue covered electrodes after 8 days was lower for 50-µm circles (70%; 8 days) as compared with 100- and 150-µm circles, which gave more or less similar results (both around 94%; 8 days). This trend in the comparison among circles was still quite similar after 15 days

Comparison of the percentage of electrically active electrodes after 8 days showed that only 0.7% of the electrodes that were microprinted with 50-µm circles became electrically active. The results with 100- and 150-µm circles were better, with 2.8% and 10.7% of the electrodes active, respectively. This positive trend

TABLE 10.1. The number of neurons N adhering on microprinted PEI circles with diameters of 50, 100, and 150 µm after 1 day.

Diameter PEI circles	50 µm	100 µm	150 µm
N(–) 1 day	16.4 ± 2.7 (n = 20)	45.6 ± 5.0 (n = 20)	81.2 ± 15.1 (n = 20)
$P_{AC\text{-}EL}$ (%) 8 days	0.7 ± 1.9 (n = 7)	2.8 ± 3.9 (n = 7)	10.7 ± 12.7 (n = 7)
$P_{COV\text{-}EL}$ (%) 8 days	70.0 ± 23.3 (n = 7)	93.6 ± 8.0 (n = 7)	93.6 ± 9.0 (n = 7)
N_C(–) Active 8 days	— (n = 0)	6.0 ± 0.0 (n = 3)	3.5 ± 2.6 (n = 7)
N_C(–) Nonactive 8 days	2.0 ± 2.1 (n = 36)	2.9 ± 2.4 (n = 34)	2.7 ± 2.5 (n = 31)
$P_{AC\text{-}EL}$ (%) 15 days	0.8 ± 2.0 (n = 6)	5.0 ± 7.7 (n = 6)	10.0 ± 13.8 (n = 6)
$P_{COV\text{-}EL}$ (%) 15 days	64.2 ± 23.8 (n = 6)	93.3 ± 8.8 (n = 6)	95.8 ± 5.9 (n = 6)
N_C(–) Active 15 days	— (n = 0)	6.0 (n = 1)	3.3 ± 2.2 (n = 4)
N_C(–) Nonactive 15 days	— (n = 0)		2.4 ± 2.3 (n = 33)

The percentage of corresponding electrically active electrodes $P_{AC\text{-}EL}$ on MEAs after 8 and 15 days, the percentage of electrodes covered with neuronal cells $P_{COV\text{-}EL}$ on MEAs after 8 and 15 days, and the number of neuronal connections with the surrounding electrodes N_C for electrically active electrodes and non-active electrodes after 8 and 15 days.

between the occurrence of electrode activity and PEI circle diameter was also seen after 15 days.

In the calculation of neuronal connectivity, a clear distinction was made between the group of active and nonactive electrodes in order to find a possible trend between neuronal connectivity between electrodes on one hand and the possibility of measuring electrical activity on the electrodes on the other hand. This was evaluated as a function of the microprinted PEI circle diameter. On 150-μm circles, active electrodes only displayed a slightly higher neuronal connectivity between electrodes as compared with nonactive electrodes after 8 and 15 days. This difference was not significant. However, on 100-μm circles, active electrodes all showed maximal connectivity ($N_C = 6$) with the surrounding six electrodes as compared with the lower connectivity ($N_C = 2.9 \pm 2.4$) on nonactive electrodes. A similar comparison for the 50-μm circles was not possible because of the fact that no electrodes with six surrounding electrodes were active.

Figure 10.3 displays an example of the morphological development of neuronal tissue on microprinted PEI circles after 8 and 15 days. Especially after 8 days, well-spread glial cells are visible on the DDS-coated ONO background, but partially disappear in the time course between 8 and 15 days. The distribution of the neuronal tissue on the PEI-coated circles changed from a still homogeneously distributed neuronal layer into centrally positioned neuronal aggregate with a high density of neurites visible at the borders of the PEI circles.

Figure 10.4 presents the average waveform of 10 recorded spikes after 8 and 15 days from an isolated microculture on electrode 49. Despite the morphological difference of the neuronal tissue after 8 and 15 days, the mean shape of the spike waveforms were quite similar and characterized by a fast negative peak followed by a slower positive peak.

Figure 10.5 shows the histogram of the interspike time interval distribution together with the average spike rate after 8 and 15 days from electrode 49 over

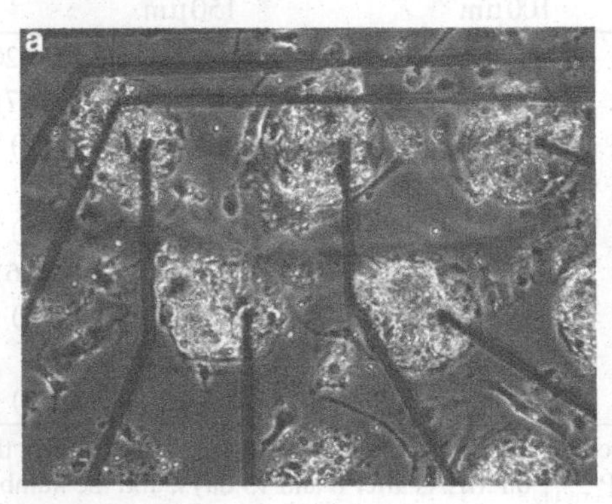

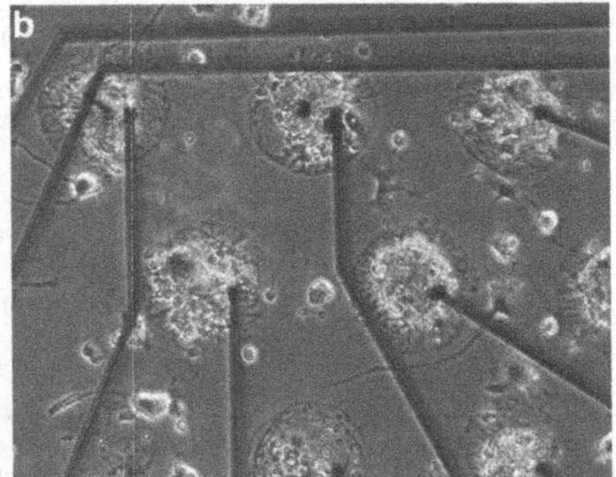

FIG. 10.3. Phase contrast images of patterned cortical neuron groups on a PEI-microprinted DDS-coated multielectrode array after 8 days (left) and 15 days (right). Electrode spacing is 240 μm. Electrodes 43, 44, 45, 49, and 50 are shown.

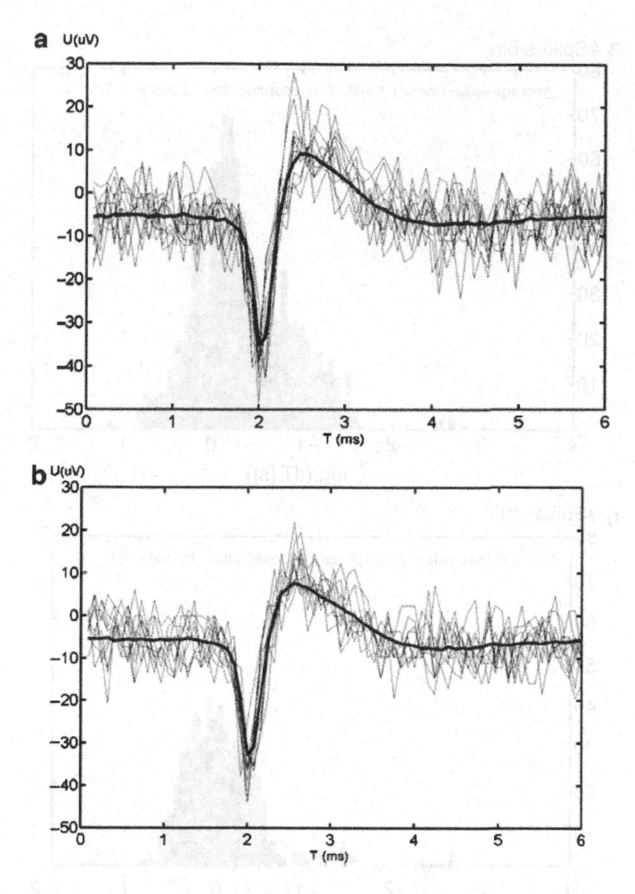

FIG. 10.4. Waveforms of 10 recorded spikes after 8 days *(left)* and 15 days *(right)* from electrode 49. Thick solid lines indicate the calculated mean from the 10 recorded waveforms.

a time course of 15 minutes. On average, the maximum of the two interspike interval histograms is situated around 1 sec. The corresponding spike rate of 1 spike/s is different from the one observed during network bursts in unpatterned large networks (Van Pelt, Wolters, Corner, Rutten, and Ramakers 2004). In those networks, although the spike rate at indvidual sites during a population burst is usually higher, it almost never comes close or exceeds 1 spike/s (Van Pelt, Wolters, Corner, Rutten, and Ramakers 2004). Therefore, it seems that the small island networks fire faster than the unpatterned large networks.

A special phenomenon, synchronized firing, was observed in connected islands at 12 DIV. Figure 10.6 gives a typical example of two connected islands at 12 DIV. They fired synchronously during periods of typically 120 sec, with regular silent intervals between these 120-sec epochs. Fig. 10.7 presents the spike rates of these islands, first as an interelectrode interval histogram (Fig. 10.7, above), then as a spike rate plot over a time interval of 120 sec (Fig. 10.7, below). It is clearly observed that the two

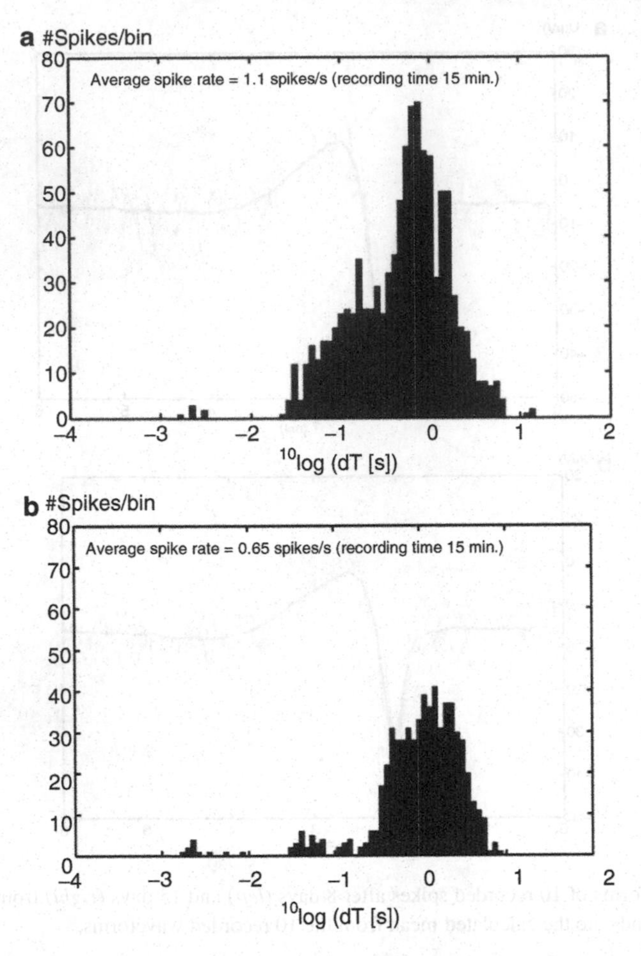

FIG. 10.5. Histograms of the interspike time interval (dT) distribution from an isolated 150-μm microculture, calculated over a time period of 15 minutes. The interval time scale is plotted on a logarithmic scale. Data from electrode 49 after 8 days (left) and 15 days (right).

clusters fire in a highly synchronized way, with periodic bursting. In the peaks, spike rates go up to 30 spikes/sec in one channel, and about 20 spikes/sec in the other.

4. DISCUSSION AND CONCLUSION

Activity in patterned MEA's will be briefly compared to unpatterned (random) network activity, as reported by Van Pelt and co-workers (Van Pelt, Wolters, Corner, Rutten, and Ramakers 2004). In the latter, activity starts approximately 1 week after seeding, i.e., 7 DIV. Spike rate, summed over all 61 electrodes, develops gradually from 0 to 32 spikes/sec between 9 and 42 DIV. This would imply a maximum of

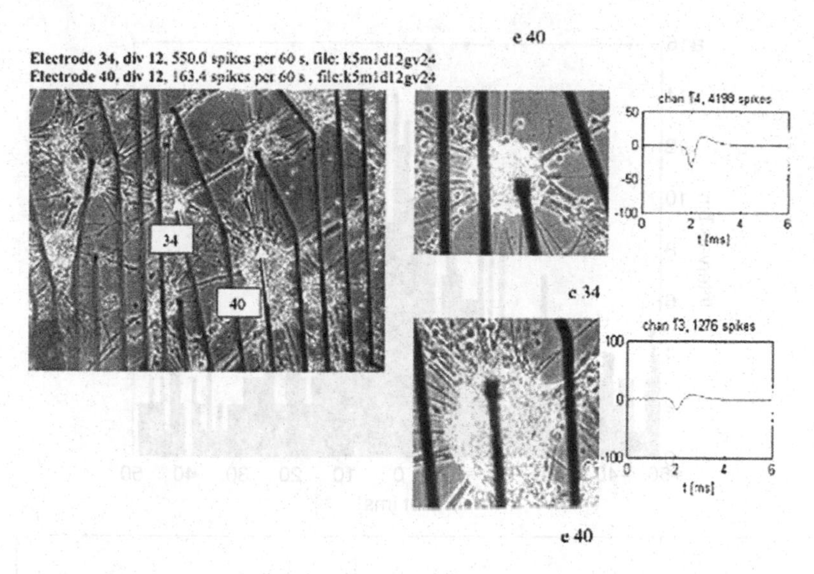

FIG. 10.6. Example of spontaneous bioelectrical activity from two connected islands at 12 DIV. Two islands at electrodes 34 and 40 were spontaneously active after a week. Electrode separation is 190 μm. Islands became interconnected later and showed synchronized behavior (see Fig. 10.7). Details on the right show the two clusters and the derived averaged spike waveform (top: electrode 40, 4198 spikes. bottom: electrode 34, 1276 spikes).

about 0.5 spikes/sec per electrode, at 42 DIV if all electrodes took part. However, the spike rate may differ considerably in time and per electrode, for example, 22 spikes/s was observed maximally in one electrode (lasting a few days).

Comparing unpatterned networks with patterned islands, activity seems to start at the same age, 7 to 8 DIV. The variability in the random networks, and absence of statistics so far in the unconnected island case, makes it hard to draw a comparison. A cautious observation may be that the spike rate in unconnected islands is in the same range as in the random network, i.e., 4 to 8 spikes/s at 22 to 28 DIV.

However, it is clear that the connected clusters can fire synchronously at a very high spike rate, much higher than observed in the unpatterned network (Van Pelt, Wolters, Corner, Rutten, and Ramakers 2004).

Another comparison can be made with regard to the probability that spontaneous firing develops in unpatterned as well as patterned large networks. In unpatterned networks (Van Pelt, Wolters, Corner, Rutten, and Ramakers 2004) 16 (each consisting of seven cultured MEAs) out of 24 experiments were selected (on several grounds, but not the presence or absence of activity) for longitudinal measurements. Forty-seven (40%) out of this set of 112 cultures (i.e., 112 MEAs with 60 electrodes each) exhibited spontaneous activity at 7 and 8 DIV, of which 16 had four or more active electrode sites. This means that 60% had no active electrodes at all, 26% (n = 31) showed activity at less than four electrodes, and 14%

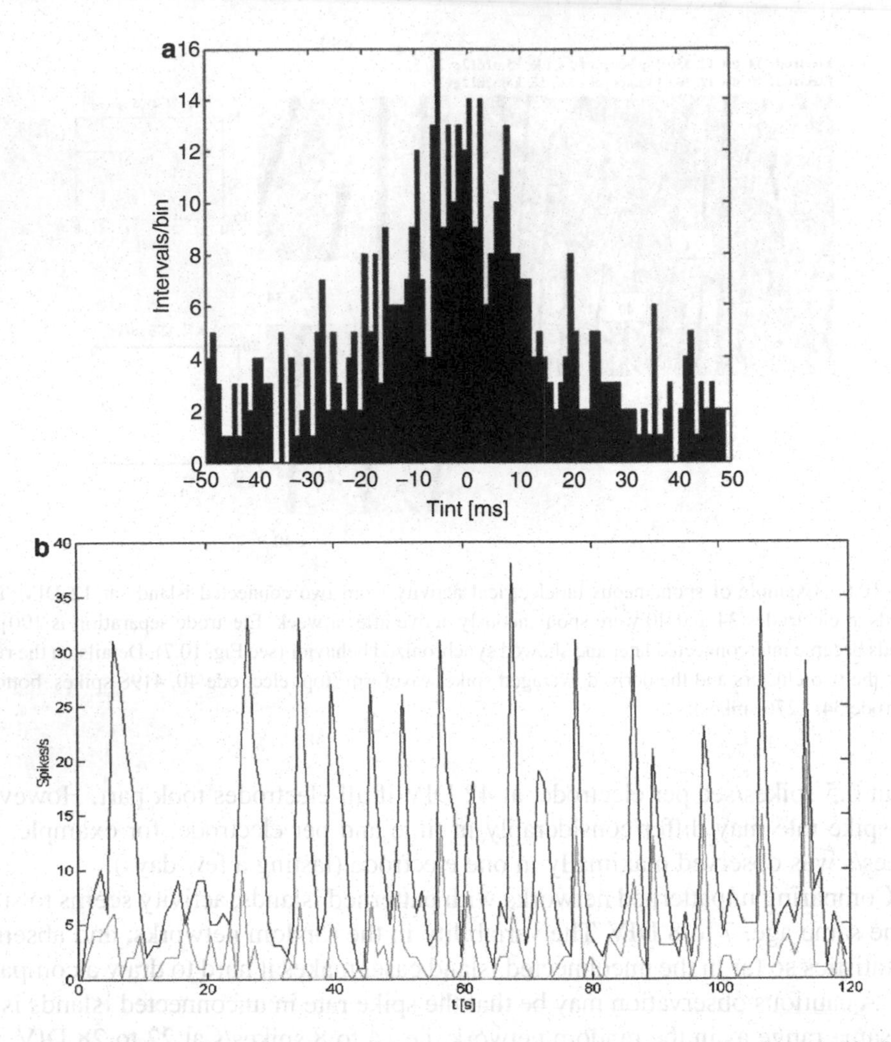

FIG. 10.7. Analysis of the spike activity of the two connected islands at electrodes e34 and e40 of Fig. 10.6. Above: Interelectrode interval time (Tint) histogram, bin width is 1 ms, of same electrodes. Below: Spikes/s during 120 sec of bursting period. Vertical scale: spikes/second. The two curves are for electrode 34 (upper, dark curve) and 40 (lower, light gray curve). The two curves clearly resemble the same oscillating pattern of bursts, pointing at synchronization between the two networks. Spearman correlation coefficient $\rho = 0.61$.

($n = 16$) at four or more electrodes. Therefore, one can estimate conservatively that the average probability that an electrode is active is in the order of 1% (under the assumption that the electrodes behave "independently," so we do not consider the bursting periods, in which many of the available electrodes show activity). Clearly, this percentage is on the same order or lower as that found for the patterned islands, as noted in Table 10.1 (i.e., 2.8% and 10.7% for the 100- and 150-μm islands, respectively) at 8 DIV. However, as the number of elements in each sample is only 7 and the standard deviations are considerable, these figures are only indicative (see Table 10.1).

The start of activity at diameter 100 micrometer and onward in circular islands probably indicates a minimum number of neurons needed to develop spontaneous activity. However, also other variables also play an essential role in that process, such as number and density of synaptic connections and cells.

Control of local density was done by Chang and co-workers (Chang, Brewer, and Wheeler 2002) on hippocampal neurons showing that alternating line patterning (40-μm wide lines) of substrates gave control over local density of neurons, 100 to 500 cells/mm^2, and did enhance the activity compared with randomly plated networks with approximately the same density.

The data give some indication that small islands (50-μm diameter) develop hardly any spontaneous activity. This may be advantageous for the use of a cultured probe as a recording device. As stated in the Introduction, for the stimulatory use of a cultured probe, spontaneous activity is unimportant as long as axonal sprouts grow from the host tissue toward the implant. Other types of cultured probes may do the reverse, that is, "send out" axons to the host tissue. In that case, spontaneous activity of the intermediate cultured networks is undesirable.

The comparison of wave shapes of Fig. 10.4 with the set of shapes from the "data base" of Fig. 10.8 (Appendix) indicates that the shape of Fig. 10.4 closely resembles type 1 of Fig. 10.8, which is by far the most frequent observed wave shape in random unpatterned networks. This gives confidence that the island populations of neurons can be regarded as subsets with identical properties.

APPENDIX

WAVE SHAPE CLASSIFICATION

Properties of the six spike waveform classes are given in Fig. 10.8 and Table 10.2: the shapes, number of measured waveforms, number of spikes per class, peak amplitude ranges, and width range. Table 10.2 shows that within each basic waveform there is a large variety of amplitude ranges and sometimes width.

From Table 10.2, it is seen that classes 4 and 5 are relatively rare. From their shape, it may be guessed that these are not action potentials from soma or axons, but field or membrane potentials. The other shapes range between 50 and 400 mv Vpp, and 1 to 2.5 ms width.

They may arise from cells or axons that cover electrodes fully or partially, or lie completely, beside an electrode. Axons may be short or long, straight, or bended. Membrane ionic channels of cells covering electrodes may be distributed homogeneously over the entire membrane surface or have a modified density in the sealed portion of the membrane (to the electrode). All these variables determine

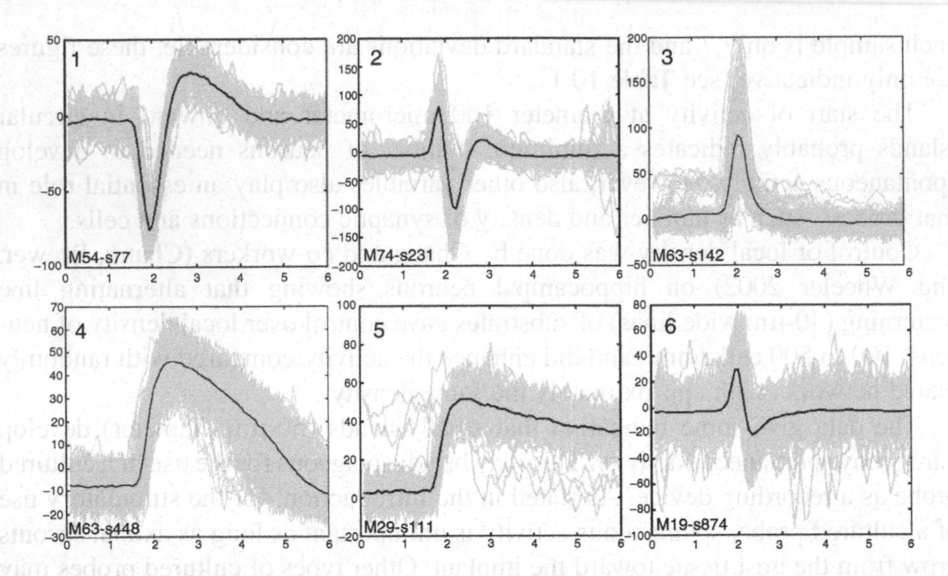

FIG. 10.8. Six basic waveform groups, obtained from 90,000 spikes in unpatterned, spontaneously active neuronal networks, age 6 to 34 days in vitro (DIV). (x-axis: ms, y-axis: mV). Solid black curves are averaged waveforms (superimposed in grey color). See also table 2.

TABLE 10.2. Summary of data of the six basic waveforms presented in Figure 10.8.

Wave form	Numberof spikes	Fist phase amplitude range (μV)	Second phase amplitude range (μV)	Third phase amplitude range (μV)	Width range (ms)
1	24115	−300 to −30	5 to 100		±2
2	2242	10 to 100	−150 to −50	5 to 50	±2
3	8998	20 to 250			1 to 1.5
4	1409	50 to 150			2.5 to 3.5
5	909	30 to 200			4+
6	3554	40 to 100	−60 to −30		±2

Waveforms can have one, two, or three phases.

the exact wave shape of the extracellularly recorded action potential and make it quite impossible to ascribe shapes to cell somata or axons.

For example, if wave shapes 1, 2, and 6 originate from axons near (but not lying over) an electrode, shape 2 is the standard triphasic shape for a straight axon, shape 1 would miss the first phase and shape 6 would miss the third phase. The latter two could then be attributed to start- and stop-phenomena in the propagation of an action potential, or to the bending of axons, as this gives rise to the missing first or third phase in a triphasic shape.

ACKNOWLEDGMENTS

The authors thank K. Zweers, R. Wiertz, and E. Marani for their contributions to the cell culturing experiments.

REFERENCES

Amiji, M. and Park, K. (1992). Prevention of protein adsorption and platelet adhesion on surfaces by PEO/PPO/PEO triblockcopolymers. Biomaterials 13, 682–692.

Blau, A. Weinl, J., Mack, S., Kienle, G., Jung and Ziegler C. (2001). Electrochemical generation of polymer films promoting neural cell adhesion on electrically conductive substrates. J. Neurosci. Meth. 112(1), 65–73.

Branch, D.W., Wheeler, B.C., Brewer, G.J. and Leckband, D. (2000). Long-term maintenance of patterns of hippocampal pyramidal cells on substrates of polyethylene glycol and microstamped polylysine. IEEE Trans. Biomed. Eng. 47, 290–300.

Branch, D.W., Wheeler, B.C., Brewer, G.J. and Leckband, D.E. (2001). Long-term stability of microstamped substrates of polylysine and grafted polyethylene glycol in cell culture conditions. Biomaterials 22, 1035–1047.

Chang, J.C., Brewer, G.J., and Wheeler, B.C. (2000). Microelectrode array recordings of patterned hippocampal neurons for four weeks. Biomed. Microdev. 2(4), 245–253.

Chang, J.C., Brewer, G.J., and Wheeler, B.C. (2002). Modulation of neural network activity by patterning. Biosens. Bioelectr. 16, 527–533.

Corey, J.M., Wheeler, B.C., and Brewer, G.J. (1996). Micrometer resolution silane-based patterning of hippocampal neurons: critical variables in photoresist and laser ablation processes for substrate fabrication. IEEE Trans. Biomed. Eng. 43, 944–955.

Jansen, H.V., Gardeniers, J.G.E., Elders, J., Tilmans, H.A.C. and Elwenspock, M. (1994). Applications of fluorocarbon polymers in micromechanics and micromachining. Sens. Actuat. A, 41–42, 136–140.

Lee, J.H., Ju, Y.M. and Kim, D.M. (2000). Platelet adhesion onto segmented polyurethane film surfaces modified by addition and crosslinking of PEO-containing block copolymers. Biomaterials 21, 683–691.

Li, J.T. and Caldwell, K.D. (1996). Plasma protein interactions with Pluronic™-treated colloids. Colloids Surfaces B: Biointerfaces 7, 9–22.

Lucas, J.H., Czisny, L.E. and Gross, G.W. (1986). Adhesion of cultured mammalian CNS neurons to flame-modified hydrophobic surfaces. In Vitro 22, 37–43.

Romein, H.J., van Huizen, F. and Wolters, P.S. (1984). Towards an improved serum-free, chemically defined medium for long-term culturing of cerebral cortex tissue. Neurosci. Biobehav. Rev. 8, 301–333.

Ruardij, T.G., Goedbloed, M.H. and Rutten, W.L.C. (2000). Adhesion and patterning of cortical neurons on polyethylenimine and fluorocarbon-coated surfaces. IEEE Trans. BioMed. Eng. 47, 1593–1599.

Ruardij, T.G., van den Boogaart, M.A.F. and Rutten, W.L.C. (2002). Adhesion and growth of electrically active cortical neurons on polyethyleneimine patterns microprinted on PEO-PPO-PEO triblockcopolymer-coated hydrophobic surfaces. IEEE Trans. Nanobiosci. 1(1), 1–8.

Van Pelt, J., Wolters, P.S., Corner, M.A., Rutten, W.L.C. and Ramakers, G.J.A. (2004). Long-term characterization of firing dynamics of spontaneous bursts in cultured neural networks. IEEE Trans. Biomed. Eng. 51, 2051–2062.

Yeung, C.K., Lauer, L., Offenhauser, A. and Knoll, W. (2001). Modulation of the growth and guidance of rat brain stem neurons using patterned extracellular matrix proteins. Neurosci Letts. 301, 147–150.

11

Nanomaterials for Neural Interfaces: Emerging New Function and Potential Applications

Allison J. Beattie, Adam S.G. Curtis, Chris D.W. Wilkinson, and Mathis Riehle

1 INTRODUCTION

One of the most potent stimuli for the control of cell adhesion, migration, orientation, and shape as well as gene expression is the topography of the substratum (Weiss 1945; Curtis and Varde 1964; Clark, Connolly, Curtis, Dow, and Wilkinson 1990; Matsuzaka, Walboomers, de Ruijter, and Jansen 2000). These stimuli appear to work on many cell types and different topographies. The scale of the topography has differing effects (den Braber, de Ruijter, Ginsel, von Recum, and Jansen 1996). Initially nearly all the evidence came from the use of microstructures made on the whole by some of the top-down techniques of the computer chip industry, but in the past 10 years nanostructured surfaces have been investigated as well. The expectation that nanometrically sized features would have profound effects on cells would have been denied in the 1990s and is still denied by those not intimately familiar with the literature. A response to this proposition should raise thoughts about the scale of cell components that might lead to reaction to nanofeatures. It may also lead to speculation as to whether a single nanofeature can affect a cell or whether quite a number of repeated nanofeatures are necessary.

A. Offenhäusser and R. Rinaldi (eds.), *Nanobioelectronics - for Electronics, Biology, and Medicine*,
DOI: 10.1007/978-0-387-09459-5_12, © Springer Science+Business Media, LLC 2009

When we consider nerve cells many of the features of these cells are organized on a small scale such dendrites, the smaller axons and presynaptic nerve terminals (boutons), post synaptic terminals, and microspikes so that it is likely that signals will be smaller in geometric size than the cellular structures that react. Furthermore, bear in mind that extensions from nerve cell bodies can extend a long way across the substratum. Such processes find their way by continuously probing, sampling, and reacting to the substratum over distances that are often in the 50s to 100s of micrometers (Fig. 11.1). Table 11.1 summarizes some of the main findings mainly for cell types other than neurons, but such data are included because they indicate that some conditions have not yet been studied for neurons, although clearly they should be.

The early work on neuron guidance by structures concentrated on linear structures such as grooves/ridges and fibers of micron sizes, but as the fabrication technology became available, a few researchers studied nanostructures (Clark, Connolly, Curtis, Dow and Wilkinson 1991; Wojciak, Crossan, Curtis, and Wilkinson 1995; Craighead, James, and Turner 2001). Initially the structures made were nanometric in only one or two of the three dimensions. Indeed, the nanoprinted tracks (see the following) were nanometric only in the thickness of the printed track, but one sparsely investigated aspect of this work is whether such tracks had nanometrically defined edges. Since neurons often align to the edges of such tracks, the question of what they were aligning to arises. Nanogrooves (Johansson, Carlberg, Danielsen, Montelius, and Kanje 2005) and nanofibers

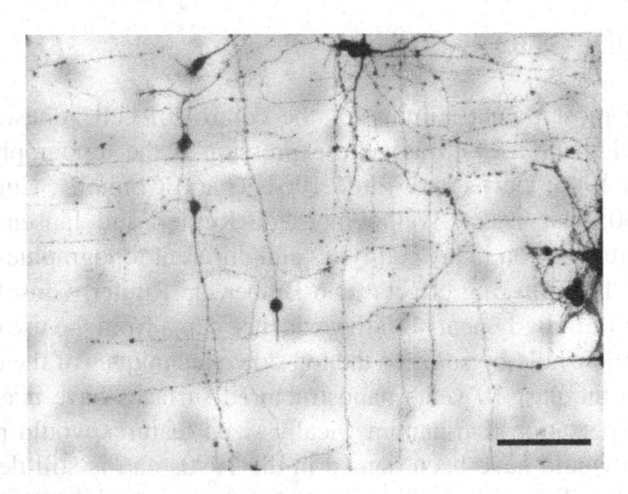

FIG. 11.1. Rat spinal cord cells growing in culture on nanopillars (150 nm diameter, 300 nm pitch, 116 nm high) in a square array made by embossing polycaprolactone. Note orthogonal pattern of the neurites and many presynaptic or postsynaptic swellings on the processes. Cell bodies partially visible at edges of the image. Cells fixed and stained with Coomassie Blue. Scale bar = 100 μm.

TABLE 11.1. Table describing the various nanotopograpies that have been reported to affect cell attachment, alignment, and growth.

Topography	Dimensions	Cell	Response	References
Nano-pits (hexagonal, orthoganol)	35 nm, 75 nm, 120 nm dia	Rat epitenon, human fibroblasts	↓ Cell adhesion (75, 120 nm) ↑ Cell alignment	(Gallagher, McGhee, Wilkinson, and Riehle 2002; Curtis, Gadegaard, Dalby, Riehle, Wilkinson, and Aitchison 2004)
Nano-colloids	20 nm, 50 nm dia	Fibroblasts	↑ Microtubule expression (20 nm), ↑ microfilament expression (50 m)	(Wood, Wilkinson, and Curtis 2006; Wood 2007)
Nano-posts, Nano-grates	"low" 50–100 nm "mid" 200–300 nm "high" 500–600 nm	Human foreskin fibroblasts	↓ Cell size and proliferation (nano-posts), ↑ cell alignment (nano-grates)	(Choi, Hagvall, Wu, Dunn, Beygui, and CJ 2007)
Nano-grooves	100–400 nm width, 200–2000 pitch, 300 nm depth	Mouse dorsal root and superior cervical ganglion explants	↑ Process alignment, groove width affected guidance	(Johansson, Carlberg, Danielsen, Montelius, and Kanje 2005)

(Schnell, Klinkhammer, Balzer, Brook, Klee, Dalton, and Mey 2007) have been used to align neurons. There is of course an alternative route to producing routing signals or cues: this is to provide nonadhesive surfaces in the regions in which cells are required to be nonadherent. Serum albumin printing and nanopit arrays can achieve this (Britland, Perez-Arnaud, Clark, McGinn, Connolly, and Moores 1992; Detrait, Lhoest, Knoops, Bertrand, and van den Bosch de Aguilar 1998; Gustavsson, Johansson, Kanje, Wallman, and Linsmeier 2007).

2 NANOFABRICATION

The lithographic tool used to define a primary pattern with less than 100 nm features is beam lithography. The pattern thus defined in resist can be used directly to make the final nanostructures by etching, or can be converted into an X-ray or optical mask for printing or into a die that can emboss a suitable polymer (as in Fig. 11.1). Although nanostructures over small areas were defined from the 1970s onward using converted scanning electron microscopes, large areas (e.g., the 1 cm^2 required for cell work) could only be obtained in the late 1980s with the advent of machines with laser-controlled stages. Such electron beam writing machines

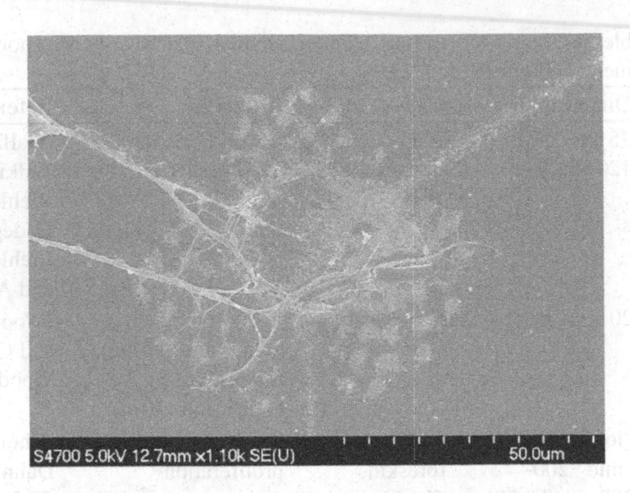

FIG. 11.2. Rat spinal cord neurons growing on chemical patterns created by microcontact printing methods.

allowed the stitching together of individual fields (the area that could be written by deflecting the beam without movement of the specimen) to form these larger areas. The accuracy of stitching is crucial in the formation of patterns for cell applications; a stitch error of only 100 nm can give rise to a linear feature to which the cells again cross. There is a requirement for the production of a large number of identical replicas of the surface, because usually biologists wish to repeat their experiments several times for each of different cell densities, growth times, proportions of different neuronal types, etc.

This replication may be done by printing through a mask using X-rays or light onto a suitable substrate covered with resist. Replication on polymeric susbstrates is conveniently done by embossing using a die.

If chemical tracks are required, the various processes of nanoprinting from an inked stamp can be used (James, Davis, Meyer, Turner, Turner, Withers, Kam, Banker, Craighead, Isaacson, Turner, and Shain 2000; Kane, Takayama, Ostuni, Ingber, and Whitesides 1999; Lauer, Klein, and Offenhausser 2001; Heller, Garga, Kelleher, Lee, Mahbubani, Sigworth, Lee, and Rea 2005). An example is shown in Fig. 11.2 Various liftoff processes are also adaptable.

2.1 MATERIALS

Although silicon and titanium with their surface oxides form good substrates for the adhesion and movement of many cell types, it may be better to add a layer of a protein over the oxide to improve cell adhesion and extension. In many cases a transparent sustratum gives the experimenter advantages in imaging the cells. The proteins used have included laminin, poly-L lysine, and poly-D lysine (Matsuzawa, Liesi, and Knoll 1996; Offenhausser, Sprossler, Matsuzawa, and Knoll 1997; Wheeler, Corey, Brewer, and Branch 1999 Chang, Brewer, and Wheeler 2003).

3 ORIENTATION, MIGRATION, AND EXTENSION

Orientation of the extensions is normally determined by the pattern of topography or chemical tracks on the substratum (Carter 1967). Chemical tracks more adhesive than surrounding areas are effective in guiding processes within the pattern. If it is possible to fabricate a substrate with a gradient in adhesion along a track, oriented movement of the cell is at least a theoretical possibility. This is haptotaxis. However, it should be appreciated that although the growth cone(s) may conform well to the substratum cue, the axon or dendrite following behind may adhere poorly to the substratum, so that if the extension is mechanically strained by the route followed, as for example by attempts to turn acute angles, the main part of the neurite may contract and pull the neurite into the shortest path. In some instances even the cell body may be detached; the cell is then left "suspended" between two or more growth cones or a growth cone and the cell body.

For these reasons the most stable patterns are orthogonally or hexagonally intersecting paths. In theory other patterns might be sufficiently stable mechanically but do not seem to have been used. A very effective pattern is shown in Fig. 11.3.

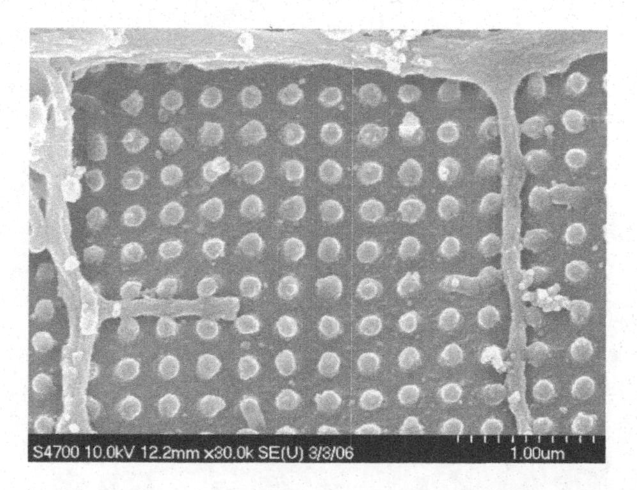

FIG. 11.3. Spinal cord cells growing in culture on nanopillars in orthogonal array on polycaprolactone, SEM. Scale bar = 1 μm.

3.1 NETWORK PATTERNS

Some work in this area has been done with simple linear tracks, which of course does not permit or at least plan for the formation of networks. Orthogonally intersecting tracks permit the building of networks, but unless the positions of synapses are very carefully specified the networks will tend to be re-entrant. Networks, or at least parts of a network, may be re-entrant in live nervous systems, but networks should "fan-in" to bring in, compare, and process different sensory inputs, and output different processed messages to a range of other networks and ultimately to motor cells.

With these matters in mind Judy Wilkinson designed a tiled set of networks that should handle fan-in and fan-out processing (Wilkinson and Curtis 1999), as shown in Fig. 11.4 For this design to be successful neurons must adhere to the circular node region, develop one process, which must split and innervate the next cells, as indicated by the arrows in Fig. 11.4 The type of network plan used recently by us can be modified slightly into a hexagonal array and it is noteworthy that cell bodies tend to locate at the junctions of three tracks with good staining for synaptic materials (required for cell interfacing) at such sites.

3.2 ORDER AND SYMMETRY

The work on the adhesion of various cell types to nanofeatures has shown that not only the size (x, y, and z dimensions) of the features affects adhesion, but also that array patterns (orthogonal, hexagonal, random, and other patterns)

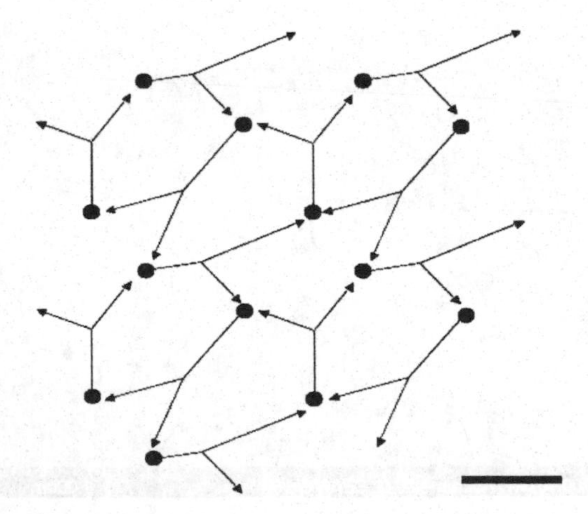

FIG. 11.4. "Fan in and fan out" Jude pattern designed by Dr. Judy Wilkinson,. Note the circular nodal areas for cell body attachment. Arrows indicate the direction of axonal growth. Scale bar = 100 µm.

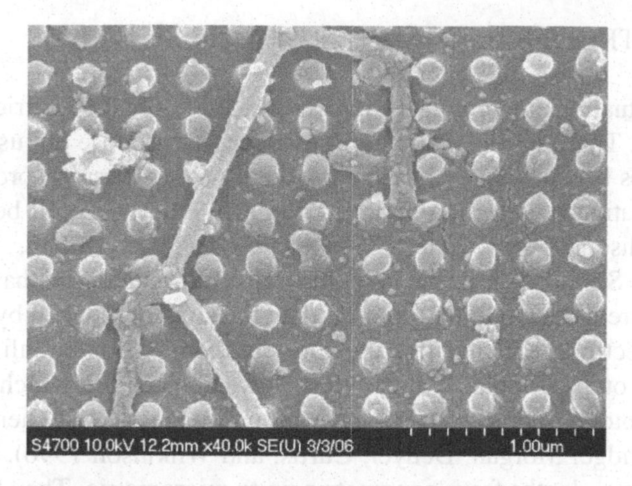

FIG. 11.5. Orientation of neuronal processes to orthogonal nanopillared array. Note the angles of process alignment. Scale bar = 1 μm.

affect adhesion (Dalby, Gadegaard, and Wilkinson 2008). Although such tests have not yet been done on the reactions of neurites, it is almost inherently obvious that there will be such effects (Fig. 11.5). This suggests that misalignment of the arrays will lead to unintended routings but it also indicates that the alignments of these nanodots and related nanofeatures could be used to construct very precise nerve networks.

3.3 GENE EXPRESSION

Although several authors have reported that growing cells on various types of nanostructure enhances gene expression most of these studies have confined their work to at most a few genes (Dalby, Giannaras, Riehle, Gadegaard, Affrossman, and Curtis 2004; Dalby, Riehle, Johnstone, Affrossman, and Curtis 2004; Dalby, Riehle, Sutherland, Agheli and Curtis 2004, 2005). However, Dalby and co-workers used large gene arrays and reported on many up and many down expressions resulting from the growth of fibroblasts on pillar nanotopography. If this finding applies to neurons, then we have a potentially very powerful system for controlling the phenotype of cells. It is possible that the nanofeatures induce changes in chromosome territories (Dalby, Biggs, Gadegaard, Kalna, Wilkinson, and Curtis 2007).

4 ELECTRODES (EXTRACELLULAR)

In many situations there will be a wish or need to obtain electrical recordings from the cells. There has been a natural tendency to propose the use of extracellular electrodes because they offer the prospect of long term recordings at many sites. Gold, platinum, and iridium and indium-tin oxide have all been used with success (Sandison, Curtis, and Wilkinson 2002; Nam, Chang, Wheeler, and Brewer 2004). Stimulation of the cells may also be desired, this has the problem that higher current densities are required than the signal produced by the cells and hence lager electrodes. Do nanofeatures present more potent stimuli for nerve cell guidance than other stimuli? A few reports have appeared in which the guidance effects of nanotopographic features have been compared with chemical patterns (Britland, Perridge, Morgan, Denyer, Curtis, and Wilkinson 1996). Britland used a range of groove depths from nanometric up to micrometric. They found that the deeper the groove the more pronounced were the effects of the topography, so that at the greatest depths the groove effect (nanotopographic) dominated the chemical track effects.

Many other cues for controlling the movement of neurons have been described (e.g., chemotaxis and mechanical forces), and it might be expected that such cues might be present in test systems. If these other cues were important it could be argued that they should have overridden topography or chemical surface patterns. They do not appear to do so.

5 SUMMARY

This chapter presents arguments and evidence that nanofeatures should have great potential for organizing nerve nets in culture.

REFERENCES

Britland, S., Perez-Arnaud, E., Clark, P., McGinn, B., Connolly, P. and Moores, G. (1992). Micropatterning proteins and synthetic peptides on solid supports: a novel application for microelectronics fabrication technology. Biotechnol. Prog. 8, 155–160.

Britland, S.T., Perridge, C., Morgan, H., Denyer, M.C., Curtis, A. and Wilkinson, C. (1996). Morphogenetic guidance cues can interact synergistically and hierarchically in steering nerve cell growth. Exp. Biol. Online 1:2.

Carter, S.B. (1967). Haptotaxis and the mechanism of cell motility. Nature 213, 256–260.

Chang, J.C., Brewer, G.J. and Wheeler, B.C. (2003). A modified microstamping technique enhances poly-lysine transfer and neuronal cell patterning. Biomaterials 24, 2863–2870.

Choi, C.H., Hagvall, S.H., Wu, B.M., Dunn, J.C., Beygui, R.E. and Kim, C.J. (2007). Cell interaction with three-dimensional sharp-tip nanotopography. Biomaterials 28, 1672–1679.

Clark, P., Connolly, P., Curtis, A.S., Dow, J.A. and Wilkinson, C.D. (1990). Topographical control of cell behaviour: II. Multiple grooved substrata. Development 108, 635–644.

Clark, P., Connolly, P., Curtis, A.S., Dow, J.A. and Wilkinson, C.D. (1991). Cell guidance by ultrafine topography in vitro. J Cell Sci. 99 (Pt 1), 73–77.

Craighead, H.G., James, C.D. and Turner, A.M.P. (2001). Chemical and topographical patterning for directed cell attachment. Curr. Opin. Solid State Mater. Sci. 5, 177–184.

Curtis, A.S., Gadegaard, N., Dalby, M.J., Riehle, M.O., Wilkinson, C.D. and Aitchison, G. (2004). Cells react to nanoscale order and symmetry in their surroundings. IEEE Trans Nanobiosci. 3, 61–65.

Curtis, A.S. and Varde, M. (1964). Control of cell behavior: topological factors. J Natl Cancer Inst. 33, 15–26.

Dalby, M.J., Biggs, M.J., Gadegaard, N., Kalna, G., Wilkinson, C.D. and Curtis, A.S. (2007). Nanotopographical stimulation of mechanotransduction and changes in interphase centromere positioning. J. Cell Biochem. 100, 326–338.

Dalby, M.J., Gadegaard, N. and Wilkinson, C.D. (2008). The response of fibroblasts to hexagonal nanotopography fabricated by electron beam lithography. J. Biomed. Mater. Res A 15;84(4):973–979.

Dalby, M.J., Giannaras, D., Riehle, M.O., Gadegaard, N., Affrossman, S. and Curtis, A.S. (2004). Rapid fibroblast adhesion to 27 nm high polymer demixed nano-topography. Biomaterials 25, 77–83.

Dalby, M.J., Riehle, M.O., Johnstone, H., Affrossman, S. and Curtis, A.S. (2004). Investigating the limits of filopodial sensing: a brief report using SEM to image the interaction between 10 nm high nanotopography and fibroblast filopodia. Cell Biol. Int. 28, 229–236.

Dalby, M.J., Riehle, M.O., Sutherland, D.S., Agheli, H. and Curtis, A.S. (2004). Fibroblast response to a controlled nanoenvironment produced by colloidal lithography. J. Biomed. Mater. Res. A. 69, 314–322.

Dalby, M.J., Riehle, M.O., Sutherland, D.S., Agheli, H. and Curtis, A.S. (2005). Morphological and microarray analysis of human fibroblasts cultured on nanocolumns produced by colloidal lithography. Eur. Cell. Mater. 9, 1–8; discussion.

den Braber, E.T., de Ruijter, J.E., Ginsel, L.A., von Recum, A.F. and Jansen, J.A. (1996). Quantitative analysis of fibroblast morphology on microgrooved surfaces with various groove and ridge dimensions. Biomaterials 17, 2037–2044.

Detrait, E., Lhoest, J.B., Knoops, B., Bertrand, P. and van den Bosch de Aguilar, P. (1998). Orientation of cell adhesion and growth on patterned heterogeneous polystyrene surface. J. Neurosci. Meth. 84, 193–204.

Gallagher, J.O., McGhee, K.F., Wilkinson, C.D. and Riehle, M.O. (2002). Interaction of animal cells with ordered nanotopography. IEEE Trans. Nanobiosci. 1, 24–28.

Gustavsson, P., Johansson, F., Kanje, M., Wallman, L. and Linsmeier, C.E. (2007). Neurite guidance on protein micropatterns generated by a piezoelectric microdispenser. Biomaterials 28, 1141–1151.

Heller, D.A., Garga, V., Kelleher, K.J., Lee, T.C., Mahbubani, S., Sigworth, L.A., Lee, T.R. and Rea, M.A. (2005). Patterned networks of mouse hippocampal neurons on peptide-coated gold surfaces. Biomaterials 26, 883–889.

James, C.D., Davis, R., Meyer, M., Turner, A., Turner, S., Withers, G., Kam, L., Banker, G., Craighead, H., Isaacson, M., Turner, J. and Shain, W. (2000). Aligned microcontact printing of micrometer-scale poly-L-lysine structures for controlled growth of cultured neurons on planar microelectrode arrays. IEEE Trans Biomed Eng. 47, 17–21.

Johansson, F., Carlberg, P., Danielsen, N., Montelius, L. and Kanje, M. (2005). Axonal outgrowth on nano-imprinted patterns. Biomaterials 27, 1251–1258.

Kane, R.S., Takayama, S., Ostuni, E., Ingber, D.E. and Whitesides, G.M. (1999). Patterning proteins and cells using soft lithography. Biomaterials 20, 2363–2376.

Lauer, L., Klein, C. and Offenhausser, A. (2001). Spot compliant neuronal networks by structure optimized micro-contact printing. Biomaterials 22, 1925–1932.

Matsuzaka, K., Walboomers, F., de Ruijter, A. and Jansen, J.A. (2000). Effect of microgrooved poly-l-lactic (PLA) surfaces on proliferation, cytoskeletal organization, and mineralized matrix formation of rat bone marrow cells. Clin. Oral Implants Res. 11, 325–333.

Matsuzawa, M., Liesi, P. and Knoll, W. (1996). Chemically modifying glass surfaces to study substratum-guided neurite outgrowth in culture. J. Neurosci. Meth. 69, 189–196.

Nam, Y., Chang, J.C., Wheeler, B.C. and Brewer, G.J. (2004). Gold-coated microelectrode array with thiol linked self-assembled monolayers for engineering neuronal cultures. IEEE Trans. Biomed. Eng. 51, 158–165.

Offenhausser, A., Sprossler, C., Matsuzawa, M. and Knoll, W. (1997). Electrophysiological development of embryonic hippocampal neurons from the rat grown on synthetic thin films. Neurosci. Lett. 223, 9–12.

Sandison, M., Curtis, A.S. and Wilkinson, C.D. (2002). Effective extra-cellular recording from vertebrate neurons in culture using a new type of micro-electrode array. J. Neurosci. Meth. 114, 63–71.

Schnell, E., Klinkhammer, K., Balzer, S., Brook, G., Klee, D., Dalton, P. and Mey, J. (2007). Guidance of glial cell migration and axonal growth on electrospun nanofibers of poly-epsilon-caprolactone and a collagen/poly-epsilon-caprolactone blend. Biomaterials 28, 3012–3025.

Weiss, P. (1945). Biological research strategy and publication policy. Science 101, 101–104.

Wheeler, B.C., Corey, J.M., Brewer, G.J. and Branch, D.W. (1999). Microcontact printing for precise control of nerve cell growth in culture. J. Biomech. Eng. 121, 73–78.

Wilkinson, C. and Curtis, A. (1999). Networks of living cells. Phys. World 12, 45–48.

Wojciak, B., Crossan, J., Curtis, A.S.G. and Wilkinson, C.D.W. (1995). Grooved substrata facilitate in vitro healing of completely divided flexor tendons. J. Mater. Sci. Mater. Med. 6, 266–271.

Wood, M.A. (2007). Colloidal lithography and current fabrication techniques producing in-plane nanotopography for biological applications. J. R. Soc. Interface 4, 1–17.

Wood, M.A., Wilkinson, C.D. and Curtis, A.S. (2006). The effects of colloidal nanotopography on initial fibroblast adhesion and morphology. IEEE Trans. Nanobiosci. 5, 20–31.

12 Interfacing Neurons and Silicon-Based Devices

Andreas Offenhäusser, Sven Ingebrandt, Michael Pabst, and Günter Wrobel

1 INTRODUCTION

The combination of biological signal processing elements such as membrane proteins, whole cells, or even tissue slices with electronic transducers for the detection of physical signals creates functional hybrid systems that bring together the living and the technical worlds. Functional coupling of physiological processes with microelectronic and nanoelectronic devices will have great impact for a wide range of applications. The high sensitivity and selectivity of biological recognition systems with a manufactured signal-detection and processing system will open up exciting possibilities for the development of new biosensors as well as for new approaches in neuroscience and computer science. This includes: (a) pharmacological as well as toxicologically lab on a chip concepts, which allows fast, high-throughput screening of potential drugs; (b) the use of the high sensitivity and selectivity of biological recognition systems with signal-amplification cascades for the development of biosensors with unprecedented detection threshold; and (c) the multisite interfacing of neuronal networks with arrays of electronic devices on the microscopic level of individual nerve cells or cell processes would facilitate spatiotemporal mapping of brain dynamics.

A. Offenhäusser and R. Rinaldi (eds.), *Nanobioelectronics - for Electronics, Biology, and Medicine*,
DOI: 10.1007/978-0-387-09459-5_13,

One of the most important reasons for the use of living cells is to obtain functional information, such as the effect of a stimulus on a living system. Information at the cellular level can yield insight into mechanisms of biochemical compound action, enabling not only detection but also classification. This has led to the development and validation of a number of cell-based biosensor concepts. In the field of biosensors cellular proteins exhibit enhanced stability when expressed in cell systems, as compared with coating these proteins on electrochemical or optical transducers. In addition, cultured excitable cells have been proposed as a cellular transducing system, in which changes in spontaneous or evoked action potentials reflect functional changes associated with the exposure to pharmacological compounds. These cells are usually cultured on multisite recording arrays to allow long-term observation from many cells at the same time. In addition, the coupling of a 2D cellular network with an extracellular recording system allows the structure–function relationship of such a network to be studied in detail. For extracellular signal recordings from electrically active cells in culture, two main concepts have been developed: multielectrode arrays (MEA) with metalized contacts on silicon or glass substrates (Pine 1980; Gross, Wen, and Lin 1985) and arrays of FETs (Bergveld, Wiersma, and Meertens 1976; Fromherz, Offenhäusser, Vetter, and Weis 1991). With these noninvasive methods, the electrical activity of single cells and networks of neurons can be observed over an extended period of time. This enables insights into long-term effects such as adaptation in neuronal networks and allows the fabrication of robust whole cell biosensors (Parce et al. 1989). Meanwhile, both concepts are growing together by designing MEAs inside a CMOS process with on chip amplification and filtering (Heer, Hafizovic, Franks, Blau, Ziegler and Hierlemann 2006; Krause, Lehmann, Lehmann, Freund, Schreiber, and Baumann 2006).

Sufficient electrical coupling between the cell and the electrode for extracellular signal recording is achieved only when a cell or a part of a cell is located directly on top of the electrode. Electrical signals recorded by these devices show lower signals and a higher noise level (owing to a weaker coupling to the [gate] electrode compared with the patch pipette), but can be monitored for weeks. In addition recordings can be done simultaneously on many sites.

The need for further manipulations of the cell adhesion and organization at the micron level becomes evident if one looks at a randomly seeded network of nerve cells. Usually neurons adhere to the substrate developing dendrites and axons and forming a totally random network structure. However, we are more interested in the formation and characterization of well-defined network architectures and, hence, have developed strategies that allow us to manipulate the adhesion sites of individual neurons and control the direction and outgrowth of dendrites and axons. This can be achieved at the level of the adhesion layer, the interface between the substrate (microelectronic device) and the adhering cell.

Using the so-called micro-contact-printing (μCP) technique gives us the required control over the network formation. In the CP approach a stamp made from poly(dimethylsiloxane) (PDMS) is "inked" with the specific peptide or with the whole extracellular matrix protein such as laminin and the pattern imprinted onto the substrate, which was precoated with the linker molecules used for the covalent attachment of the adhesion promoter.

2 THEORETICAL CONSIDERATIONS

The electrical activity of biological cells in vitro can be recorded by measuring changes in the local extracellular voltage with planar metal electrodes or integrated planar field effect transistors (FETs). These changes in local extracellular voltage result from the flux of ions across the membrane of the cell, mainly sodium (Na^+), potassium (K^+), and calcium (Ca^{2+}) ions. The distribution and strength of the potential near the sensor depends on the geometry of the cleft between attached cell and sensor surface and the specific electrical behavior of the cell membrane.

For a quantitative understanding of the extracellular signals it is necessary to describe the experimental situation. A schematic picture of a typical setup is shown in Fig. 12.1. Biological cells (human embryonic kidney [HEK293] cells expressing a voltage gated K^+ channel) (Wrobel, Seifert, Ingebrandt, Enderlein, Ecken, Baumann, Kaupp, and Offenhäusser 2005) are cultured on top of the gate of a FET. Outside the cell and inside the cleft, the region between cell and sensor surface, there is extracellular electrolyte solution. By electrical excitation, the ion channels in the cell's membrane open and ions can flow from the inside to the

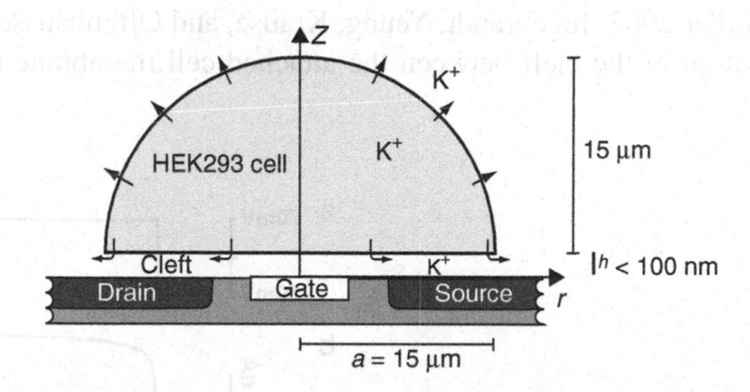

FIG. 12.1. Schematic drawing of the cell–sensor interface between a cell and a field-effect transistor (not to scale). The cell is located on the top of a FET gate. The cell is approximated by a half-sphere with a radius $a = 15\,\mu m$. The cleft height is $h < 100\,nm$ (Lambacher and Fromherz 1996; Wrobel, Höller, Ingebrandt, Dieluweit, Sommerhage, Bochem, and Offenhäusser 2007). The gate area of the FET is about $50\,\mu m^2$. Smaller letter-size corresponds to a smaller concentration of K^+.

outside. Although in the upper part of the cell (free membrane) these ions just enter the surrounding electrolyte bath directly, it is different at the attached membrane. Here, the ions have to pass the cleft before entering the bath. The flux of ions into the cleft causes an increase of charge. Because of the connected electric field other charged ions will move: Ions with similar charge leave the cleft and move into the surrounding bath, whereas ions of opposite charge are attracted and enter the cleft. The change of charge inside the cleft influences the source drain current under the gate and is sensed by the FET. It can be directly correlated to the membrane current into the cleft.

Figure 12-2 shows the comparison of whole-cell membrane currents (I_M) and extracellular signal shapes (V_{FET}) recorded with a p-channel FET for HEK293 cells transfected with a K^+ channel: (A) Rectangular stimulation pulses from the holding voltage led to (B) whole-cell current I_M and (C) corresponding FET signals V_{FET} for activation of the K^+ channel. The FET signal shape resembles a combination of different signal components, each with distinct kinetics depending on the respective cell type: (a) capacitive transients, caused by the capacitive coupling of the stimulation pulse (Schätzthauer and Fromherz 1998; Sprössler, Denyer, Britland, Curtis, Knoll, and Offenhäusser 1999), (b) an increase of V_{FET} to a steady state amplitude, (c) a partially instantaneously decline of V_{FET} at the end the pulse, and (d) a slow relaxation of V_{FET}, that is absent in I_M.

For a quantitative description of the FET signal shapes the ion flux in the cleft has to be understood: Regehr and co-workers (Regehr, Pine, Cohan, Mischke, and Tank 1989) developed the "point contact model," a simple equivalent circuit that describes the extracellular voltage waveforms of invertebrate neurons measured with planar metal electrodes. This model was also used to describe FET signal shapes of different types of electrogenic cells (Fig. 12-3) (Schätzthauer and Fromherz 1998; Sprössler, Denyer, Britland, Curtis, Knoll, and Offenhäusser 1999; Kind, Issing, Arnold, and Müller 2002; Ingebrandt, Yeung, Krause, and Offenhäusser 2005). The electrolyte solution in the cleft between the attached cell membrane and the FET

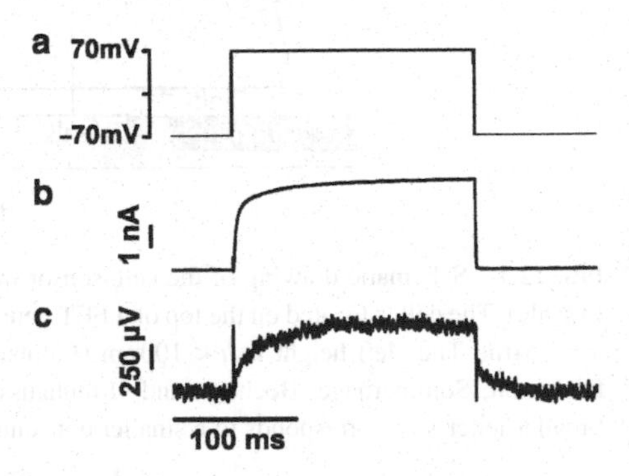

FIG. 12.2. Exemplary electrical coupling of a HEK293 cells transfected with K^+ channels to a FET gate. **A.** Voltage-clamp stimulation pulse. **B.** Transmembrane K^+ current measured by patch-clamp. **C.** Corresponding FET recording.

surface led to a seal resistance R_J between cleft and bath. The magnitude of R_J is typically in the order of several 100 kΩ up to MΩ. The model considers the current components in the cleft between the FET surface and the attached cell membrane (see Fig. 12-3). Incorporated inside the membrane are various channels of specific ion conductances that contribute to the current through the membrane. Based on the theory of (Hodgkin and Huxley 1952) the membrane current consists of several components, a capacitive one characterized by the membrane capacitance C_{AM}, resistive parts R^i depending on the particular ion species i, and a leakage current through $R_{AM}{}^L$.

The current through the cleft is determined by the seal resistance R_J and the voltage V_J. C_G and R_G are the capacitance and the resistance of the gate, respectively. By neglecting the leakage current through the membrane and assuming $R_G >> R_J$ one receives by application of Kirchhoff's law to the current components in the cleft a quantitative description of the point contact model:

$$C_{AM}\frac{d(V_M - V_J)}{dt} + \sum_i I^i_{AM} - \frac{V_J}{R_J} - C_G\frac{dV_J}{dt} = 0 \qquad (1)$$

Here V_M is the homogeneous intracellular membrane voltage.

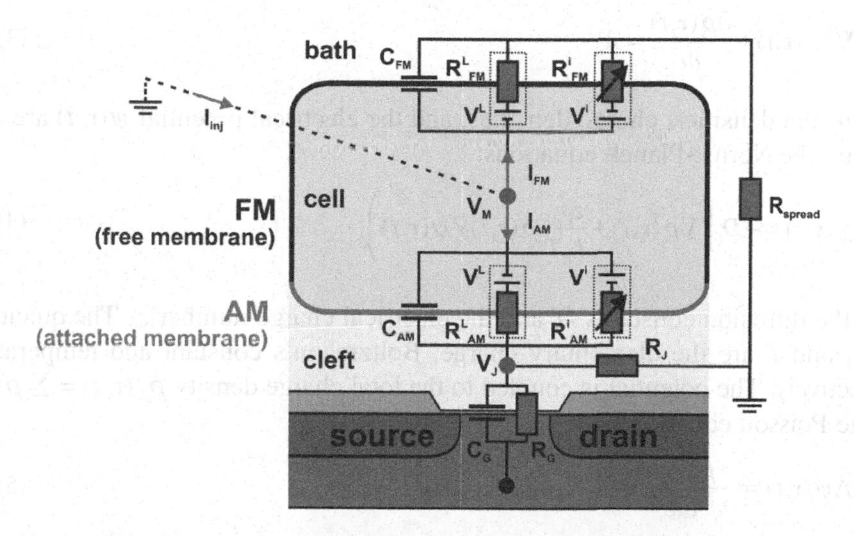

FIG. 12.3. Schematics of the point contact model of cell-transistor coupling. The cell membrane is divided into free (*FM*) and attached membrane (*AM*) with the respective values of membrane area (A_{FM}, A_{JM}) and membrane capacitance (C_{FM}, C_{JM}). Hodgkin-Huxley elements (Hodgkin and Huxley 1952) describe the properties of the cell membrane and the different ion channels. V_L and R_L represent the leakage voltage and leakage resistance of the membrane. C_G and R_G are the capacitance and the resistance of the gate, respectively. The seal resistor R_J represents the electrical properties of the cleft between the membrane and the transistor. In case of patch-clamp experiments, a current I_{inj} can be injected into the cell.

The Point contact model gives a global description and does not provide any local information of the potential inside the cleft. An improvement of this description is the use of the two-dimensional cable equation (Weis and Fromherz 1997).

$$\frac{1}{r_J}\Delta V_J = -c_{AM}\frac{d(V_M - V_J)}{dt} - \sum_i j_{AM}^i + c_G\frac{dV_J}{dt} \tag{2}$$

with V_J depending at the local coordinates inside the cleft. In analogy to R_J, C_{AM}, I_{AM} and C_G the parameters r_J, c_{AM}, j_{AM}, and c_G are local quantities depending on the coordinates.

Equations 1 and 2 consider the electrical properties of the cell-cleft-FET system only. No electrodiffusive effects are included. However in the complete description electrodiffusion has to be included. A fundamental approach is the description of the electrodiffusion of the ions in the cleft by a differential equation system, consisting out of the continuity equation, the Nernst-Planck equation end the Poisson equation, the Poisson-Nernst-Planck (PNP) equation system. For each ion species i let $\rho_i(\mathrm{r}, t)$ and $\mathrm{j}_i(\mathrm{r}, t)$ present the electrical charge densities and the electrical current densities as functions of position r and time t. If λ_i defines the source rate of the different ions, the continuity equations are given by:

$$\nabla \cdot \mathbf{j}_i(\mathbf{r},t) + \frac{\partial \rho_i(\mathbf{r},t)}{\partial t} = \lambda_i \tag{3}$$

Current densities, charge densities, and the electrical potential $\psi(\mathrm{r}, t)$ are coupled by the Nernst-Planck equations:

$$\mathbf{j}_i(\mathbf{r},t) = -D_i\left(\nabla \rho_i(\mathbf{r},t) + \frac{z_i e_0}{k_B T}\rho_i(\mathbf{r},t)\nabla \psi(\mathbf{r},t)\right) \tag{4}$$

with the diffusion constants D_i and the electrical charge number z_i. The quantities e_0, k_B and T are the elementary charge, Boltzmann's constant and temperature, respectively. The potential is coupled to the total charge density $\rho_{tot}(\mathrm{r}, t) = \sum \rho_i(\mathrm{r}, t)$ by the Poisson equation:

$$\Delta \psi(\mathbf{r},t) = -\frac{\rho_{tot}(\mathbf{r},t)}{\varepsilon_0 \varepsilon_r} \tag{5}$$

with the dielectric constants ε_0 and ε_r. The disadvantage of this system is that it results in a nonlinear integro-differential equation which usually cannot be solved analytically. However, under certain boundary conditions and simplifications the PNP system can be linearized and analytical solutions are possible.

3 METHODS

3.1 FIELD EFFECT TRANSISTORS FOR EXTRACELLULAR RECORDINGS

FETs used for extracellular recordings had either nonmetalized transistor gates with cells growing directly on the gate oxide (Bergveld, Wiersma, and Meertens 1976; Fromherz, Offenhäusser, Vetter, and Weis 1991; Offenhäusser et al. 1997; Krause, Lehmann, Lehmann, Freund, Schreiber, and Baumann 2006) or metalized gates. The latter were in direct contact with the electrolyte (Jobling et al. 1981) or they were electrically insulated, so-called floating gates (Offenhäusser et al. 1995; Cohen, Spira, Yitshaik, Borghs, Shwartzglass, and Shappir 2004; Meyburg, Goryll, Moers, Ingebrandt, Böcker-Meffert, Lüth, and Offenhäusser 2006) (Fig. 12-4).

These semiconductor devices are based on concepts developed for the fabrication of ion-sensitive FETs (ISFETs), which combine the operational principle of a high-input impedance FET and an ion-sensitive electrode. ISFETs are usually made from a simplified metal gate FET process (Matsuo et al., 1981). However, various groups have proposed the fabrication of ISFETs on the basis of a modified standard CMOS process, which allows integration of measurement circuitry on the same chip and achieves a cost reduction in fabrication. Extended gate ISFET structures were developed (van der Spiegel et al. 1983; Katsube et al. 1986) in which a chemically sensitive layer was deposited over a metal that was connected by a metal line to the gate of a MOSFET.

Bousse and co-workers (Bousse, Scott, and Meindl 1988) fabricated ISFETs with an electrically floating polysilicon (PS) gate in a modified CMOS process. The PS was used to define the source and drain regions, and was covered by a silicon nitride layer used as pH-sensitive material. Previously, a similar device structure had been considered by Smith and co-workers (1984) for electrostatic

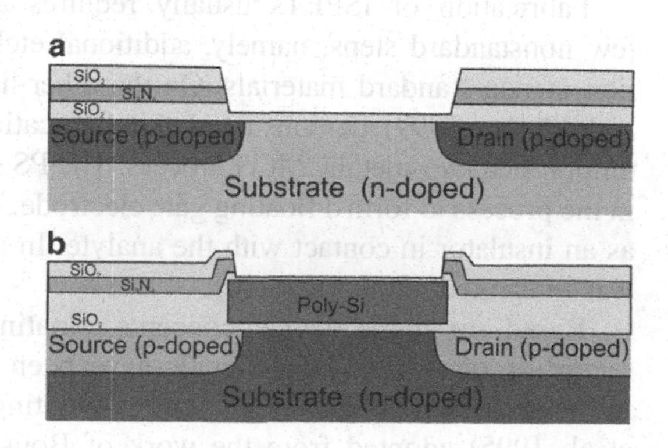

FIG. 12.4. Cross-section of the transistor for extracellular recording of neuronal signals (not to scale). (left) Open gate transistor (right) Floating gate transistor.

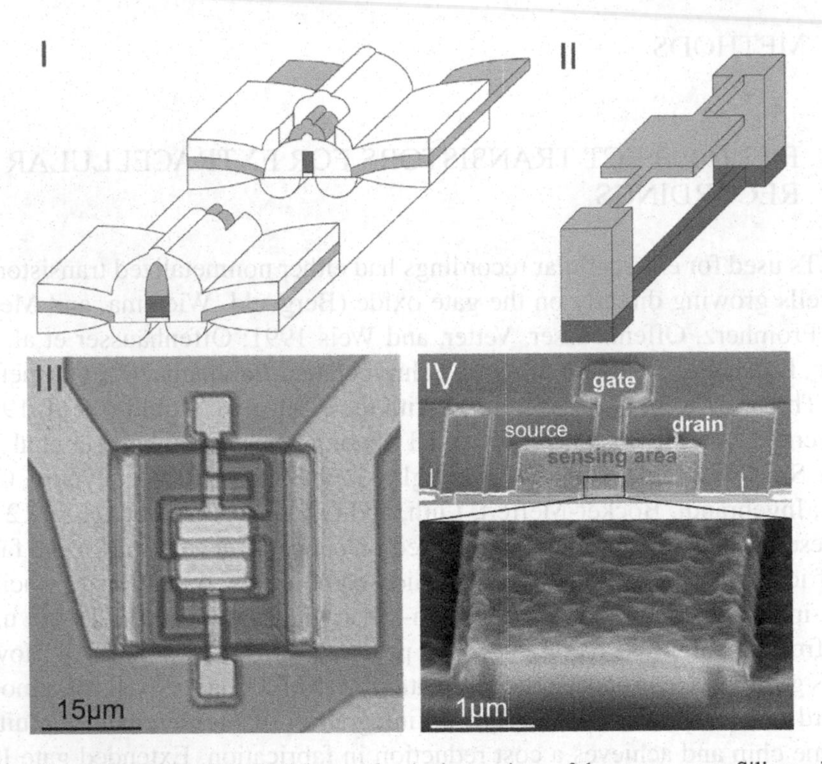

FIG. 12.5. Floating gate transistor. **I.** Scheme of a single transistor of the sensor array. Silicon nitride of the ONO stack is omitted for clarity. **II.** Exposed scheme of the compound floating gate. Notice the distance between the top and bottom gate in the middle of the transistor of silicon oxide. The top and bottom gate were electrically connected at both ends. The schemes are not to scale. **III.** Differential interference contrast (DIC) micrograph of a meander-shaped transistor (L_g = 2 μm, exposed area 100 μm^2). **IV.** SEM micrograph of a cross-section of a linear transistor (L_g = 2 μm, exposed area 460 μm^2). The inset shows a magnification of the transistor gate region. (Adapted from Meyburg, Goryll, Moers, Ingebrandt, Böcker-Meffert, Lüth, and Offenhäusser 2006).

protection of the ISFET gate. However, it was not completely floating, as it was connected via a MOSFET switch to the external measurement circuitry.

Fabrication of ISFETs usually requires a CMOS-like process including a few nonstandard steps, namely, additional etching steps, passivation and deposition of non-standard materials. On the other hand (Bausells, Carrabina, Errachid, and Merlos 1999) demonstrated the fabrication of a pH-sensitive ISFET in an unmodified two-metal CMOS process. The PS gate is connected to the metal layers in the process to form a floating gate electrode. The CMOS passivation layer is used as an insulator in contact with the analyte. In addition, an ISFET amplifier circuit was integrated in the same chip.

Based on these device concepts, floating gate transistor designs for the recording of extracellular signals have been developed. In an earlier work we fabricated an n-channel FET with a floating PS gate structure (Offenhäusser et al. 1995) adopted from the work of Bousse and co-workers (Bousse, Scott, and Meindl 1988). A thinner dielectric layer was used in order to increase the input capacitance. However, this implementation had limitations because of the almost identical geometry of the sensing and the gate area. Cohen and co-workers (Cohen, Spira, Yitshaik, Borghs, Shwartzglass, and Shappir 2004) reported a fabrication process of depletion-type p-channel floating gate transistors with octagonal sensing area realized in a shortened 0.5-μm CMOS technology with double PS layers. The sensing area was exposed by post-CMOS processing steps, which removed the second PS layer. We developed a floating gate concept with two PS layers in which an n-channel MOS transistor gate is connected to a sensing area exposed to the cell culture (Meyburg, Goryll, Moers, Ingebrandt, Böcker-Meffert, Lüth, and Offenhäusser 2006). The sensing area was isolated from the electrolyte by a thin oxide. This concept combined an n-channel transistor optimized for large transconductance with a sensing area that was independently adapted to the size of an adhering biological cell. In addition, the separation between the sensing area and PS gate should prevent cations such Na^+ and K^+ from penetrating the gate oxide and thus altering the properties of the transistor, a problem especially familiar with nonmetalized n-channel FETs.

3.2 CHARACTERIZATION OF THE CELL–DEVICE INTERFACE

The signal transfer between cells and electronic devices depends primarily on the interface between cell and sensor surface (Regehr, Pine, Cohan, Mischke, and Tank 1989; Schätzhauer and Fromherz 1998; Ingebrandt, Yeung, Krause, and Offenhäusser 2005). As a consequence, the characterization of the cell–device interface is of high interest. Optical methods are very well suited to study the attachment of the cells. Here, in particular reflection interference contrast microscopy (RICM) (Curtis 1964; Izzard and Lochner 1976), total internal reflection fluorescence (TIRF) microscopy (Axelrod 1981; Gingell et al. 1985), and fluorescence interference contrast (FLIC) microscopy (Lambacher and Fromherz 1996; Braun and Fromherz 1998) are used. However, RICM and TIRF microscopy can not be used with opaque substrates. Although these optical microscopy techniques provide excellent in-depth information, their lateral resolution is restricted by the wavelength of the light. For example, FLIC microscopy can resolve 1-nm differences in depth at a lateral resolution of 400 nm (Lambacher and Fromherz 1996; Braun and Fromherz 1998). For the characterization of the cell–device interface at a higher lateral resolution, transmission electron microscopy (TEM) can be employed and has been used with a different focus by several groups before (Meyle et al. 1993; Singhvi et al. 1994; den Braber et al. 1998, Walboomers et al. 1998; Pfeifer et al. 2003; Khakbaznejad et al. 2004; Baharloo et al. 2005).

Both—optical as well as electron microscopy—require a detailed knowledge of the protein layers between cell membrane and solid substrate. These protein layers can influence cell–substrate measurements and may lead to wrong interpretations. To reduce these errors, protein layer thickness can be determined by scanning probe microscopy or surface sensitive optical techniques. Here, imaging ellipsometry was used to determine the protein layer thickness on opaque silicon/silicon dioxide substrates.

We used substrates with surface properties similar to those of our FET recording devices: smooth silicon from the border of a wafer used in the FET fabrication, as well as FET chips. A thin layer of SiO_2, on which the cells were growing, was the final passivation layer for all substrates. Before cell culturing, the substrates were cleaned and coated with different protein solutions. HEK cells were cultured on these substrates that expressed a voltage-gated K^+ channel as previously described (Wrobel, Seifert, Ingebrandt, Enderlein, Ecken, Baumann, Kaupp, and Offenhäusser 2005). The cells were kept for 3 to 5 days at 37°C and 5% CO_2. The attached cells were fixed with glutaraldehyde and postfixed with OsO_4 in and uranyl acetate. Following dehydration with a graded acetone series, the specimens were incubated in propylene oxide. The specimens were then embedded in epoxy resin.

We removed the substrates by liquid N_2 freeze fracture followed by sonication. By this procedure the cells remained in the epoxy resin and the interface became uncovered. We preserved that interface by sputtering a thin gold or platinum/coal layer onto the surface. The specimens were then re-embedded in epoxy resin and ultrathin sections (approx. 70 nm) prepared.

The detailed TEM study provided clear differences for the ultrathin sections of HEK cells grown on the protein coatings. Figure 12-6 shows a typical TEM

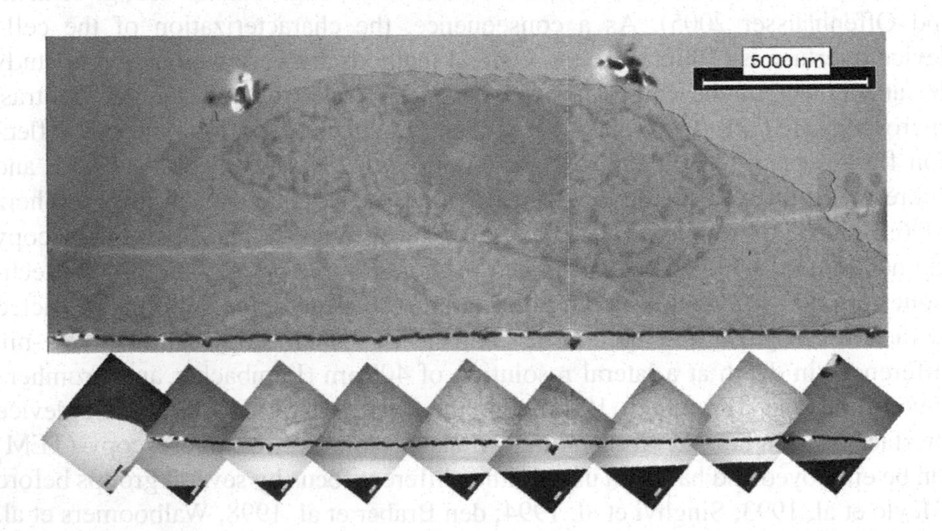

FIG. 12.6. Typical TEM images of HEK cells grown on smooth substrates. **A.** HEK cell grown for 4 days on a PLL-coated substrate. **B.** Higher magnification of the upper image.

image of a cell grown for 4 days on a PLL-coated smooth substrate. We found no distinct shrinking artefacts, and the whole cross-section and prominent cellular structures were well preserved. The cell attached tightly to the substrate surface, which is marked by the sputtered layer *(black line)*. Small defects became visible by a partially slight roughness of the sputtered layer and by small gaps in this layer. Higher magnification of a part of this interface showed the differences in the membrane attachment in detail: areas of close adhesion with $d_j \leq 10$ nm and areas with enhanced distance of 100 to 150 nm in this particular case.

We found that contact geometry between attached cell membrane and substrate was dependent on the type of protein coating used. In the presence of polylysine, the average distance of the membrane–substrate interface was in the range of 35 to 40 nm. However, the cell membrane was highly protruded in the presence of other proteins such as fibronectin, laminin, or concanavalin-A.

4 NEURON TRANSISTOR HYBRID SYSTEMS

The electrical activity of neurons, the action potential (AP), can be monitored by a classical micropipette recording unit and shows the well known voltage-time profile given in Fig. 12-7A, resulting from (time-dependent) contributions of Na^+-, K^+-, and Ca^{2+}-currents across the cell membrane. Alternatively, the local extracellular voltage caused by action potentials can be recorded using planar metal electrodes or integrated planar field effect transistors (FETs). The patch pipette recording shows an excellent signal-to-noise ratio; however, the cell survives that treatment for not much longer than a few hours, and multielectrode recording beyond a few electrodes/cells is barely possible. Recordings by FET devices show lower signals and a higher noise level (owing to a weaker coupling to the gate electrode compared to the patch pipette), however, can be monitored for much longer times, and with the current array design on thousands of channels simultaneously!

In some of the recording experiments, FET recordings had signal-to-noise (S/N) levels smaller than one and signals had to be averaged. In some cases the signals are strong enough to overcome the noise level. One example, which was recorded from a locust neuron, is given in Fig. 12-7B, which is the simultaneously measured event shown in Fig. 12-7A. By stimulation with 1 nA in current-clamp mode, the neuron generated APs, which were in parallel recorded by the FET *(left traces)*.

The recorded extracellular signal shapes caused by APs show a biphasic signal with negative transient followed by a weaker positive peak. The time courses of the extracellular signals differed from the patch-clamp recordings. The FET response to an AP is dominated by a negative peak during the rising phase of an AP.

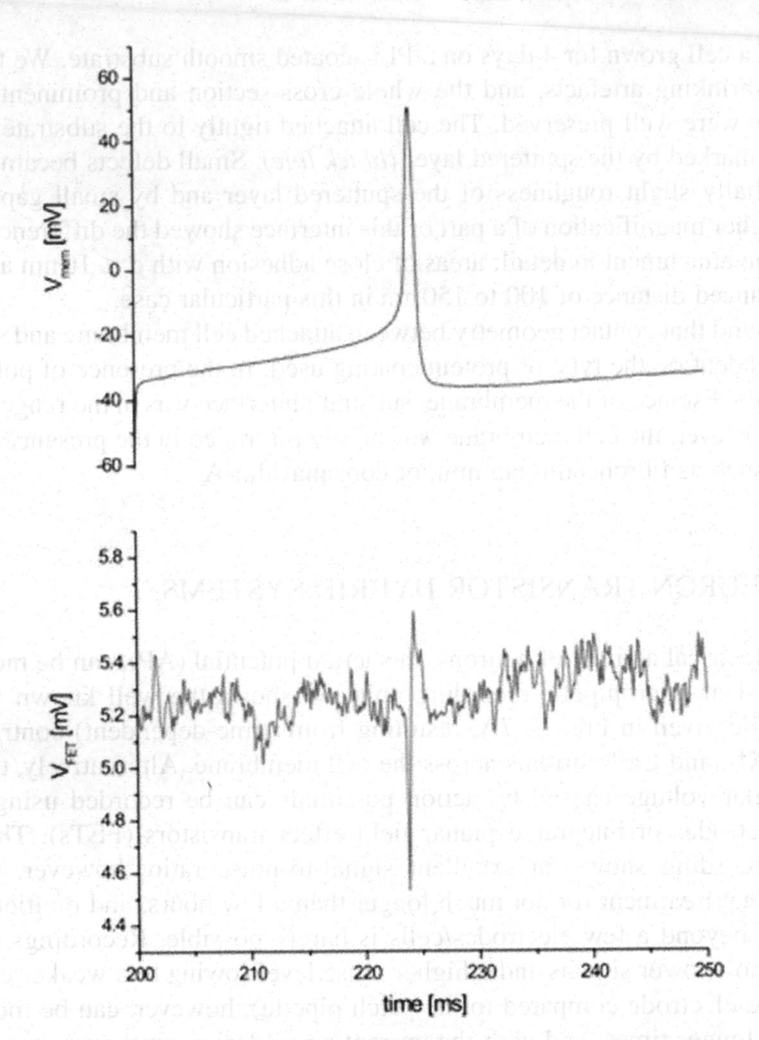

FIG. 12.7. Comparison of recordings of neuronal signal by the FET and patch-clamp method: action potentials in current-clamp mode of a locust neuron (Vrest –58 mV; DIV 3). During the rising phase of an AP (recorded by a patch clamp electrode) *(A)* a negative transient was recorded by the FET *(B)* followed by a positive peak of smaller amplitude during the falling phase of the AP.

In the falling phase of the AP a weaker positive peak was recorded. Comparing the FET signal to the time course of ionic currents in voltage-clamp mode, revealed a similar time shift of the FET signal peaks and the peaks in the patch-clamp signal. The FET registered a fast negative signal during the inward current. The extracellular signal reversed and reached a positive peak parallel to the upstroke of the patch-clamp signal.

The extracellular signal shape recorded from the locust neuron was similar to the C-type response described by Schätzthauer and Fromherz (1998). In a C-type response, the FET signal is dominated by a high Na^+ and K^+ conductance in the junction membrane compared to the capacitive current over the attached membrane.

5 CONCLUSIONS

Among the many possible visions for future directions in research and development on the basis of the presented research directions the integration with elements of microfluidics are very obvious and actually already down the road: Any practical use of a cell-based sensor will need a liquid-handling system that has to implement microfluidic design principles. In as much as we will see "lab-on-a-chip"–based devices there will be concepts developed that also give us the tools and modules to support our cell-based bio-electronic hybrid system by the required buffer supply, temperature control units, etc.

Another direction to go is in the developing of electronic devices that are designed to the specific cytoarchitecture of the brain region in which they will be implanted. This "neuromorphic" approach should enable the implant to blend in with neural tissue in a functionally seamless and biocompatible manner.

Finally, we expect more profits from this biomimetic engineering effort than a cochlea, retina, or a hippocampal prosthesis. The process of studying the input–output relationships of electrical signals interacting with a neuron over space and time will lead to computational devices that incorporate true neural network concepts. One can envisage that biomimetic or neural-silicon hybrid processors will be able to accomplish far more than the first generation of neural network systems that emerged in the last two decades based on von Neumann computer architectures.

ACKNOWLEDGMENTS

It is our pleasure to acknowledge the very competent and dedicated help of our colleagues: S. Britland, M. Denyer, St. Eick, J. Eschermann, M. Höller, M. Krause, L. Lauer, S. Meyburg, M. Prömpers, M. Schindler F. Sommerhage, Ch. Sprössler, R. Stockmann, P. Thiebaud, S. Weigel, C.-K. Yeung, and Y. Zhang.

REFERENCES

Axelrod, D. (1981). Cell–substrate contacts illuminated by total internal-reflection fluorescence. J. Cell Biol. 89, 141–145.

Bausells, J., Carrabina, J., Errachid, A. and Merlos A. (1999). Ion-sensitive field-effect transistors fabricated in a commercial CMOS technology. Sens. Actuators B: Chem. 57, 56–62.

Baharloo, B., Textor, M., & Brunette, D. M. (2005). Substratum roughness alters the growth, area, and focal adhesions of epithelial cells, and their proximity to titanium surfaces. J. Biomed. Mater. Res. A 74A, 12–22.

Bergveld P., Wiersma J., and Meertens H. (1976). Extracellular potential recordings by means of a field-effect transistor without gate metal, called OSFET. IEEE Trans. Biomed. Eng. 23, 136–144.

Bousse L., Scott J., and Meindl J.D. (1988). A process for the combined fabrication of ion sensors and CMOS circuits. IEEE Electron Dev. Letts. 9, 44–46.

Braun, D. & Fromherz, P. (1998). Fluorescence interferometry of neuronal cell adhesion on microstructured silicon. Phys. Rev. Lett. 81, 5241–5244.

Cohen A., Spira M.E., Yitshaik S., Borghs G., Shwartzglass O. and Shappir J. (2004). Depletion type floating gate p-channel MOS transistor for recording action potentials generated by cultured neurons. Biosens. Bioelectr. 19, 1703–1709.

Curtis, A. S. G. (1964). Mechanism of adhesion of cells to glass—study by interference reflection microscopy. J. Cell Biol. 20, 199–215.

den Braber, E. T., Jansen, H. V., de Boer, M. J., Croes, H. J. E., Elwenspoek, M., Ginsel, L. A. & Jansen, J. A. (1998) Scanning electron microscopic, transmission electron microscopic, and confocal laser scanning microscopic observation of fibroblasts cultured on microgrooved surfaces of bulk titanium substrata. J. Biomed. Mater. Res. 40, 425–433.

Fromherz P., Offenhäusser A., Vetter T. and Weis J. (1991). A neuron-silicon junction: a Retzius cell of the leech on an insulated-gate field-effect transistor. Science 252, 1290–1293.

Gingell, D., Todd, I., & Bailey, J. (1985). Topography of cell glass apposition revealed by total internal-reflection fluorescence of volume markers. J. Cell Biol. 100, 1334–1338.

Gross G.W., Wen W.Y. and Lin J.W. (1985). Transparent indium tin oxide electrode patterns for extracellular, multisite recording in neuronal cultures. J. Neurosci. Meth. 15, 243–252.

Heer F., Hafizovic S., Franks W., Blau A., Ziegler C. and Hierlemann A. (2006). CMOS microelectrode array for bidirectional interaction with neuronal networks. IEEE J. Solid-State Circuits 41, 1620–1629.

Hodgkin A.L. and Huxley A.F. (1952). A quantitative description of membrane current and its application to conduction and excitation in nerve. J. Physiol. 117, 500–544.

Ingebrandt S., Yeung C.K., Krause M. and Offenhausser A. (2005). Neuron-transistor coupling: interpretation of individual extracellular recorded signals. Eur Biophys J. 34, 144–154.

Izzard, C. S. & Lochner, L. R. (1976). Cell-to-substrate contacts in living fibroblasts—interference reflection study with an evaluation of technique. J. Cell Sci. 21, 129–159.

Jobling, D. T., Smith, J. G., & Wheal, H. V. (1981). Active microelectrode array to record form the mammalian central nervous system in vitro. Med. Biol. Eng. & Comput. 19, 553–560.

Katsube, T., Araki, T., Hara, M., Yaji, T., Kobayashi, S., Suzuki, K. (1986). A multi-species biosensor with extended-gate field effect transistors. In: Proceedings of the Sixth Sensor Symposium, pp. 211–214.

Kind T., Issing M., Arnold R., and Müller B. (2002). Electrical coupling of, single cardiac rat myocytes to field-effect and bipolar transistors. IEEE Trans Biomed Eng. 49, 1600–1609.

Lambacher A. and Fromherz P. (1996). Fluorescence interference-contrast microscopy on oxidized silicon using a monomolecular dye layer. Appl. Phys. A 63, 207–216.

Khakbaznejad, A., Chehroudi, B. & Brunette, D. M. (2004). Effects of titanium-coated micromachined grooved substrata on orienting layers of osteoblast-like cells and collagen fibers in culture. J. Biomed. Mater. Res. A 70A, 206–218.

Krause G., Lehmann S., Lehmann M., Freund I., Schreiber E. and Baumann W. (2006). Measurements of electrical activity of long-term mammalian neuronal networks on semiconductor neurosensor chips and comparison with conventional microelectrode arrays. Biosens. Bioelectron. 21, 1272–1282.

Matsuo, T., Esashi, M. (1981). Methods of ISFET fabrication. Sens. Actuators 1, 77–96

Meyle, J., Gultig, K., Wolburg, H., & Von Recum, A. F. (1993). Fibroblast anchorage to microtextured surfaces. J. Biomed. Mater. Res. 27, 1553–1557.

Meyburg S., Goryll M., Moers J., Ingebrandt S., Böcker-Meffert S., Lüth H., and Offenhäusser A. (2006). N-Channel field-effect transistons with floating gates for extracellular recordings Biosens. Bioelectron. 21, 1037–1044.

Offenhäusser, A., Rühe, J., Knoll, W. (1995). Neuronal Cells Cultured on Modified Microelectronic Device Surfaces. Journal of Vacuum Science & Technology A-Vacuum Surfaces and Films 13(5), 2606–2612.

Pabst M., Wrobel G., Ingebrandt S., Sommerhage F., and Offenhäusser A. (2007). Solution of the Poisson-Nernst-Planck equations in the cell-substrate interface. Eur. Phys. J. E 24, 1–8.

Parce, J.W., Owicki, J.C., Kercso, K.M., Sigal, G.B., Wada, H.G., Muir, V.C., Bousse, L.J., Ross, K.L., Sikic, B.I., & McConnell, H.M. (1989). Detection of Cell-Affecting Agents with a Silicon Biosensor. Science 246, 243–247.

Pfeiffer, F., Herzog, B., Kern, D., Scheideler, L., Geis-Gerstorfer, J., & Wolburg, H. (2003). Cell reactions to microstructured implant surfaces. Microelectron. Eng. 67-8, 913–922.

Pine J. (1980). Recording action-potentials from cultured neurons with extracellular micro-circuit electrodes. J. Neurosci. Meth. 2, 19–31.

Regehr W.G., Pine J., Cohan C.S., Mischke M.D., and Tank D.W. (1989). Sealing cultured invertebrate neurons to embedded dish electrodes facilitates long-term stimulation and recording, J. Neurosci. Meth. 30, 91–106.

Smith, R., Huber, R.J., Janata, J. (1984). Electrostatically protected ion sensitive field effect transistors. Sens. Actuators 5, 127–136.

Schätzthauer R. and Fromherz P. (1998). Neuron-silicon junction with voltage-gated ionic currents, Eur. J. Neurosci. 10, 1956–1962.

Singhvi, R., Stephanopoulos, G. & Wang, D. I. C. 1994 Effects of substratum morphology on cell physiology—review. Biotechnol. Bioeng. 43, 764–771.

Sprössler C., Denyer M., Britland S., Curtis A., Knoll W., and Offenhäusser A. (1999). Electrical recordings from rat cardiac muscle cells using field-effect transistors. Phys. Rev. E. 60, 2171–2176.

Van der spiegel, J., Lauks, I., Chan, P., Babic, D. (1983). The extended gate chemically sensitive field effect transistor as multi-species microprobe. Sens. Actuators 4, 291–298.

Walboomers, X. F., Croes, H. J. E., Ginsel, L. A. & Jansen, J. A. (1998). Growth behavior of fibroblasts on microgrooved polystyrene. Biomaterials 19, 1861–1868.

Weis R. and Fromherz P. (1997). Frequency dependent signal-transfer in neuron-transistors. Phys. Rev. E. 55, 877–889.

Wrobel G., Seifert R., Ingebrandt S., Enderlein J., Ecken H., Baumann A., Kaupp U.B. and Offenhäusser A. (2005). Cell-transistor coupling: Investigation of potassium currents recorded with p- and n-channel FETs. Biophys. J. 89, 3628–3638.

Wrobel G., Höller M., Ingebrandt S., Dieluweit S., Sommerhage F., Bochem H.P. and Offenhäusser A. (2007). Cell-transistor coupling: transmission electron microscopy study of the cell-sensor interface. J. R. Soc. Interface 5(19), 213–222.

Electronics for Cellomics

13
Hybrid Nanoparticles for Cellular Applications

Franco Calabi

1 INTRODUCTION

Nanoparticles can be defined as supramolecular entities that combine linear dimensions below 100 nm (at which quantum mechanical effects become significant) with self-assembly from smaller building blocks (atoms or molecules), i.e., bottom-up synthesis rather than top-down fabrication. Nanoparticles show remarkably enhanced optical, magnetic, and electronic properties compared with bulk materials, and these properties can be finely tuned by controlling their size, shape, and composition.

Although humans' ability to manipulate and characterize matter at the nanoscale is very recent, nanoparticles are of widespread occurrence in nature and in particular in living systems, where they have been diversified and refined by evolution. Being comparable in size to pieces of the cell machinery, artificial nanoparticles are attracting an increasing interest as cell interaction tools, leading to the development of new approaches to the detection and control of biological processes.

This chapter focuses on hybrid inorganic/organic nanoparticles that are primarily designed to exploit the electronic properties of their constituents. Quite apart from the latter, however, hybrid nanoparticles may be used as scaffolds for the

A. Offenhäusser and R. Rinaldi (eds.), *Nanobioelectronics - for Electronics, Biology, and Medicine*,
DOI: 10.1007/978-0-387-09459-5_14,

assembly of multifunctional complexes in defined geometric arrangements. By assembling multivalent ligands, or combinations of ligands, on the surface of a single nanoparticle, binding affinity and specificity can be dramatically increased (Mammen, Choi, and Whitesides 1998).

Based on the nature of the inorganic component, hybrid nanoparticles may be grouped into three main classes: (a) semiconductor colloidal nanocrystals (quantum dots); (b) metal; and (c) superparamagnetic (iron oxide) nanoparticles. The following briefly summarizes some essential properties of hybrid nanoparticles, discusses strategies by which nanoparticles are brought to interact with cells, both in vitro and in vivo, and finally reviews some of the most significant results up to date. Only applications to live cells are considered, with emphasis on the most recent results. The reader is also referred to an excellent recent review (Alivisatos, Gu, and Larabell 2005).

2 PROPERTIES OF HYBRID NANOPARTICLES FOR CELLULAR APPLICATIONS

The preparation of biocompatible hybrid nanoparticles and their characteristics are discussed in detail in several recent reviews (Daniel and Astruc 2004; Medintz, Uyeda, Goldman, and Mattoussi 2005; Michalet, Pinaud, Bentolila, Tsay, Doose, Li, Sundaresan, Wu, Gambhir, and Weiss 2005; Morawski, Lanza, and Wickline 2005; Neuberger, Schopf, Hofmann, Hofmann, and von Rechenberg 2005; Gupta and Gupta 2005b; Hirsch, Gobin, Lowery, Tam, Drezek, Halas, and West 2006) and in other chapters of this book. Only the features most relevant to cellular applications are briefly recapitulated.

2.1 SEMICONDUCTOR COLLOIDAL NANOCRYSTALS (QUANTUM DOTS)

Up to now, QD have been used essentially as fluorochromes although applications have been described based on electron transfer with organic molecules (Clarke, Hollmann, Zhang, Suffern, Bradforth, Dimitrijevic, Minarik and Nadeau 2006). The essential optical features of fluorescent QD are:

- Bright luminescence (mainly resulting from high one- and two-photon capture cross sections), enabling single nanoparticle detection
- High photostability, with low decrease in luminescence upon prolonged excitation
- Narrow, symmetric emission spectra, with peak wavelength tunable based on composition and size, enabling high-level multiplexing

- Broad absorption spectra, which allows excitation of different emitters with a single source
- Comparatively slow radiative decay (>10 ns), enabling time-gated detection to improve the signal to background ratio
- Multiple imaging modalities (e.g., combined optical and electron microscopy)

Most used quantum dots are made from II – VI elements (e.g., CdSe) and emit in the visible range. The synthesis has recently been reported of novel QD with emission in the NIR (Kim, Fisher, Eisler, and Bawendi 2003; Gao, Cui, Levenson, Chung, and Nie 2004). These are attractive for in vivo imaging, since the range between 700 and 900 nm is one of two "clear" spectral windows (the second being between 1,200 and 1,600 nm) that are well separated from major absorption peaks of blood and water, thus enabling deep tissue imaging (millimeters to centimeters) (Lim, Kim, Nakayama, Stott, Bawendi, and Frangioni 2003).

The risk of cytotoxicity from II-VI nanocrystals has raised concerns, because of the highly toxic nature of the constituents. Toxicity of CdSe-core QD on primary rat hepatocytes was found to correlate with surface oxidation and a decrease of QD size with release of free cadmium ions in solution. Surface coatings with ZnS and BSA significantly reduced, but did not eliminate toxicity. However, red-emitting CdSe QD capped with ZnS, coated with PEG to prevent nonspecific binding and conjugated to EGF (a hepatocyte ligand), could be used for long-term (7 days) in vitro cell labeling without deleterious effects on viability, migration, or differentiated functions (Derfus, Chan, and Bhatia 2004).

Mercaptopropionic acid–coated cadmium telluride (CdTe) QD were shown within 24 hours to trigger plasma membrane, mitochondrial, and nuclear damage, and a decrease in metabolic activity followed by a reduction in cell number of MCF-7 breast cancer cells. Cell death was by nonclassical (caspase-independent) apoptosis and correlated with QD-induced ROS generation and cytoplasmic release of mitochondrial cytochrome c. N-Acetylcysteine reduced QD-induced cytotoxicity by ROS scavenging (Lovric, Cho, Winnik, and Maysinger 2005b). Similar results were reported with cysteamine-coated cationic CdTe QD in PC12 and N9 cells. Moreover, higher toxicity correlating with a unique nuclear localization was found for green- than for red-emitting QD (Lovric, Bazzi, Cuie, Fortin, Winnik, and Maysinger 2005).

Surface functionalization plays a key role in nanoparticle toxicity. The cytotoxicity of CdSe/ZnS nanoparticles with different coatings was tested by examining their effects on the adhesion of NRK fibroblasts, known to be highly sensitive to toxic metal ions (Kirchner, Liedl, Kudera, Pellegrino, Javier, Gaub, Stolzle, Fertig, and Parak 2005). The presence of a ZnS shell around CdSe particles was reported to increase the maximum nontoxic concentration nearly by a factor of 10. Cross-linked silica was found to reduce dramatically the release of Cd^{2+} from the particle surface, and addition of PEG groups to the silica shell further decreased nanoparticle uptake and cytotoxicity. Apart from the release of Cd^{2+} atoms, some coats (e.g., an amphiphilic polymer) were found to induce cell poisoning by causing nanoparticle precipitation on cells.

Recently, a detailed analysis has been carried out at the cellular and molecular level of the impact of short-term (48-h) treatment with two doses of PEG silanized QD on in vitro cultured human lung and skin epithelial cells (Zhang, Stilwell, Gerion, Ding, Elboudwarej, Cooke, Gray, Alivisatos, and Chen 2006). These QD did not cause any significant change in the total cell number, apoptosis/necrosis or cell cycle profile, but for a possible G2/M block in skin fibroblasts. Transcriptional profiling with the Affymetrix HG-U133A Gene Chip showed significant changes (>twofold) in only 46 out of ~22,0000 genes (~0.2%). Most changes consisted in downregulation, and involved genes controlling M-phase progression, spindle formation, and cytokinesis, as well as FOXM1 and BHLHB2/Dec1 transcription factors. Upregulated genes were related to carbohydrate binding, intracellular vesicle localization, and cell membrane–associated and intracellular vesicular proteins involved in the cellular response to stress. Genes associated with heavy metal exposure were not induced by treatment. PEG-silica-QD were concluded to have negligible short-term toxicity at least in epithelial cells.

2.2 GOLD NANOPARTICLES

Although gold nanoparticles have been extensively used as label in electron microscopy, noble metal nanoparticles show strong interactions with radiations at optical frequencies, from the UV through to the infrared. This is due to collective resonance of the conduction electrons with the incident electromagnetic field, causing highly efficient absorption and/or scatter of radiation. The magnitude, peak wavelength, and spectral bandwidth of the plasmon resonance associated with a nanoparticle are dependent on the particle size,, shape, and material composition, as well as the local environment. Hence gold nanoparticles can be used both for optical imaging (by scattering as in dark field microscopy, confocal reflectance microscopy, and optical coherence tomography) and for cell ablation (by heating resulting from conversion of absorbed electromagnetic energy). The main advantages of gold nanoparticles are:

- Large optical cross-sections that exceed conventional dyes by many orders of magnitude and are readily tuned from the visible to approximately 3 μm in the infrared
- High photostability resulting from a rigid structure
- High threshold for optical saturation
- Easy coupling to peptides or other biomolecules (e.g., via sulfhydryl chemistry)
- Biocompatibility

Apart from gold nanospheres, nanorods and nanoshells have been most studied, the latter consisting of a dielectric silica core covered by a thin metal (typically gold) shell (Loo, Lowery, Halas, West, and Drezek 2005). The calculated optical properties (resonance wavelength, extinction cross-section, and the ratio of scattering to absorption) of these three types of gold nanoparticles have recently been compared (Jain, Lee, El-Sayed, and El-Sayed 2006). With gold nanospheres, the magnitude of extinction as well as the relative contribution of scattering to the extinction rapidly increases with size. The absorption cross-section of 40 nm gold nanospheres was estimated at 5 orders of magnitude higher than that of conventional dyes, and light scattering by 80-nm gold nanospheres as 5 orders of magnitude stronger than the emission from the best fluorescing organic dyes. The variation in peak plasmon wavelength with the size of nanospheres is, however, restricted to 520 to 550 nm.

The optical cross-section of gold nanoshells is comparable to or higher than that of nanospheres, with total extinction linearly dependent on overall size, whilst independent of the core/shell radius ratio, and optical resonance in the NIR. The relative scattering contribution to the extinction increases rapidly with the nanoshell size or a decrease in the ratio of core/shell radius. The resonance wavelength can be increased by either increasing the total nanoshell size or increasing the ratio of the core/shell radius.

Gold nanorods are estimated to have optical cross-sections comparable to nanospheres and nanoshells at a much smaller effective size, or an order of magnitude higher at a similar size. Optical resonances can be linearly tuned across the NIR by changing either the effective size or the aspect ratio. The total extinction as well as the relative scattering contribution increase rapidly with the effective size, independently of the aspect ratio. Nanorods with a high aspect ratio and a small effective radius were predicted to be the strongest photoabsorbing nanoparticles, and nanorods with a high aspect ratio and large effective radius to give the highest scattering contrast for imaging applications.

2.3 SUPERPARAMAGNETIC NANOPARTICLES

Superparamagnetic iron oxide nanoparticles (SPION) consist of a <20-nm core of magnetite (Fe_3O_4) or maghemite (γ-Fe_2O_3). Although SPION have a much larger magnetic moment than individual paramagnetic atoms (e.g., Gd), they do not show magnetization outside a magnetic field at ambient temperature since their magnetic moment fluctuates randomly, the energy required to change its direction being comparable to thermal energy due to their size.

SPION are used primarily as contrast agents for MRI. However, they have also been exploited for cell ablation or drug delivery (by heating resulting from conversion of alternating magnetic fields, or by physical targeting with magnetic fields). Although the spatial resolution of magnetic resonance imaging (MRI) has substantially increased over the past few years, achieving voxel resolutions of ~10 µm³ in vitro and about 50 µm³ in vivo, its sensitivity remains well lower than optical microscopy. The MRI signal is mostly generated by water and lipid protons, and depends both on proton density and the magnetic relaxivity of the surrounding medium. Contrast agents modify this relaxivity, resulting in changes of the longitudinal (R1), or transverse (R2) relaxation rates that can be visualized as positive or negative enhancement.

Although SPION are generally regarded as biocompatible and noncytotoxic, exposure of human fibroblasts for 24 hours to ~40- to 45-nm nanoparticles with a magnetite inner core was found to result in a dose-dependent reduction in cell viability and cell adhesion, and in a disruption of actin filament and microtubule networks. These effects were prevented by coating the nanoparticles with pullalan, a nonionic, nonantigenic, highly hydrophilic polysaccharide (Gupta and Gupta 2005a).

3 NANOPARTICLE–CELL INTERACTIONS

3.1 CELL LABELING IN VITRO

Several different methods have been used to label cells in vitro with nanoparticles. Direct delivery to the intracellular compartment of semiconductor nanoparticles embedded in micelles formed from PEG-conjugated lipids has been achieved by microinjection in *X. laevis* embryos (Dubertret, Skourides, Norris, Noireaux, Brivanlou, and Libchaber 2002). Nanoparticles (up to 10^9 per cell) did not show aggregation up to 4 days after injection and were restricted to the progeny of the injected cells.

Particles up to 100 nm in size can be non-specifically ingested by most eukaryotic cells by fluid phase endocytosis, an active, temperature-sensitive process, and delivered to endosomal compartments. Thus, upon incubation with HeLa cells at 37°C, dihydroxylipoic acid-capped quantum dots were shown to be internalized and to localize to a large number of juxtanuclear vesicles, suggestive of late endosomes. The nature of these vesicles was confirmed by colocalization with an endosomal marker (Jaiswal, Mattoussi, Mauro, and Simon 2003). In the case of gold nanoshells, high-density binding to cells occurs via the protein-adsorbent gold surface (Hirsch, Stafford, Bankson, Sershen, Rivera, Price, Hazle, Halas, and West 2003). It has been suggested that endocytosis efficiency is optimal for ~50-nm particles, as shown with QD-conjugated sugar balls (Osaki, Kanamori, Sando, Sera, and Aoyama 2004) or colloidal gold nanoparticles (Chithrani, Ghazani, and Chan 2006). However, the kinetics of intracellular uptake is inversely related to size (Chithrani, Ghazani, and Chan 2006). Nonspecific QD or SPION uptake can be massive and is particularly efficient when the nanoparticles are positively charged at physiological pH, allowing them to bind to negatively charged cell-surface glycoproteins and glycolipids (Lovric, Bazzi, Cuie, Fortin, Winnik, and Maysinger 2005a; Parak, Pellegrino, and Plank 2005; Song, Choi, Huh, Kim, Jun, Suh, and Cheon 2005).

Transduction efficiency of negatively charged nanoparticles can be increased by the addition of cationic lipids, which are known to promote DNA translocation into cells. Labeling of up to 85–95% has been reported with Lipofectamine 2000 for B16F10 melanoma cells (Voura, Jaiswal, Mattoussi, and Simon 2004). Protein transduction domains (PTD) are short arginine-rich peptide sequences found in natural proteins (e.g., the HIV-1 transcriptional activator Tat protein, the HSV VP22 structural protein and the *Drosophila* homeotic factor Antennapoedia), or their synthetic analogues, that can mediate intracellular delivery of a variety of macromolecules. Monocrystalline SPION coated with cross-linked aminated dextran (~45 nm) and derivatized with short HIV-Tat peptides are efficiently internalized by cell lines (such as HeLa; Josephson, Tung, Moore, and Weissleder 1999) or primary cells (such as lymphocytes and CD34+ hematopoietic or neural progenitor cells; Josephson, Tung, Moore, and Weissleder 1999; Lewin, Carlesso, Tung, Tang, Cory, Scadden, and Weissleder 2000; Lewin, Carlesso, Tung, Tang, Cory, Scadden, and Weissleder 2000) in quantities up to 10 to 30 pg of superparamagnetic iron per cell, localizing to the cytoplasm and nuclear compartment, with no deleterious effect on cell viability, differentiation, or proliferation. Similar results were reported with QD conjugated to Tat or a poly-arginine peptide in a human prostate cancer cell line, C4-2 (Gao, Cui, Levenson, Chung, and Nie 2004; Stroh, Zimmer, Duda, Levchenko, Cohen, Brown, Scadden, Torchilin, Bawendi, Fukumura, and Jain 2005) (Fig. 13.1) and in a variety of mammalian cell types (Lagerholm, Wang, Ernst, Ly, Liu, Bruchez, and Waggoner 2004), and with Tat-gold nanoparticle conjugates in HeLa (human cervical carcinoma), NIH-3T3 (murine fibroblastoma), and HepG2 (human hepatocarcinoma) cells (Tkachenko, Xie, Liu, Coleman, Ryan, Glomm, Shipton, Franzen, and Feldheim 2004). Once again, cellular uptake was temperature dependent and the nanoparticles were concentrated in intracellular vesicles characteristic of endosomes and lysosomes. Targeting has also been achieved by using gold nanoparticles coupled to an adenovirus vector. Viral infectivity was preserved up to a ratio of 100 nanoparticles per virus. Specific in vitro targeting to a tumour cell line expressing high levels of CEA was shown by using an adapter protein, sCAR-MFE, binding both to the adenoviral capsid and to CEA (Everts, Saini, Leddon, Kok, Stoff-Khalili, Preuss, Millican, Perkins, Brown, Bagaria, Nikles, Johnson, Zharov, and Curiel 2006).

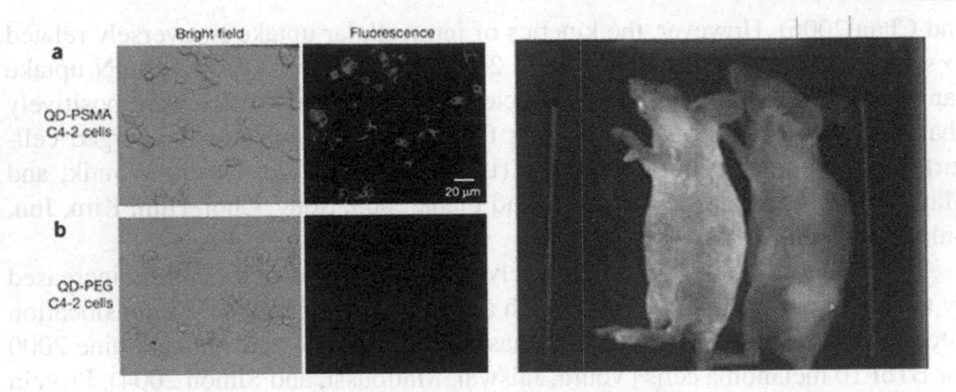

FIG. 13.1. **Left.** Binding of QD-antiPSMA to cultured C4-2 prostate cancer cells. **Right.** Imaging of C4-2 tumor xenografts targeted by QD-antiPSMA conjugates in a live animal *(right mouse)*. A control animal *(left mouse)* injected with the same amount of conjugate showed no localized fluorescence signal. Images were obtained with a spectral unmixing algorithm to remove background autofluorescence (Reprinted by permission from Macmillan Publishers Ltd: Gao, X. H., Cui, Y. Y., Levenson, R. M., Chung, L. W. K., and Nie, S. M. (2004). In vivo cancer targeting and imaging with semiconductor quantum dots, Nat. Biotechnol. 22, 969–976, copyright 2004.)

Nuclear targeting of hybrid nanoparticles in living cells has required further modifications, because of the presence of the double-membrane nuclear envelope with pore complexes ranging in size between 20 and 50 nm and impermeable to several classes of molecules. These problems have been partly circumvented by using compact QD complexes bound to a peptide spanning the NLS from the SV40 virus and introduced in HeLa cells by electroporation (Chen and Gerion 2004). However, only 15% of the cells showed nuclear nanoparticle localization, with the remaining 85% preferentially perinuclear. This was explained by invoking cell-cycle–dependent variations in nuclear pore permeability. NLS-mediated nuclear trafficking could be directly visualized. Labeled nuclei were still visible after a week, at which time cell viability was comparable to unlabeled controls, as shown by colony-forming capability. In a similar approach, a synthetic 21-residue amphipathic peptide (Pep-1) spanning a hydrophilic lysine-rich sequence from the SV40 large T antigen NLS fused via a spacer to a hydrophobic tryptophan-rich sequence (for QD binding) was reported to mediate efficient intracellular QD delivery (Mattheakis, Dias, Choi, Gong, Bruchez, Liu, and Wang 2004). Nuclear targeting in HepG2 cells with peptide-derivatized gold nanoparticles was explored by video-enhanced color differential interference contrast microscopy. It was found that combined derivatization with both a NLS and a peptide inducing receptor-mediated endocytosis was required for nuclear targeting, and that this was more efficient when the two signals were attached to nanoparticles as separate peptides rather than a single, fusion peptide (Tkachenko, Xie, Coleman, Glomm, Ryan, Anderson, Franzen, and Feldheim 2003).

Generalized QD labeling of the cell surface in living cells has been achieved with avidin-conjugated nanoparticles following non-specific chemical biotinylation of membrane proteins (Jaiswal, Mattoussi, Mauro, and Simon 2003). More recently, specific QD labeling has been demonstrated of proteins engineered to span a 15-amino acid peptide that can be biotinylated by biotin ligase, the product of the *E. coli BirA* gene (Howarth, Takao, Hayashi, and Ting 2005). Advantages of cell labeling with avidin-conjugated QD over methods relying on primary plus secondary antibodies are the reduced size of the labeling complex and the high stability of the biotin–streptavidin interaction.

Highly specific and efficient labeling of cell surface components has been achieved with hybrid nanoparticles conjugated to ligands for membrane receptors. Notable among the latter have been growth factors: EGF (Lidke, Nagy, Heintzmann, Arndt-Jovin, Post, Grecco, Jares-Erijman, and Jovin 2004; Lidke, Lidke, Rieger, Jovin, and Arndt-Jovin 2005) (Figs. 13.1 and 13.2), folate (Bharali, Lucey, Jayakumar, Pudavar, and Prasad 2005; Sonvico, Mornet, Vasseur, Dubernet, Jaillard, Degrouard, Hoebeke, Duguet, Colombo, and Couvreur 2005; Dixit, Van den Bossche, Sherman, Thompson, and Andres 2006), serum proteins (Chan and Nie 1998; Gupta and Curtis 2004), neurotransmitters (Rosenthal, Tomlinson, Adkins, Schroeter, Adams, Swafford, McBride, Wang, DeFelice, and Blakely 2002; Dahan, Levi, Luccardini, Rostaing, Riveau, and Triller 2003; Groc, Heine,

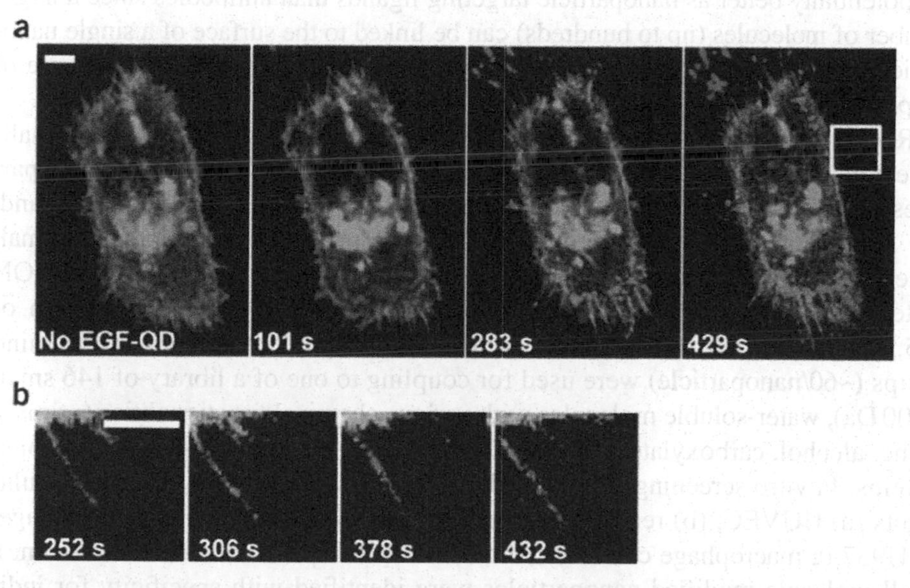

FIG. 13.2. Filopodial binding and retrograde migration of single EGF-QD nanoparticles *(orange)*. The green color corresponds to erbB3/mCitrine. Images in (**b**) correspond to magnifications of the boxed area in (**a**). Scale bars: 5 μm (Reprinted by permission from Macmillan Publishers Ltd: Lidke, D. S., Nagy, P., Heintzmann, R., Arndt-Jovin, D. J., Post, J. N., Grecco, H. E., Jares-Erijman, E. A., and Jovin, T. M. (2004). Quantum dot ligands provide new insights into erbB/HER receptor-mediated signal transduction, Nat. Biotechnol. 22, 198–203, copyright 2004.)

Cognet, Brickley, Stephenson, Lounis, and Choquet 2004; Clarke, Hollmann, Zhang, Suffern, Bradforth, Dimitrijevic, Minarik, and Nadeau 2006), ion-channel blockers (Beaurepaire, Buissette, Sauviat, Giaume, Lahlil, Mercuri, Casanova, Huignard, Martin, Gacoin, Boilot, and Alexandrou 2004), and TNFα (Paciotti, Kingston, and Tamarkin 2006). Conjugation to folate (essential in mammalian cells for the synthesis of nucleic acids and some amino acids) has become a widely used strategy for targeting nanoparticles to tumor cells, since its receptor has high affinity for folic acid ($K_d \sim 0.1$ nM), is upregulated in many epithelial cancers, and becomes expressed on cell surfaces accessible from the circulation as a consequence of the loss of cell polarity upon transformation. Alternatively, cell surface components such as differentiation markers (CD antigens) or growth factor receptors have been targeted by antibodies, such as against leukemia-specific CD90 (Bakalova, Ohba, Zhelev, Nagase, Jose, Ishikawa, and Baba 2004), human endothelial specific E-selectin/CD62E (Kang, Josephson, Petrovsky, Weissleder, and Bogdanov 2002), EGFR homologue/protooncogene product Her2/neu (Wu, Liu, Liu, Haley, Treadway, Larson, Ge, Peale, and Bruchez 2003; Huh, Jun, Song, Kim, Choi, Lee, Yoon, Kim, Shin, Suh, and Cheon 2005). At 37°C, binding of nanoparticles to cell surface receptors is followed by receptor-mediated endocytosis and hence ultimately results in intracellular labeling. It has been claimed that peptides or peptidomimetics are potentially better as nanoparticle targeting ligands than antibodies since a larger number of molecules (up to hundreds) can be linked to the surface of a single nanoparticle, yielding stronger binding affinity and better targeting efficacy because of the polyvalency effect (Mammen, Choi, and Whitesides 1998).

Recently, a novel approach relying on multivalent binding mediated by small-molecules has been proposed for the generation of cell surface–specific nanoparticles irrespective of known molecular targets (Weissleder, Kelly, Sun, Shtatland, and Josephson 2005). Proof of principle was provided by creating a small molecule/nanoparticle library starting from aminated monocrystalline SPION coated with cross-linked dextran and fluorescently labeled with fluorescein or Cy5.5 in a molar ratio of 2 to 3 fluorochromes/nanoparticle. Residual free amino groups (~60/nanoparticle) were used for coupling to one of a library of 146 small (<500 Da), water-soluble molecules with various chemical functionalities (primary amine, alcohol, carboxylate, anhydride, sulfhydryl), not previously known to bind proteins. In vitro screening was carried out for differential uptake by five cellular targets (a) HUVEC; (b) resting macrophage; (c) GM-CSF activated macrophage; (d) U937 (a macrophage cell line); and (e) CaPa2 (a pancreatic cancer cell line). Small molecule-modified nanoparticles were identified with specificity for individual cell types, and even for closely related functional states of the same cell type (resting macrophage versus macrophage activated by ox-LDL, LPS, or GM-CSF). The potential of the strategy has been shown not only for targeting novel cell types (e.g., pancreatic carcinoma), but also for improving the target-to-macrophage ratio conferred by other ligands (VCAM-1) by concurrent coupling to a small molecule (iodoacetate) giving low macrophage uptake.

3.2 IN VIVO TARGETING

The use of hybrid nanoparticles for in vivo imaging, or payload delivery, faces a number of problems additional to those encountered in vitro (Jain 1999; Moghimi, Hunter, and Murray 2001). In order to reach an extravascular cell target, a nanoparticle systemically injected in an organism must generally undergo long-range transport via hematic and/or lymphatic networks, cross the microvascular wall, and navigate through the interstitial compartment via a combination of diffusion and convection. Blood clearance and metabolism occur in competition with these processes. Although small molecules injected in the bloodstream undergo rapid extravasation (within <1 min), nanoparticles can have much longer circulation or retention times, and clearance is generally by the RES (midzonal and periportal liver Kupffer cells, spleen marginal zone, and red pulp macrophage). Thus, macrophage targeting (at least in some anatomical/functional compartments) occurs naturally and is one of the mechanisms of passive targeting. The process is influenced by both nanoparticle size and surface chemistry, either directly or through deposition of blood factors (fibronectin, immunoglobulins, and complement proteins) that mediate interaction with specific macrophage receptors. A second mechanism of passive targeting relies on regional differences in vascular permeability. Endothelial fenestrations can reach up to 150 nm in the liver and 250 nm in the spleen. In contrast, exchanges between the blood and the brain are severely restricted by the existence of a specialized barrier (blood–brain barrier, BBB). A substantial increase in endothelial permeability (with openings >300 nm) occurs at foci of infection or inflammation and in many tumors, where it reflects dysregulated angiogenesis from released vascular endothelial growth factors and is often coupled to reduced lymphatic drainage of interstitial fluid (Matsumura and Maeda 1986), a phenomenon termed enhanced permeation and retention (EPR) (Fig. 13.3). However, nanoparticle extravasation may be hindered by a high interstitial pressure, which may also occur in tumors.

Active targeting refers to the use of ligands or homing/recognition sequences that specifically bind with high affinity to cell surface epitopes or receptors at the target sites, mediating retention and/or rapid and efficient uptake (see Fig. 13.3).

Thus, specific in vivo targeting to either lung or tumour vasculature was reported by using ZnS-capped CdSe quantum dots conjugated to synthetic peptides isolated through phage display technology and specific respectively for a membrane dipeptidase on endothelial cells in lung blood vessels, for tumor cells and tumor blood vessels, or for tumor cells and tumor lymphatic vessels. Accumulation in the RES was prevented by QD coating with PEG (see the

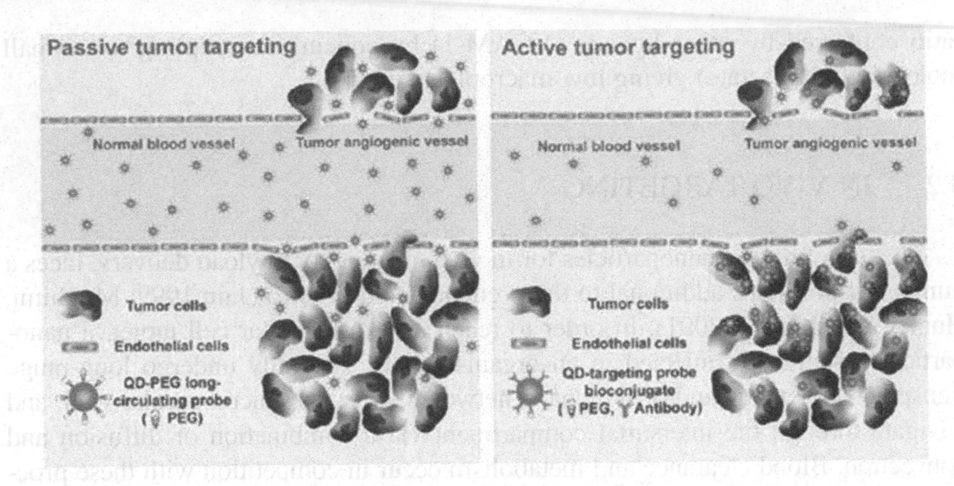

FIG. 13.3. Mechanisms of in vivo targeting (Used with permission from Gao, Cui, Levenson, Chung, and Nie 2004.) (Reprinted by permission from Macmillan Publishers Ltd: Gao, X. H., Cui, Y. Y., Levenson, R. M., Chung, L. W. K., and Nie, S. M. (2004). In vivo cancer targeting and imaging with semiconductor quantum dots, Nat. Biotechnol. 22, 969-976, copyright 2004.)

following). Based on the subcellular pattern of luminescence, it was concluded that peptide-coated nanoparticles are internalized after binding to the cell surface (Akerman, Chan, Laakkonen, Bhatia, and Ruoslahti 2002).

Active targeting competes with passive. Moreover, the active moiety may be recognized as foreign and cause activation of a host immune response. For this reasons, active targeting is generally used in conjunction with "stealthing" technologies, aimed at shielding the nanoparticle from both the RES and the immune system, whilst preserving the targeting interaction. PEG is the most widely used polymer because of its low toxicity and cost, and the many molecular weight variants that are available commercially. PEGylation has been shown to prolong plasma half-lives of nanoparticles by preventing rapid renal clearance and uptake by the RES, to reduce immunogenicity, and to increase solubility and stability (Moghimi, Hunter, and Murray 2001; Orive, Gascon, Hernandez, Dominguez-Gil, and Pedraz 2004).

The effect of different PEG coatings on QD circulation half-lives was compared in mice by both venipuncture and in vivo imaging (Ballou, Lagerholm, Ernst, Bruchez, and Waggoner 2004). QD coated by amphiphilic coats, short-chain (750 Da) methoxy-PEG or 3400 Da carboxy-PEG were cleared from the circulation within 1 hour from injection, whereas long-chain (5000 Da) methoxy-PEG QD persisted over 3 hours. This correlated with significant differences in uptake in lymph nodes and liver. Albeit at a reduced level, fluorescence due to QD was still macroscopically visible in lymph nodes and bone marrow >4 months after injection. It has been suggested that, although modification with long PEG may prolong the circulation time and reduce the washout rate, it may also reduce the nanoparticle probability to extravasate and target tissues because of the substantially increased particle size. Moreover, a longer

PEG linker may also hamper the binding of ligands to cell surface receptors because of steric hindrance (Cai, Shin, Chen, Gheysens, Cao, Wang, Gambhir, and Chen 2006; Sunderland, Steiert, Talmadge, Derfus, and Barry 2006).

PEG passivation of gold nanoshells was found to prevent colloidal aggregation after suspension in saline, and thereby restore the nanoshell peak absorbance (Hirsch, Stafford, Bankson, Sershen, Rivera, Price, Hazle, Halas, and West 2003). Similarly, incorporation of a phospholipid conjugated to methoxy-PEG-2000 in magnetic fluid–loaded liposomes increased their physical stability both in vitro and in vivo, as shown by MRI contrast persisting in mouse vessels over 24 h after intravenous injection (Martina, Fortin, Menager, Clement, Barratt, Grabielle-Madelmont, Gazeau, Cabuil, and Lesieur 2005). Association with macrophages was significantly reduced in the absence of cell toxicity.

4 CELL/ANIMAL BIOLOGICAL APPLICATIONS OF HYBRID NANOPARTICLES

4.1 DYNAMICS OF CELLULAR RECEPTORS

Because of their size and photostability, QD can be expected to access the synapse and be tracked for long periods of time. Hence, they are well suited to study the mobility of neurotransmitter receptors, which is central to the development and plasticity of synapses. A labeling complex consisting of primary and secondary antibodies and streptavidin-QD has been used for the GlyR, the main inhibitory neurotransmitter receptor in the adult spinal cord (Dahan, Levi, Luccardini, Rostaing, Riveau, and Triller 2003). Movement of individual GlyR in living spinal neurons was followed by single QD tracking, which relies on the random intermittency (blinking) of individual QD fluorescence. Trajectories of single QD-GlyR were visualized for >20 min, against <5 sec for Cy3, and with a signal-to-noise ratio almost an order of magnitude higher. This enabled lateral resolution of 5 to 10 nm, well below the 40 nm achieved with Cy3. Distinct diffusion coefficients were estimated for different domains of the neuronal somatodendritic membrane (extrasynaptic, perisynaptic, and synaptic), related to the distribution of perisynaptic adhesion molecules and/or intrasynaptic scaffolding molecules. These coefficients were approximately four times as large as those previously measured with 500-nm latex beads and comparable to those obtained with a smaller labeling complex (a Cy3-labeled Fab fragment). Moreover, QD-labeled samples could also be used for electron microscopy upon fixation and silver enhancement, enabling the correlation of high-resolution temporal dynamics and cellular localization. Correlated light and electron microscopic imaging has also been reported in other systems (Giepmans, Deerinck, Smarr, Jones, and Ellisman 2005).

However, in similar single-molecule fluorescence microscopy on culture of hippocampal neurons, although diffusion coefficients for AMPA (a subtype of glutamate) receptors in the extrasynaptic membrane were comparable using QD- and Cy3-coupled antibodies, QD-estimated synaptic diffusion was approximately five times lower, likely due to steric hindrance (Groc, Heine, Cognet, Brickley, Stephenson, Lounis, and Choquet 2004).

QD have also been used to study the cellular dynamics of receptor tyrosine kinases. Biotinylated EGF in combination with VFP-tagging of receptors was used to track the EGF receptor (erbB1) and study the dependence of its fate on interactions with related family members, erbB2 and erbB3 (Lidke, Nagy, Heintzmann, Arndt-Jovin, Post, Grecco, Jares-Erijman, and Jovin 2004; Lidke, Lidke, Rieger, Jovin, and Arndt-Jovin 2005). EGF-QD were found to be completely endocystosed in 5 min at 37°C in CHO cells via clathrin-coated pits (as shown by colocalization with transferrin). A single ligand, as present in complexes prepared at a low (0.1:1) EGF:QD molar ratio, was sufficient to induce binding and endocytosis. EGF-QD binding led to erbB1 activation. Single QD tracking (as revealed by blinking and intensity comparison with known single QD samples) revealed a novel form of cellular transport, whereby EGF binding on filopodia is followed by retrograde transport toward the cell body at a constant rate of ~10 nm/s, with endocytosis occurring at the base of the filopodia (see Fig. 13.2). Initiation of transport requires the interaction and concerted activation of at least two liganded receptors and is blocked both by specific inhibitors of the erbB1 kinase (PD153035) and cytochalasin D, which disrupts the actin cytoskeleton. Extended temporal observations spanning up to 30 min revealed the dynamics of endocytosed ligand–receptor complexes (Brownian motion, directed motion associated with microtubules, vesicular fusion). An estimate was also made of the kinetics of EGF surface binding (exponential) vs. internalization (linear, rate-limiting). By determining differences in EGF-QD-induced internalization rates among different members of the erbB RTK family, it was concluded that a substantial fraction of erbB1 occurs in heterodimers with erbB2 that are not readily internalized, whereas no association occurs with erbB3.

Binding and activation of cellular receptors by QD-conjugated ligands has also been reported in other systems, and allows combined imaging and control of cell functions. A βNGF-QD conjugate was internalized by TrkA receptor positive PC12 cells and induced a switch to a nonproliferative neural phenotype, as demonstrated by changes in the shape of cell somata and extensive neurite outgrowth and elongation. Although the bioactivity of the conjugate was reduced compared to that of free βNGF, largely due to the QD moiety, it allowed monitoring of βNGF trafficking over up to 4.5 days (Vu, Maddipati, Blute, Nehilla, Nusblat, and Desai 2005). However, serotonin-labeled QD were ineffective in eliciting currents in *X. laevis* oocytes expressing the serotonin receptor, although it was unclear whether this was due to lack of binding to the receptor, to ineffective gating or to channel block (Rosenthal, Tomlinson, Adkins, Schroeter, Adams, Swafford, McBride, Wang, DeFelice, and Blakely 2002). Binding did occur to serotonin transporters, albeit at EC_{50} values 1 to 2 orders of magnitude higher than free serotonin.

QD have also been used in combination with TIRF to study chondrocyte adhesion to substrates, and in particular the dynamics of the transition from the initial hyaluronan-mediated stage to the formation of integrin-mediated focal adhesions (Cohen, Kam, Addadi, and Geiger 2006). This transition requires substantial thinning of the hyaluronan layer or its removal from the cell–substrate interface. The approach was based on the use of biotinylated hyaluronan-binding protein (bHABP) in conjunction with streptavidin-conjugated QD. Time-lapse images were collected at 15-sec intervals for up to 75 min. For the first 20 min after cell seeding onto a glass surface, QD (i.e., hyaluronan) were found homogeneously distributed under the cells, whereas at later stages, concomitantly with cell spreading and increased focal adhesion formation, they became progressively clustered in confined pockets and their movement in the x-y plane was restrained. These pockets resemble "close contacts," i.e., broad areas that were previously identified by interference reflectance microscopy and that may greatly facilitate the nucleation of focal adhesions.

4.2 SENSING/SENSITIZING

QD conjugated to an organic electron donor (dopamine) have been used as intracellular redox probes (Clarke, Hollmann, Zhang, Suffern, Bradforth, Dimitrijevic, Minarik, and Nadeau 2006). Electron transfer in a dopamine–QD_{560} conjugate results in QD fluorescence quenching. In the presence of a reducing agent, dopamine quinone formation is prevented, and the conjugate can be internalized by dopamine receptor–expressing cells without inducing cell damage. Unquenching of QD fluorescence occurs on exposure to an oxidized intracellular environment (e.g., the lysosomes). Thus, the QD conjugate can be used as an alternative, or complement, to redox-sensitive dyes, with much greater photostability. A further advantage is the ability to target cell populations expressing specific receptors.

On exposure to UV light, the dopamine–QD_{560} conjugate generates singlet oxygen, causing cytotoxicity preceded by alterations in mitochondrial morphology typical of oxidative damage. Similarly, a CdSe antiCD90–QD conjugate, targeted to human leukemia cells, caused in vitro UV-mediated killing of leukemia cells and potentiated the effect of conventional photosensitizers, whilst sparing normal cells (Bakalova, Ohba, Zhelev, Nagase, Jose, Ishikawa, and Baba 2004). Since, upon UV-irradiation the QD were shown to induce free radical oxidation of liposomes, although the release of toxic amount of cadmium could be excluded, it was concluded that the effect was a result of the nanocrystals mediating UV-induced local generation of free radicals and ROS.

4.3 MOLECULAR INTERACTIONS

The use of SPION has been proposed to detect molecular interactions in living cells (Won, Kim, Yi, Kim, Jung, and Kim 2005). Streptavidin-conjugated SPION (50 nm) are derivatized with a biotinylated "bait" molecule. Transduction of the nanoparticles into cultured cells is achieved either by simultaneous conjugation to a Tat-HA2 peptide (fusing HIV-1 Tat to a sequence from the influenza virus hemagglutinins, in order to mediate endosome release) or microinjection. Application of a permanent magnet (1.1 T) against the tissue culture slide causes the SPION to concentrate to a narrow zone within the cells that is closest to the magnet, dragging with it any interacting component. When the latter is fluorescent (e.g., by fusion to a VFP, generated by recombinant DNA techniques), its changed distribution within the cell due to the molecular interaction can be visualized by confocal fluorescent microscopy. The method has been validated by identifying clones from a cDNA library that encode proteins binding to the immunosuppressant drug FK506, and by detecting TNFα-induced phosphorylation of IkB and the TNFα-regulated interaction between IkB and RelA or TcRP. It can be applied to the identification of both biological interactions and protein modifications within living cells in a variety of tissues and hosts, and can be readily adapted to genome-wide interaction screens. It also allows dynamic, single-cell analysis of interactions, whilst artifacts are reduced by the clear readout signal and by the use of a physiologically relevant context.

4.4 GENE CONTROL

Femtosecond NIR laser irradiation of a DNA/gold nanorod conjugate has been shown to result in shape transformation and release of DNA (Chen, Lin, Wang, Tzeng, Wu, Chen, Chen, Chen, and Wu 2006). Upon transduction of the conjugate in HeLa cells and NIR irradiation, expression of the DNA encoded protein (EGFP) was observed, indicating that gold nanorods and NIR irradiation can be used for the remote control of gene expression in living cells.

4.5 IN VIVO IMAGING

Imaging of tumor vasculature in a murine engraft model was obtained with a QD emitting in the NIR (705 nm) conjugated at a high ratio to a cyclic c(RGDyK) peptide, which is potent antagonist of $\alpha_v\beta_3$ integrin (Cai, Shin, Chen, Gheysens, Cao, Wang, Gambhir, and Chen 2006; Sunderland, Steiert, Talmadge, Derfus, and Barry 2006). The QD was coated with amine-modified PEG-2000 covalently attached to the surface. Twenty minutes after injection through the tail vein in mice harboring a tumor xenograft, a fluorescence signal was recorded from the tumor, reaching a peak at 6 hours. Vasculature targeting was suggested by the heterogeneity of the fluorescence on ex vivo tumor imaging, and confirmed by colocalization with the CD31 marker. Hence it was concluded that these QD, with a size estimated at 15 to 30 nm, did not extravasate into the tumor. Specificity was confirmed by the lack of specific staining with unconjugated QD. Incubation in serum resulted in a 20% loss of QD fluorescence. Compared with a Cy5.5 conjugate, the QD conjugate was claimed to provide a stronger signal, better photostability, and the possibility of multiplexing. However, uptake by the RES was higher, localization slower, and imaging restricted to the intravascular space, due to a larger size.

In vivo active tumor targeting was achieved by using a multifunctional QD probe, consisting of an encapsulating high-molecular-weight triblock copolymer, PEG, and an antibody directed against prostate-specific membrane antigen (PSMA) (Gao, Cui, Levenson, Chung, and Nie 2004). After a single tail vein administration, the nanoparticles were delivered and retained by a tumor xenograft (see Fig. 13.1). However nonspecific uptake was apparent mainly in the liver and spleen. Passive targeting of PEGylated QD unconjugated to anti-PSMA required a 15-fold higher dose, although no targeting was observed with QD devoid of both anti-PMSA and PEG coating, which was attributed to the excess of negative charges on the nanoparticle surface. In vivo fluorescence imaging was achieved. To improve contrast, a spectral unmixing algorithm was used to separate autofluorescence from the QD signal. It was suggested that QD are more suitable for in vivo imaging than traditional fluorochromes, because of their large absorption coefficient, and hence higher luminescence, under photon-limited excitation conditions, as is found in deep tissues.

In a similar approach, 9-nm water-soluble SPION were conjugated to anti-Her2/neu antibody and injected into mice implanted with a Her2/neu overexpressing cell line (Huh, Jun, Song, Kim, Choi, Lee, Yoon, Kim, Shin, Suh, and Cheon 2005). A significant drop of T2 values was recorded within 10 min, which doubled by 20 min postinjection. Dynamic monitoring of T2 signal changes at higher magnetic fields revealed a heterogeneous pattern of tumour infiltration, resulting from nanoparticle localization mainly at highly vascular regions, as confirmed by CD31 expression (Fig. 13.4).

Targeting of SPION to apoptotic tumor cells was shown by both flow cytometry and confocal microscopy upon conjugation to the C_2 domain of synaptotagmin I, which binds to anionic phospholipids in cell membranes and hence to the plasma membrane of apoptotic cells (Zhao, Beauregard, Loizou, Davletov, and Brindle 2001). Using these SPION as contrast agent allowed detection of apoptosis by MRI both in vitro and in vivo, in a mouse tumour model treated with chemotherapeutic drugs. Similarly, dextran-coated SPION tagged with $A\beta_{1-40}$, an Aβ peptide with high binding affinity for Aβ plaques,

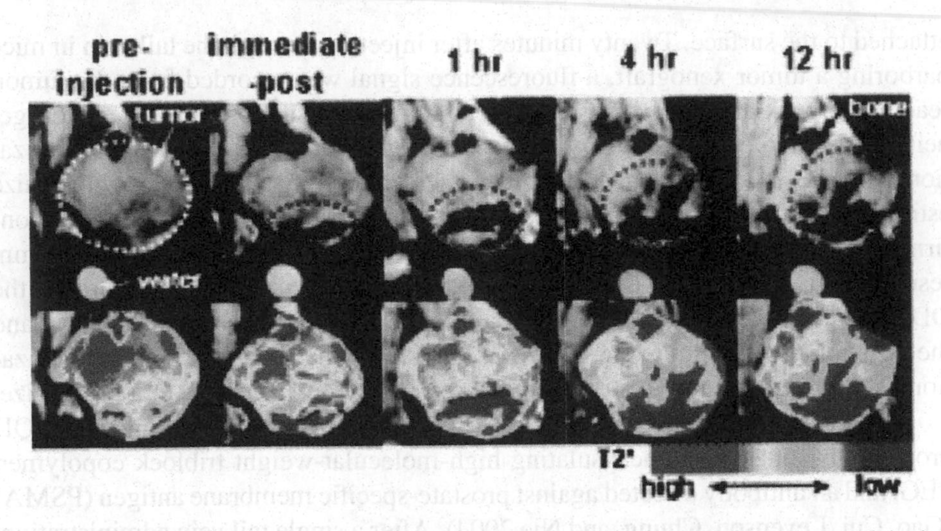

FIG. 13.4. **Top.** Sequential T2*-weighted MR images at 9.4 T of tumor implants after intravenous injection of SPION conjugated to an anti-Her2 (erbB2) antibody. The tumor area is circled with a *white dotted line*. A dark MR image *(red dotted circle)* first appears near the bottom region then gradually diffuses to the central and upper regions of the tumor. **Bottom.** Color mapping of the same images shows greater details (Reprinted with permission from Huh, Y. M., Jun, Y. W., Song, H. T., Kim, S., Choi, J. S., Lee, J. H., Yoon, S., Kim, K. S., Shin, J. S., Suh, J. S., and Cheon, J. (2005). In vivo magnetic resonance detection of cancer by using multifunctional magnetic nanocrystals. J. Am. Chem. Soc. 127, 12387–12391. Copyright 2005 American Chemical Society.)

allowed micro MR imaging of Aβ plaques in the brains of mice transgenic for either APP or APP and presenilin 1. Successful imaging required opening of the BBB by coinjection of the nanoparticles with mannitol (Wadghiri, Sigurdsson, Sadowski, Elliott, Li, Scholtzova, Tang, Aguinaldo, Pappolla, Duff, Wisniewski, and Turnbull 2003). Monocrystalline SPION (3 nm) coated with low-molecular-weight dextran and conjugated to human holo-transferrin have been used to image transgene expression in vivo by MRI technology (Weissleder, Moore, Mahmood, Bhorade, Benveniste, Chiocca, and Basilion 2000). The transgene encoded a mutant form of the transferrin receptor, engineered to be expressed constitutively at high levels. Upon intravenous injection of the superparamagnetic transferrin conjugate, substantial differences in MRI signal-to-noise ratios were recorded at the implantation site of a transgene-positive tumor with respect to the controlateral site implanted with a transgene-negative tumor.

4.6 CELL TRACKING

Bone-marrow lineage-negative stem cells have been QD tagged in vitro (λ_{em}: 660 nm) by using PEG-lipid coated QD derivatized with HIV-1 Tat, and co-injected in tumor-bearing mice with a QD-based plasma marker (λ_{em}: 470 nm) in order to track their fate in the circulation by two-photon in vivo imaging (Fig. 13.5).

The QD-based plasma marker was reported to provide sharper boundaries of vascular networks compared with high molecular weight dextrans, because of the greater hindrance to interstitial transport of spherical, rigid nanoparticles compared to flexible molecules. Moreover, the QD allowed clear simultaneous 2-photon imaging of GFP-labeled perivascular cells and SHG imaging of collagen fibers, due to the wide spacing of emission wavelengths and to the QD resistance to photobleaching at the higher laser powers required for SHG imaging (Stroh, Zimmer, Duda, Levchenko, Cohen, Brown, Scadden, Torchilin, Bawendi, Fukumura, and Jain 2005).

Tracking of dendritic cells following intranodal injection as a cancer vaccine in melanoma patients was accomplished by in vitro labeling with SPION, taking advantage of the naturally high endocytotic activity of immature dendritic cells (de Vries, Lesterhuis, Barentsz, Verdijk, van Krieken, Boerman, Oyen, Bonenkamp, Boezeman, Adema, Bulte, Scheenen, Punt, Heerschap, and Figdor 2005). Comparison of gradient-echo images before and 2 days after injection revealed significant differences in signal intensity both at the injected lymph node and at neighboring sites, indicating dendritic cell migration to downstream nodes. As few as 5×10^5 cells could be readily visualized by MRI. Comparison to scintigraphy

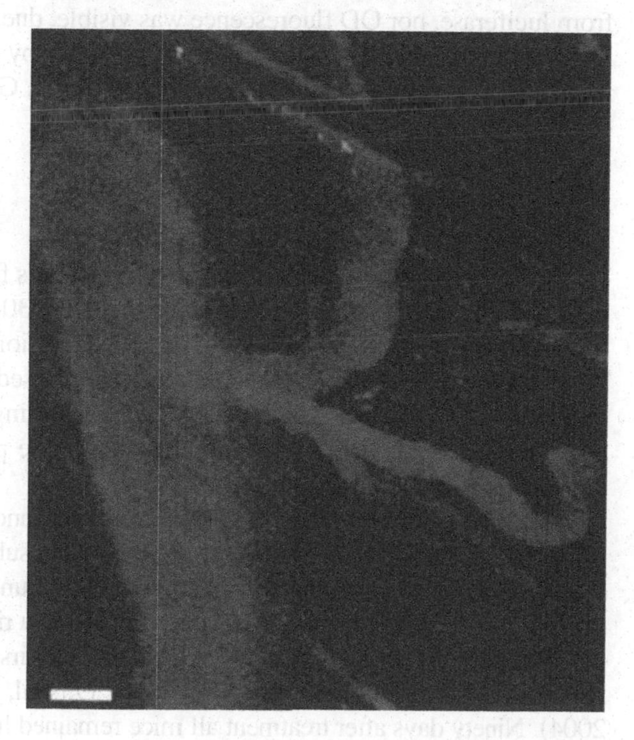

FIG. 13.5. Superimposition of seven sequential in vivo images of a single bone-marrow stem cell labeled with QD_{590}-Tat *(orange)* navigating tumor vessels highlighted with QD_{470} *(blue)*. Scale bar: 50 mm. (Reprinted by permission from Macmillan Publishers Ltd: Stroh, M., Zimmer, J. P., Duda, D. G., Levchenko, T. S., Cohen, K. S., Brown, E. B., Scadden, D. T., Torchilin, V. P., Bawendi, M. G., Fukumura, D., and Jain, R. K. (2005). Quantum dots spectrally distinguish multiple species within the tumor milieu in vivo, Nat. Med. 11, 678–682, copyright 2005.)

revealed much greater anatomical detail, resulting from a combination of higher resolution and better soft tissue contrast.

Tissue deposition of in vitro QD-labeled B16 melanoma cells has been analyzed following injection in the bloodstream in the mouse and imaging by multiphoton microscopy (Voura, Jaiswal, Mattoussi, and Simon 2004). By using multicolor imaging (i.e., labeling of different B16 cohorts with QD emitting at 510 or 570 nm), it was found that lung seeding is not entirely random, since some locations were infiltrated by more cells than others. The conclusions were validated by showing that QD do not have deleterious effects on tumor cells since both the fraction of QD-labeled cells seeded to the lung shortly (5 h) after injection, and the fraction of QD-labeled tumor nodules 40 days after injection were similar to the fraction of QD-labeled cells in the injected sample.

In vivo fluorescence imaging is hampered by autofluorescence of ubiquitous endogenous fluorochromes (collagen, porphyrins, flavins) and by the poor penetration of exciting radiation, due to scattering and absorption of light by tissues. An alternative strategy makes use of self-illuminating QD, consisting of QD conjugated to luciferase and excited by BRET in the presence of coelenterazine (a luciferase substrate). When C6 glioma cells were labeled with such QD conjugates, prepared from QD emitting at > 650 nm (a wavelength at which tissues are essentially transparent), and injected in mice, followed by injection of coelenterazine, a BRET signal was visualized corresponding to the lung. In contrast, neither direct luminescence from luciferase, nor QD fluorescence was visible, due to the absorption/scattering by intervening tissues of the radiation directly emitted by luciferase (480 nm) or required for efficient direct QD excitation (So, Xu, Loening, Gambhir, and Rao 2006).

4.7 TARGETED THERAPY

Cell death was induced in $CD8^+$ T lymphocytes from peripheral blood by incubation with anti-CD8-coated gold nanoparticles (30-nm) followed by pulsed laser irradiation (λ_{exc}: 565 nm, 100 pulses of 20 ns duration at 0.5 J/cm^2) (Pitsillides, Joe, Wei, Anderson, and Lin 2003). Cell death increased with the number of nanoparticles bound per cell and was preceded by swelling. Irradiation conditions were also identified that led to transient cell membrane permeabilization (up to 2 min) without inducing significant cell lethality.

Similarly, passive targeting of PEG-coated gold nanoshells (~130 nm diameter), with peak optical absorption in the NIR, was achieved at subcutaneous sites of implantation of murine colon carcinoma cells (CT26.WT) in immunocompetent mice. Photothermal treatment (λ_{exc}: 808 nm, 4 W/cm^2, 3 min) resulted in a rise of the temperature of the targeted regions and in tumor killing in all animals (Hirsch, Stafford, Bankson, Sershen, Rivera, Price, Hazle, Halas, and West 2003; O'Neal, Hirsch, Halas, Payne and West 2004). Ninety days after treatment all mice remained healthy and tumor free.

Gold/silica nanoshells with a shell/core ratio optimized for both peak optical scattering and absorption efficiencies have been suggested for dual imaging and treatment applications. When derivatized with anti-Her2 (erbB2) antibody, these nanoparticles were shown to provide increased scatter-based optical contrast of SKBr3 breast cancer cells and cell death upon laser irradiation in the NIR (λ_{exc}: 820 nm, 0.8 W/cm^2 for 7 min) (Loo, Lowery, Halas, West, and Drezek 2005). Gold nanorods tuned to give absorption in the NIR (i.e., with an aspect ratio of 3.9, obtained by titrating the silver concentration during growth) and conjugated to anti-EGFR (erbB1) antibodies, were used both to visualize human carcinoma cells by dark-field microscopy and kill them upon irradiation with an 800-nm continuous power laser (Huang, El-Sayed, Qian, and El-Sayed 2006) (Fig. 13.6).

Unlike gold nanospheres, however, cell binding did not result in diagnostic wavelength maxima and absorption bandwidths. The higher sensitivity of malignant cells to nanorod-mediated photothermal destruction (corresponding to an energy threshold for killing approximately half of that required by nonmalignant cells) was attributed to the higher amount of bound gold nanorods due to EGFR overexpression. Nanorods were claimed to be more efficient at inducing cell killing than nanoshells, possibly because of a higher absorption cross-section.

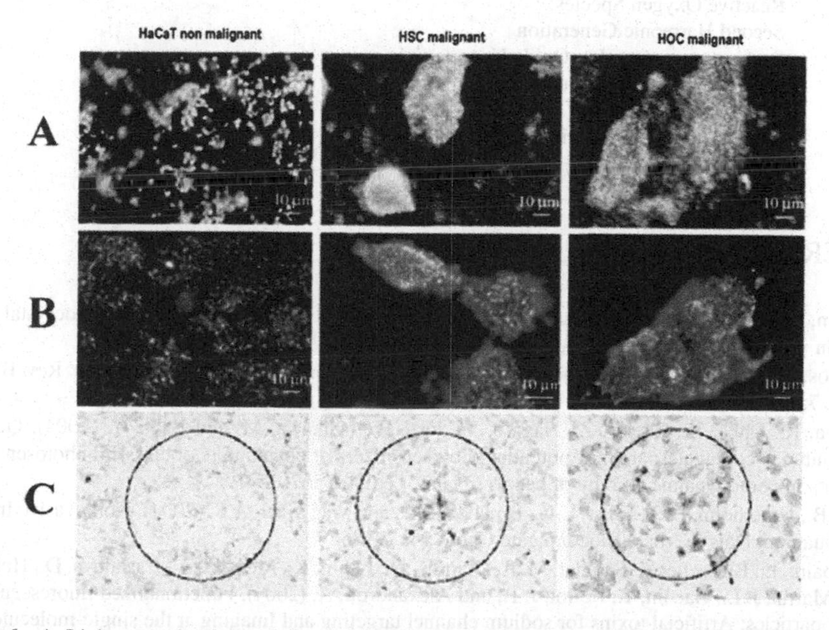

FIG. 13.6. **A.** Light scattering images of anti-EGFR/Au nanospheres after incubation with cells for 30 min at room temperature. **B.** Light scattering images of anti-EGFR/Au nanorods after incubation with cells for 30 min at room temperature. **C.** Irradiation of anti-EGFR/Au nanorod-targeted cells (10 W/cm^2) results in selective killing of malignant cells (Reprinted with permission from Huang, X. H., El-Sayed, I. H., Qian, W., and El-Sayed, M. A. (2006). Cancer cell imaging and photothermal therapy in the near-infrared region by using gold nanorods. J. Am. Chem. Soc. 128, 2115-2120. Copyright 2006 American Chemical Society.)

ABBREVATION

App: Amyloid Precursor Protein
BBB: Blood-Brain Barrier
BRET: Bioluminescence Resonant Energy Transfer
CEA: Carcino Embryonic Antigen
EGF: Epidermal Growth Factor
EGFR: Epidermal Growth Factor Receptor
EPR: Enhanced Permeation and Retention
GM-CSF: Granulocyte/Macrophage Colony Stimulating Factor
GlyR: Glycine Receptor
HIV-1: Human Immunodeficieny Virus 1
HSV: Herpes Simples Virus
HUVEC: Human Umbilical Vein Endothelial Cells
LPS: Lipopolysaccharide
MRI: Magnetic Resonance Imaging
NIR: Near Infrared Region
NLS: Nuclear Localization Signal
ox-LDL: oxidised Low Density Lipoprotein
PEG: Poly(Ethylene Glycol)
PSMA: Prostate-Specific Membrane Antigen
PTD: Protein Transduction Domain
QD: Quantum Dots
RES: Reticulo-Endothelial System
ROS: Reactive Oxygen Species
SHG: Second Harmonic Generation
SPION: Superparamagnetic Ion Oxide Nanoparticle
TIRF: Total Internal Reflection Microscopy
VCAM: Vascular Cell Adhesion Molecule 1
VFP: Visible Flurescent Protein

REFERENCES

Akerman, M.E., Chan, W.C.W., Laakkonen, P., Bhatia, S.N. and Ruoslahti, E. (2002). Nanocrystal targeting in vivo. Proc. Natl. Acad. Sci. U S A 99, 12617–12621.

Alivisatos, A.P., Gu, W.W. and Larabell, C. (2005). Quantum dots as cellular probes. Annu. Rev. Biomed. Eng. 7, 55–76.

Bakalova, R., Ohba, H., Zhelev, Z., Nagase, T., Jose, R., Ishikawa, M. and Baba, Y. (2004). Quantum dot anti-CD conjugates: are they potential photosensitizers or potentiators of classical photosensitizing agents in photodynamic therapy of cancer? Nano. Lett. 4, 1567–1573.

Ballou, B., Lagerholm, B.C., Ernst, L.A., Bruchez, M.P. and Waggoner, A.S. (2004). Noninvasive Imaging of Quantum Dots in Mice. Bioconjugate Chem 15, 79–86.

Beaurepaire, E., Buissette, V., Sauviat, M. P., Giaume, D., Lahlil, K., Mercuri, A., Casanova, D., Huignard, A., Martin, J. L., Gacoin, T., Boilot, J. P., and Alexandrou, A. (2004). Functionalized fluorescent oxide nanoparticles: Artificial toxins for sodium channel targeting and Imaging at the single-molecule level. Nano Lett. 4, 2079–2083.

Bharali, D.J., Lucey, D.W., Jayakumar, H., Pudavar, H.E. and Prasad, P.N. (2005). Folate-receptor–mediated delivery of InP quantum dots for bioimaging using confocal and two-photon microscopy. J. Am. Chem. Soc. 127, 11364–11371.

Cai, W.B., Shin, D.W., Chen, K., Gheysens, O., Cao, Q.Z., Wang, S.X., Gambhir, S.S. and Chen, X.Y. (2006). Peptide-labeled near-infrared quantum dots for imaging tumor vasculature in living subjects. Nano. Lett. 6, 669–676.

Chan, W.C.W. and Nie, S.M. (1998). Quantum dot bioconjugates for ultrasensitive nonisotopic detection. Science 281, 2016–2018.

Chen, C.C., Lin, Y.P., Wang, C.W., Tzeng, H.C., Wu, C.H., Chen, Y.C., Chen, C.P., Chen, L.C. and Wu, Y.C. (2006). DNA-gold nanorod conjugates for remote control of localized gene expression by near infrared irradiation. J. Am. Chem. Soc. 128, 3709–3715.

Chen, F.Q. and Gerion, D. (2004). Fluorescent CdSe/ZnS nanocrystal-peptide conjugates for long-term, nontoxic imaging and nuclear targeting in living cells. Nano. Lett. 4, 1827–1832.

Chithrani, B.D., Ghazani, A.A. and Chan, W.C.W. (2006). Determining the size and shape dependence of gold nanoparticle uptake into mammalian cells. Nano. Lett. 6, 662–668.

Clarke, S.J., Hollmann, C.A., Zhang, Z.J., Suffern, D., Bradforth, S.E., Dimitrijevic, N.M., Minarik, W.G. and Nadeau, J.L. (2006). Photophysics of dopamine-modified quantum dots and effects on biological systems. Nat. Mater. 5, 409–417.

Cohen, M., Kam, Z., Addadi, L. and Geiger, B. (2006). Dynamic study of the transition from hyaluronan- to integrin-mediated adhesion in chondrocytes. Embo J 25, 302–311.

Dahan, M., Levi, S., Luccardini, C., Rostaing, P., Riveau, B. and Triller, A. (2003). Diffusion dynamics of glycine receptors revealed by single-quantum dot tracking. Science 302, 442–445.

Daniel, M.C. and Astruc, D. (2004). Gold nanoparticles: assembly, supramolecular chemistry, quantum-size–related properties, and applications toward biology, catalysis, and nanotechnology. Chem. Rev. 104, 293–346.

de Vries, I. J. M., Lesterhuis, W. J., Barentsz, J. O., Verdijk, P., van Krieken, J. H., Boerman, O. C., Oyen, W. J. G., Bonenkamp, J. J., Boezeman, J. B., Adema, G. J., Bulte, J. W. M., Scheenen, T. W. J., Punt, C. J. A., Heerschap, A., and Figdor, C. G. (2005). Magnetic resonance tracking of dendritic cells in melanoma patients for monitoring of cellular therapy. Nat. Biotechnol. 23, 1407–1413.

Derfus, A.M., Chan, W.C.W. and Bhatia, S.N. (2004). Probing the cytotoxicity of semiconductor quantum dots. Nano. Lett. 4, 11–18.

Dixit, V., Van den Bossche, J., Sherman, D.M., Thompson, D.H. and Andres, R.P. (2006). Synthesis and grafting of thioctic acid-PEG-folate conjugates onto Au nanoparticles for selective targeting of folate receptor-positive tumor cells. Bioconjugate Chem. 17, 603–609.

Dubertret, B., Skourides, P., Norris, D.J., Noireaux, V., Brivanlou, A.H. and Libchaber, A. (2002). In vivo imaging of quantum dots encapsulated in phospholipid micelles. Science 298, 1759–1762.

Everts, M., Saini, V., Leddon, J. L., Kok, R. J., Stoff-Khalili, M., Preuss, M. A., Millican, C. L., Perkins, G., Brown, J. M., Bagaria, H., Nikles, D. E., Johnson, D. T., Zharov, V. P., and Curiel, D. T. (2006). Covalently linked au nanoparticles to a viral vector: Potential for combined photothermal and gene cancer therapy. Nano Lett. 6, 587–591.

Gao, X.H., Cui, Y.Y., Levenson, R.M., Chung, L.W.K. and Nie, S.M. (2004). In vivo cancer targeting and imaging with semiconductor quantum dots. Nat. Biotechnol. 22, 969–976.

Giepmans, B.N.G., Deerinck, T.J., Smarr, B.L., Jones, Y.Z. and Ellisman, M.H. (2005). Correlated light and electron microscopic imaging of multiple endogenous proteins using Quantum dots. Nat Meth. 2, 743–749.

Groc, L., Heine, M., Cognet, L., Brickley, K., Stephenson, F.A., Lounis, B. and Choquet, D. (2004). Differential activity-dependent regulation of the lateral mobilities of AMPA and NMDA receptors. Nat. Neurosci. 7, 695–696.

Gupta, A.K. and Curtis, A.S.G. (2004). Lactoferrin and ceruloplasmin derivatized superparamagnetic iron oxide nanoparticles for targeting cell surface receptors. Biomaterials 25, 3029–3040.

Gupta, A.K. and Gupta, M. (2005a). Cytotoxicity suppression and cellular uptake enhancement of surface modified magnetic nanoparticles. Biomaterials 26, 1565–1573.

Gupta, A.K. and Gupta, M. (2005b). Synthesis and surface engineering of iron oxide nanoparticles for biomedical applications. Biomaterials 26, 3995–4021.

Hirsch, L.R., Gobin, A.M., Lowery, A.R., Tam, F., Drezek, R.A., Halas, N.J. and West, J.L. (2006). Metal nanoshells. Ann. Biomed. Eng. 34, 15–22.

Hirsch, L.R., Stafford, R.J., Bankson, J.A., Sershen, S.R., Rivera, B., Price, R.E., Hazle, J.D., Halas, N.J. and West, J.L. (2003). Nanoshell-mediated near-infrared thermal therapy of tumors under magnetic resonance guidance. Proc. Natl. Acad. Sci. U S A 100, 13549–13554.

Howarth, M., Takao, K., Hayashi, Y. and Ting, A.Y. (2005). Targeting quantum dots to surface proteins in living cells with biotin ligase. Proc. Natl. Acad. Sci. U S A 102, 7583–7588.

Huang, X.H., El-Sayed, I.H., Qian, W. and El-Sayed, M.A. (2006). Cancer cell imaging and photothermal therapy in the near-infrared region by using gold nanorods. J. Am. Chem. Soc. 128, 2115–2120.

Huh, Y.M., Jun, Y.W., Song, H.T., Kim, S., Choi, J.S., Lee, J.H., Yoon, S., Kim, K.S., Shin, J.S., Suh, J.S. and Cheon, J. (2005). In vivo magnetic resonance detection of cancer by using multifunctional magnetic nanocrystals. J. Am. Chem. Soc. 127, 12387–12391.

Jain, P.K., Lee, K.S., El-Sayed, I.H. and El-Sayed, M.A. (2006). Calculated absorption and scattering properties of gold nanoparticles of different size, shape, and composition: Applications in biological imaging and biomedicine. J. Phys. Chem. B 110, 7238–7248.

Jain, R.K. (1999). Transport of molecules, particles, and cells in solid tumors. Ann. Rev. Biomed. Eng. 1, 241–263.

Jaiswal, J.K., Mattoussi, H., Mauro, J.M. and Simon, S.M. (2003). Long-term multiple color imaging of live cells using quantum dot bioconjugates. Nat. Biotechnol. 21, 47–51.

Josephson, L., Tung, C.H., Moore, A. and Weissleder, R. (1999). High-efficiency intracellular magnetic labeling with novel superparamagnetic-tat peptide conjugates. Bioconj. Chem. 10, 186–191.

Kang, H.W., Josephson, L., Petrovsky, A., Weissleder, R. and Bogdanov, A. (2002). Magnetic resonance imaging of inducible e-selectin expression in human endothelial cell culture. Bioconj. Chem. 13, 122–127.

Kim, S., Fisher, B., Eisler, H.J. and Bawendi, M. (2003). Type-II quantum dots: CdTe/CdSe (core/shell) and CdSe/ZinTe (core/shell) heterostructures. J. Am. Chem. Soc. 125, 11466–11467.

Kirchner, C., Liedl, T., Kudera, S., Pellegrino, T., Javier, A. M., Gaub, H. E., Stolzle, S., Fertig, N. and Parak, W. J. (2005). Cytotoxicity of colloidal CdSe and CdSe/ZnS nanoparticles. Nano. Lett. 5, 331–338.

Lagerholm, B.C., Wang, M.M., Ernst, L.A., Ly, D.H., Liu, H.J., Bruchez, M.P. and Waggoner, A.S. (2004). Multicolor coding of cells with cationic peptide coated quantum dots. Nano. Lett. 4, 2019–2022.

Lewin, M., Carlesso, N., Tung, C.H., Tang, X.W., Cory, D., Scadden, D.T. and Weissleder, R. (2000). Tat peptide-derivatized magnetic nanoparticles allow in vivo tracking and recovery of progenitor cells. Nat. Biotechnol. 18, 410–414.

Lidke, D.S., Lidke, K.A., Rieger, B., Jovin, T.M. and Arndt-Jovin, D.J. (2005). Reaching out for signals: filopodia sense EGF and respond by directed retrograde transport of activated receptors. J Cell Biol 170, 619–626.

Lidke, D.S., Nagy, P., Heintzmann, R., Arndt-Jovin, D.J., Post, J.N., Grecco, H.E., Jares-Erijman, E.A. and Jovin, T.M. (2004). Quantum dot ligands provide new insights into erbB/HER receptor-mediated signal transduction. Nat. Biotechnol. 22, 198–203.

Lim, Y., Kim, S., Nakayama, A., Stott, N.E., Bawendi, M.G. and Frangioni, J.V. (2003). Selection of quantum dot wavelengths for biomedical assays and imaging. Mol. Imag. 2, 50–64.

Loo, C., Lowery, A., Halas, N., West, J. and Drezek, R. (2005). Immunotargeted nanoshells for integrated cancer imaging and therapy. Nano. Lett. 5, 709–711.

Lovric, J., Bazzi, H.S., Cuie, Y., Fortin, G.R.A., Winnik, F.M. and Maysinger, D. (2005a). Differences in subcellular distribution and toxicity of green and red emitting CdTe quantum dots. J. Mol. Med. 83, 377–385.

Lovric, J., Cho, S.J., Winnik, F.M. and Maysinger, D. (2005b). Unmodified cadmium telluride quantum dots induce reactive oxygen species formation leading to multiple organelle damage and cell death. Chem. Biol. 12, 1227–1234.

Mammen, M., Choi, S.K. and Whitesides, G.M. (1998). Polyvalent interactions in biological systems: implications for design and use of multivalent ligands and inhibitors. Angew Chem.-Int. Ed. 37, 2755–2794.

Martina, M.S., Fortin, J.P., Menager, C., Clement, O., Barratt, G., Grabielle-Madelmont, C., Gazeau, F., Cabuil, V. and Lesieur, S. (2005). Generation of superparamagnetic liposomes revealed as highly efficient MRI contrast agents for in vivo imaging. J. Am. Chem. Soc. 127, 10676–10685.

Matsumura, Y. and Maeda, H. (1986). A new concept for macromolecular therapeutics in cancer chemotherapy: mechanism of tumoritropic accumulation of proteins and the antitumor agent smancs. Cancer Res. 46, 6387–6392.

Mattheakis, L.C., Dias, J.M., Choi, Y.J., Gong, J., Bruchez, M.P., Liu, J.Q. and Wang, E. (2004). Optical coding of mammalian cells using semiconductor quantum dots. Anal. BioChem. 327, 200–208.

Medintz, I.L., Uyeda, H.T., Goldman, E.R. and Mattoussi, H. (2005). Quantum dot bioconjugates for imaging, labeling and sensing. Nat. Mater. 4, 435–446.

Michalet, X., Pinaud, F.F., Bentolila, L.A., Tsay, J.M., Doose, S., Li, J.J., Sundaresan, G., Wu, A.M., Gambhir, S.S. and Weiss, S. (2005). Quantum dots for live cells, in vivo imaging, and diagnostics. Science 307, 538–544.

Moghimi, S.M., Hunter, A.C. and Murray, J.C. (2001). Long-circulating and target-specific nanoparticles: theory to practice. Pharmacol. Rev. 53, 283–318.

Morawski, A.M., Lanza, G.A. and Wickline, S.A. (2005). Targeted contrast agents for magnetic resonance imaging and ultrasound. Curr. Opin. Biotechnol. 16, 89–92.

Neuberger, T., Schopf, B., Hofmann, H., Hofmann, M. and von Rechenberg, B. (2005). Superparamagnetic nanoparticles for biomedical applications: possibilities and limitations of a new drug delivery system. J. Magn. Magn. Mater. 293, 483–496.

O'Neal, D.P., Hirsch, L.R., Halas, N.J., Payne, J.D. and West, J.L. (2004). Photo-thermal tumor ablation in mice using near infrared-absorbing nanoparticles. Cancer Lett. 209, 171–176.

Orive, G., Gascon, A.R., Hernandez, R.M., Dominguez-Gil, A. and Pedraz, J.L. (2004). Techniques: New approaches to the delivery of biopharmaceuticals. Trends Pharmacol. Sci. 25, 382–387.

Osaki, F., Kanamori, T., Sando, S., Sera, T. and Aoyama, Y. (2004). A quantum dot conjugated sugar ball and its cellular uptake on the size effects of endocytosis in the subviral region. J. Am. Chem. Soc. 126, 6520–6521.

Paciotti, G.F., Kingston, D.G.I. and Tamarkin, L. (2006). Colloidal gold nanoparticles: a novel nanoparticle platform for developing multifunctional tumor-targeted drug delivery vectors. Drug Dev. Res. 67, 47–54.

Parak, W.J., Pellegrino, T. and Plank, C. (2005). Labeling of cells with quantum dots. Nanotechnology 16, R9–R25.

Pitsillides, C.M., Joe, E.K., Wei, X., Anderson, R.R., and Lin, C.P. (2003). Selective cell targeting with light-absorbing microparticles and nanoparticles. Biophys. J. 84, 4023–4032.

Rosenthal, S. J., Tomlinson, I., Adkins, E. M., Schroeter, S., Adams, S., Swafford, L., McBride, J., Wang, Y., DeFelice, L.J. and Blakely, R.D. (2002). Targeting cell surface receptors with ligand-conjugated nanocrystals. J. Am. Chem. Soc. 124, 4586–4594.

So, M.K., Xu, C.J., Loening, A.M., Gambhir, S.S. and Rao, J.H. (2006). Self-illuminating quantum dot conjugates for in vivo imaging. Nat. Biotechnol. 24, 339–343.

Song, H.T., Choi, J.S., Huh, Y.M., Kim, S., Jun, Y.W., Suh, J.S. and Cheon, J. (2005). Surface modulation of magnetic nanocrystals in the development of highly efficient magnetic resonance probes for intracellular labeling. J. Am. Chem. Soc. 127, 9992–9993.

Sonvico, F., Mornet, S., Vasseur, S., Dubernet, C., Jaillard, D., Degrouard, J., Hoebeke, J., Duguet, E., Colombo, P. and Couvreur, P. (2005). Folate-conjugated iron oxide nanoparticles for solid tumor targeting as potential specific magnetic hyperthermia mediators: synthesis, physicochemical characterization, and in vitro experiments. Bioconj. Chem. 16, 1181–1188.

Stroh, M., Zimmer, J.P., Duda, D.G., Levchenko, T.S., Cohen, K.S., Brown, E.B., Scadden, D.T., Torchilin, V.P., Bawendi, M.G., Fukumura, D. and Jain, R.K. (2005). Quantum dots spectrally distinguish multiple species within the tumor milieu in vivo. Nat. Med. 11, 678–682.

Sunderland, C.J., Steiert, M., Talmadge, J.E., Derfus, A.M. and Barry, S.E. (2006). Targeted nanoparticles for detecting and treating cancer. Drug Dev. Res. 67, 70–93.

Tkachenko, A.G., Xie, H., Coleman, D., Glomm, W., Ryan, J., Anderson, M.F., Franzen, S. and Feldheim, D.L. (2003). Multifunctional gold nanoparticle-peptide complexes for nuclear targeting. J. Am. Chem. Soc. 125, 4700–4701.

Tkachenko, A.G., Xie, H., Liu, Y.L., Coleman, D., Ryan, J., Glomm, W.R., Shipton, M.K., Franzen, S. and Feldheim, D.L. (2004). Cellular trajectories of peptide–modified gold particle complexes: comparison of nuclear localization signals and peptide transduction domains. Bioconj. Chem. 15, 482–490.

Voura, E.B., Jaiswal, J.K., Mattoussi, H. and Simon, S.M. (2004). Tracking metastatic tumor cell extravasation with quantum dot nanocrystals and fluorescence emission-scanning microscopy. Nat. Med. 10, 993–998.

Vu, T.Q., Maddipati, R., Blute, T.A., Nehilla, B.J., Nusblat, L. and Desai, T.A. (2005). Peptide-conjugated quantum dots activate neuronal receptors and initiate downstream signaling of neurite growth. Nano. Lett. 5, 603–607.

Wadghiri, Y. Z., Sigurdsson, E. M., Sadowski, M., Elliott, J. I., Li, Y. S., Scholtzova, H., Tang, C. Y., Aguinaldo, G., Pappolla, M., Duff, K., Wisniewski, T., and Turnbull, D. H. (2003). Detection of Alzheimer's amyloid in Transgenic mice using magnetic resonance microimaging. Magn. Reson. In Med. 50, 293–302.

Weissleder, R., Kelly, K., Sun, E.Y., Shtatland, T. and Josephson, L. (2005). Cell-specific targeting of nanoparticles by multivalent attachment of small molecules. Nat. Biotechnol. 23, 1418–1423.

Weissleder, R., Moore, A., Mahmood, U., Bhorade, R., Benveniste, H., Chiocca, E.A. and Basilion, J.P. (2000). In vivo magnetic resonance imaging of transgene expression. Nat. Med. 6, 351–354.

Won, J., Kim, M., Yi, Y.W., Kim, Y.H., Jung, N. and Kim, T.K. (2005). A magnetic nanoprobe technology for detecting molecular interactions in live cells. Science 309, 121–125.

Wu, X., Liu, H., Liu, J., Haley, K.N., Treadway, J.A., Larson, J.P., Ge, N., Peale, F. and Bruchez, M.P. (2003). Immunofluorescent labeling of cancer marker Her2 and other cellular targets with semiconductor quantum dots. Nat. Biotech. 21, 41–46.

Zhang, T.T., Stilwell, J.L., Gerion, D., Ding, L.H., Elboudwarej, O., Cooke, P.A., Gray, J.W., Alivisatos, A.P. and Chen, F.F. (2006). Cellular effect of high doses of silica-coated quantum dot profiled with high throughput gene expression analysis and high content cellomics measurements. Nano. Lett. 6, 800–808.

Zhao, M., Beauregard, D.A., Loizou, L., Davletov, B. and Brindle, K.M. (2001). Non-invasive detection of apoptosis using magnetic resonance imaging and a targeted contrast agent. Nat. Med. 7, 1241–1244.

Index